To my (already) good friends
the Gourlie's

Robert H S Robertson
Dunmore
Pitlochry
PH16 5ED

1st June 1988

FULLER'S EARTH:

A HISTORY OF CALCIUM MONTMORILLONITE

By the same author

MINERAL USE GUIDE
(Cleaver-Hume Press, London, 1960, 1961)

With A. G. Clement:
SCOTLAND'S SCIENTIFIC HERITAGE
(Oliver and Boyd, Edinburgh, 1961)

This book is dedicated to

Joan M. Eyles and Robert C. Mackenzie

MINERALOGICAL SOCIETY
OCCASIONAL PUBLICATION

FULLER'S EARTH:

A HISTORY

of Calcium Montmorillonite

by

ROBERT H.S. ROBERTSON
The Resource Use Institute, Ltd.,
Dunmore, Pitlochry, PH16 5ED,
Scotland

VOLTURNA PRESS

Published 1986 by Volturna Press, Hythe, Kent, UK

BRITISH LIBRARY CATALOGUING IN PUBLICATION DATA

Robertson, R.H.S.
Fuller's Earth: A History of Calcium Montmorillonite

1. Fuller's Earth
2. Cloth industry
3. Pharmaceutical clays
4. Localities
5. Modern uses – history

I Title

First publication, 1986

ISBN 0 85606 070 4

Typeset in Baskerville by
Mathematical Composition Setters Ltd, Salisbury, UK
and printed and bound by
Billing and Sons Ltd, Worcester

ACKNOWLEDGMENTS TO SPONSORS

I thank the Council of The Resource Use Institute Ltd for allowing me to complete this work as their Project No. 132 (1983–85).

This book could not have been completed without Research Awards from **The Leverhulme Trust** and from **Laporte Industries Ltd** nor could it have been published without a grant from the latter and a grant and a loan from **The Royal Society** of London. I am most grateful for the help and encouragement given by these bodies and also by the **Clay Minerals Group** of the **Mineralogical Society of Great Britain**, who agreed to accord the book the status of an **Occasional Publication** of the Group.

I also wish to thank Dr Douglas M. C. MacEwan, an X-ray physicist and a founder of the Clay Minerals Group in 1947, for editing the book as well as for publishing it; Mathematical Composition Setters Ltd for typesetting direct from manuscript; and Miss Muffy Robertson, graphic artist, for improving many of the illustrations. Following suggestions made by three referees in the Clay Minerals Group and five readers in Laporte, many paragraphs and some chapters were rewritten. I thank them all for giving so generously of their time. Nevertheless I should perhaps emphasize that many of the opinions expressed in this book are my own.

Thanks to individual scholars and for permission to publish illustrations are given at the end before the bibliography and index.

CONTENTS

LIST OF TABLES

Principal Tables

Other Tables in Text

LIST OF ILLUSTRATIONS

TECHNICAL TERMS

Units

1 lbf/in^2 = 1 pound (force) per square inch
= 6695 Newtons per square meter
= 6695 Pascals
1 meq./100 g = 1 milliequivalent per 100 grams
1 ton = 1 long ton = 2240 lb = 1016 Kg
1 tonne = 1000 Kg

In pharmacy

1 gr a grain is 1/7000 of a pound
1 dr a dram or drachm is (approx.) 60 grains
1 oz an ounce is 480 grain

Avoirdupois

1 cwt hundredweight or 112 pounds
1 qr quarter or 28 pounds
1 st stone or 14 pounds
1 oz ounce (av.) is 1/16th of a pound

Abbreviations

cec cation exchange capacity
DTA differential thermal analysis
fob free on board (a ship)
LL liquid limit
LOI Loss on ignition
tpa tonnes per annum, nowadays
XRD X-Ray diffraction

Glossary

absorption the drawing in of liquids into the pores of a solid
adsorption the attraction of a substance from a solution or a vapour to the surface of a particle

chemisorption	adsorption by chemical bonds
endotherm	heat absorbed in a reaction and in graphic form in DTA
exotherm	heat evolved in a reaction and in graphic form in DTA
hygroscopic	having the capacity to absorb water vapour
sorption	the attraction of a gas or vapour into the pores of solid; it may be either absorption or adsorption, or it may be by both

Greek texts

The Greek texts in this book are given in MS, because the author prefers a bold upright hand to the thin sloping characters of conventional Greek typography.

INTRODUCTION

1

It was obvious at a glance that the story of the fuller's earth was the merest fabrication.
SIR ARTHUR CONAN DOYLE, The Engineer's Thumb.

THE present book, on the other hand, traces the history of that particular sort of clay which was used for cleaning woollen cloth for nearly seven thousand years. It was begun in 1934 when Lt. Col. W. T. Raikes of The Fullers' Earth Union Ltd and his son Duncan at Radley helped by collecting classical references to fuller's earth. In 1938 copies of the accumulated notes were handed to John Wills who made a profession of writing Company Histories. All that could then be done was to abandon hopes of authorship, to hand over the whole of the rest of the notes and quotations and to assist Wills in his writing page by page.

Events than took a dramatic turn. After a few weeks and the completion of a few short chapters, Wills left to make a start with another company's history. The pattern was repeated until he was recalled to Redhill. On the way back to London I realised he was unwell and unlikely ever to complete the book. He returned all the notes and books which he had borrowed, went to France and died not long after. Only one rare book was lost in this misadventure.

An opportunity arose in 1958 for making progress. Marian J. McKellar was engaged as a ghost writer as I was stricken by rheumatoid arthritis and could not wield a pen. The draft manuscript was written on a long continuous sheet and new material was inserted by scissors and transparent tape. To find the right place Marian would unroll the "Deadly Scroll" across the lawn, as charming a sight as Nausikaa of whom we shall read.

The History of Fuller's Earth has been written with the unattainable model of Charles Singer's *The Earliest Chemical Industry* in mind. Derek Spence conceived the plan of this volume and brought great scholarship to bear on the history of alum. As Peter Spence and Sons Ltd is now a component part of Laporte Industries Ltd, a few more references to alum may be quoted.

London: The Folio Society, (1948).

Christopher Columbus' 218-ton caravel *Santa Maria*, which left Palos 3 August 1492 and reached San Salvador – probably Watling Island – 12 December 1492, carried a cargo which included mercury, alum, cloth, ivory, velvet, charms, bracelets, etc., for trading with the natives.

Illustrated London News, (11 June 1949), 611.

1

○ *or* ⊏⊐

John Harris, in his *Lexicon Technicum* (the first technical dictionary, 1704) gives a list of "characters chymical". Among them was Allum.

The last business of the Scottish Parliament (1707), before passing the Act of Union with England and Wales, was consideration of a proposed Scottish alum industry.

History of Whitby and Streoneshalh Abbey, Whitby (1817), 806–817, and 199.

Michael Drayton, *Polyolbion, or a chorographicall description . . . of Great Britain* (1612, 1622) (also 1876).

Then, in a *History of Whitby* (1817), the Rev. George Young gives a long account of the alum industry, though he may have copied from Lionel Charlton, quoted by Singer; and quotes Drayton's *Poly-olbion: Her earth with Allome veines most richly intermin'd.*

Fuller's earth cannot compete with alum pictorially. Alum, for long a papal monopoly and source of great wealth, lent itself to magnificent illustrations, but the humble fuller's earth was seldom noticed by artists and most of the people connected with the industry are but shadowy figures.

References to fuller's earth are so scattered, often in obscure and ancient documents, that the original words are often quoted here, making this a source book as well as a history. Often an old translation has a charm of its own; who can resist Holland's translation of Pliny's *Natural History* (1601)?

The history of sheep breeds, of the woollen industry and of fulling machines are only lightly touched on in this book because these subjects have been treated elsewhere; but I have referred to such works and quoted enough to set the scene for the story of fuller's earth.

All that need be said here is that the washing of woollen cloth with fuller's earth was certainly practised in Cyprus about 5000 BC. Pure soft water has some detergent properties; but some river muds were found to be more effective. The washing of cloth and clothes is traced from Ancient Egypt, the Middle East, Homeric and Classical Greece, Rome, Gaul and onwards to the 20th century when reliable modern accounts of the industry can be referred to.

As English fuller's earth was for so many centuries of great importance to the wool trade, its history has been described at some length.

Calcium montmorillonite: a clay mineral, the essential constituent of fuller's earth.

In the short accounts of the rise of the calcium montmorillonite industries in many other countries, I have tried to show how they came about. I have also tried to show how each of the modern uses of these clays was discovered or invented, having described the early work on the nature and properties of fuller's earth.

The commoner usage of *fuller's earth* has been adopted throughout, except where the apostrophe is differently placed or absent in a quotation or in the Company's name, The Fullers' Earth Union Ltd. I plead accuracy and not discourtesy: indeed an essay on the apostrophe is given in chapter 70 i.

A definition

1i

Fuller's earth is that kind of material which is used in applications where fuller's earth is traditionally used.
U.S. BUREAU OF MINES when asked by James J. Fallon (formerly of Engelhard) for a precise definition.

THIS book aims to be more precise than this bureaucratic reply. It is a history of the discovery, properties and uses of calcium montmorillonite, a clay mineral of defined structure.

R. E. Grim, *Clay Mineralogy*, Mc-Graw Hill Publishing Co Ltd (1953), 55.

The word *fuller* and the French *foulon* are derived from the Latin word *fullo*, whose trade was that of scouring and felting cloth. In Roman times pieces of cloth were first soaked in stale urine or some other alkaline solution and then trampled on in a trough or vat containing a slurry of fuller's earth (*creta fullonia*). Several rinses were given with clean water to remove the earth, which carried away with it much of the grease and dirt remaining in the wool. The processes of making cloth were clearly well-developed and of great antiquity.

In prehistoric times whole fleeces were sewn together with leather thongs to make warm garments as insulation against the cold. These and the sheepskin bedding were smelly, especially when wet. The need was felt to clean the fleece in some way so that better quality garments could be cut and tailored. This was no simple problem because a sheep's fleece contains many sorts of impurity.

(**Nature of Wool**, below)
A note on the materials of ancient textiles and baskets. Julius Grant in *A History of Technology*, ed. by C. Singer, E.J. Holmyard and A.R. Hall, Oxford Clarendon Press, (1954), *1*, 451.

The nature of wool

1ii

ACCORDING to Julius Grant's "note on the materials of ancient textiles" (note above) in that invaluable reference book *A History of Technology*, edited by Charles Singer, E. J. Holmyard and A. R. Hall (1954), *wool is microscopically distinct from hair and is more elastic. It is similar in chemical composition to horn, but contains some forty per cent of natural grease. The principal microscopic feature of wool fibres, which distinguishes them from hair, is their scaly surface-structure, which accounts for their felting properties.*

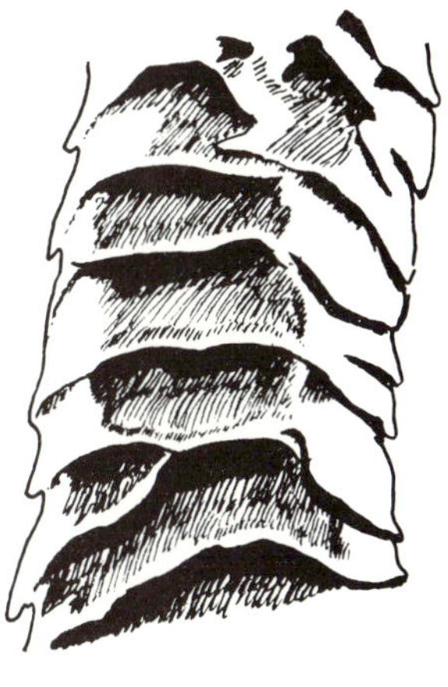

Fig. 1. Scale edges of a wool fibre from a microphotograph of CIBA-GEIGY Ltd, Basle

A wool fibre consists of three parts: (a) the central medulla, usually best seen near the root, and, in the fine grades, only there; (b) the cuticular tissue, which forms the bulk of the fibre; and (c) the outer layer of highly characteristic flat, horny tubular cells forming overlapping scales.

We are accustomed to think of wool as essentially a product of the sheep, but there are many other ungulates which grow wool, usually

1ii

under a coat of hair. The sheep of certain domestic breeds are perhaps unique in carrying wool alone, but there are wild and domesticated breeds of sheep that carry both wool and hair. It is said that there are sheep which, like their kindred goats, carry only hair. At any rate, sheep have been kept for their wool since Neolithic times.

1iii

Impurities in wool

A FLEECE contains fecal matter, mineral dirt (mainly sand and clay), leaves, plant stems and burs, plant stains, and, in recent times, ownership marks and tar now replaced by synthetic pesticides, soluble impurities rich in potash (dried sweat or suint) and most importantly wool grease. Knots had also to be removed in later times, but primitive man was more concerned with removing some of the suint and wool-grease, but more particularly the adventitious dirt or filth.

Only when the wool was plucked from the sheep and spun into yarn and woven was it discovered that the grease interfered with the weaving. In modern sheep there may be 2 to 12 per cent of suint and as much as 35 per cent of grease in the wool; ideally there should be only one half to 3 per cent of residual grease for spinning. Actually, coarse wool contains 7 to 10 per cent of grease, hairy fleeces very little, and, as early as 1890, fine wool of the type imported from the Cape, Australia and New Zealand, 30 to 40, or even 50, per cent.

Industrial Detergency, Wm. W. Niven, Jr., Reinhold Publishing Corp., (1955), 120.

Woolfat and processes of obtaining it with special relation to wool-washing, H. W. Langbeck, *J. Soc. Chem. Ind.*, (1890), 356–359.

As has been said, very soft water is a mild detergent; for it is the lime and magnesia in hard water which make grease removal more difficult through the formation of insoluble soaps. If a suint solution is clarified and heated, it can be used in a first wash known as "suint scouring", but its presence in the wool would in any case contribute to the detergency of soft water.

This history will therefore begin by some examples of fresh water scouring; and then proceed to the earliest known use of fuller's earth in Cyprus. Other scouring agents of antiquity include stale—and therefore ammoniacal—urine; naturally occurring alkalis scraped from the surface of the desert or washed out of the ashes of plants; and the juices of saponaceous plants.

As the technology developed it was discovered that the scouring of woollen fabrics can be combined with fulling or felting, which is a shrinkage process. Since wind penetrates unfulled cloth, a means had to be found to make the web thicker, denser and less permeable. Fuller's earth was used for absorbing the grease and helping to lubricate the fibres in the fulling operation for nearly 7000 years. Almost comparable in its antiquity is the use of alum in the dying of leather and wool with madder.

The Earliest Chemical Industry, Charles Singer, London: The Folio society, (1948), note 3 and page 3.

In the last few centuries soap and synthetic detergents have replaced fuller's earth in fulling and milling as the process came to be called when it was mechanised. Indeed the word "mill" came to be used to describe a textile factory as a whole, whether it was darkly Satanic or not.

William Blake "Jerusalem".

The type material, cimolian earth, contains a high proportion of what mineralogists call calcium montmorillonite. When earths of different composition were used for want of a true fuller's earth, they will be cautiously mentioned.

In so far as fuller's earth was used for millenia in the preparation of woollen cloth, and clothing is of great social and practical value, we see in this commodity a greater importance than has hitherto been accorded to it. *Society* [and may one say civilisation] declared Carlyle *is founded upon cloth.*

> *Of the three primary necessities of life, clothing is a far more potent civilising factor than either food or shelter. In food and wine we have sustenance, contentment and inspiration; and in stone and timber, refuge from the elements; but in clothing there is more. For clothes not only cover man's nakedness; they give him dignity, confidence and courage; and a peculiarly agreeable means of self-expression.*

J. Wills, draft, page 19.

In this century calcium montmorillonite has been "activated" by acid treatment to make highly adsorptive oil-decolorising earths and by alkali, usually soda ash, treatment to make swelling clays similar to the naturally occurring "bentonite", which is substantially sodium montmorillonite. Bentonite is a rockname and is often used in commerce to name both sodium and calcium montmorillonites, sometimes with the prefix sodium- or calcium-.

I have not written a history of sodium montmorillonite, of which the best known example is "Genuine Wyoming Bentonite" in the trade, for this deserves a separate account of its own; nor have I traced the history of attapulgite, the commercial name for a clay containing a high proportion of the fibrous clay-mineral palygorskite. In some countries such clays have been called "floridins" from the trade name of the Floridin Company of Georgia, USA. Unfortunately this kind of clay has often been called "American fuller's earth" or simply "fuller's earth". Here attapulgite is mentioned only when a distinction has to be made from the subject of this history. Moreover, the definition that a clay which decolorises oil is a fuller's earth must be abandoned.

Especially the USSR.

ORIGINS

2

Washing clothes in pure soft water

2i

Athene, desiring to see Nausicaa meet Odysseus, exhorted her to bestir her father to lend her a waggon and mules to take her raiment to clean it: "for thy marriage is at hand," she urged, "when thou must needs be clad in fair garments."

HOMER, Odyssey, 6, 87

Fig. 2. Greek laundrywomen. The shipwrecked Ulysses surprises Nausicaa and her companions by the sea-shore. Right: a woman wringing a cloth. Left: laundry hanging on a tree to dry. Between Ulysses and Nausicaa is the goddess Athene. The women are wearing chitons caught up at the waist. 5th century BC.

CIBA Review. p. 286, *after E. Pfuhl.*

AND in due course Nausicaa and her maidens reached the noble river with its never-failing pools, in which there was enough clear water always bubbling up and swirling by to clean the dirtiest clothes. Here they turned the mules loose from under the yoke and drove them along the eddying stream to graze on the sweet grass. Then they lifted the clothes by armfuls from the cart, dropped them into the dark water and trod them down briskly in the troughs, competing with each other in the work. When they had rinsed them till no dirt was left, they spread them out in a row along the sea-shore, just where the waves washed the shingle clean when they came tumbling up the beach. Next, after bathing and rubbing themselves with olive-oil, they took their meal at the riverside, waiting for the sunshine to dry the clothes.

Homer, *Odyssey*, vi, 87, translated by E.V. Rieu, Harmondsworth: Penguin Books, (1946) 104. . .

The etymology of **πλυνός**, a washing trough, is given later.

2i

They then played with a ball; but when it fell into the river, their shrieks awoke the shipwrecked and naked Odysseus who had swum ashore nearby.

No divine prompting was available in a similar episode in Tasso, the model for Sir Walter Scott's scene in *Waverley*.

Sir Walter Scott, *Waverley*, Ch.9.

Upon the green aforesaid two bare-legged damsels, each standing in a spacious tub, performed with their feet the office of a patent washing machine. These did not, however, like the maidens of Armida, remain to greet with their harmony the approaching guest, but, alarmed at the appearance of a handsome stranger on the opposite side, dropped their garments (I should say garment, to be quite correct) over their limbs, which their occupation exposed somewhat too freely, and with a shrill exclamation of "Eh, sirs!" uttered with an accent between modesty and coquetry, sprang off like deer in different directions.

(sprung in the original)

D.H. Lawrence reassumed an idyllic note in his retelling of the Odysseus story in *The Man of Tyre*.

2ii Scouring agents

Concerning the history of sheep, Michael L Ryder's *Sheep and Man*, Duckworth (1983), pp 846 is "a seemingly inexhaustible mine of information about sheep and their exploitation during prehistoric and historic times."

LONG before the Dark Age described by Homer, two detergents were discovered, no doubt accidentally, river mud and stale urine. No doubt when women folk took fleeces and later woven cloth to be washed in a river, they would have discovered that the woollen goods were cleaned more easily and effectively when they had been trampled into the mud. There might have been some competition then in finding the muddiest river bottom or river bank, the clay from which washed wool whiter than their neighbours. In the course of time different sorts of clay would have been tried and eventually fuller's earth would have been discovered.

One can also imagine that the cleansing power of stale urine would have been discovered before cloth was invented. A child, wrapped in skins, wetted his bed; his mother angrily cast them aside, gave him clean skins and forgot the soiled ones for several days. The urine having gone stale and alkaline, polluted the atmosphere of the cave. So when the mother tried to wash the skin, she noticed that the soiled part washed whiter than the rest.

Probably mineral alkalis and plant ashes were used at a much later date.

2iii Felting

IN THE process of trampling to cleanse fleeces or clothes the effect of felting the fibres or thickening the cloth was bound to have been noticed in prehistoric times. Indeed the wool trade has its own version of the origin of felting. The story runs that Noah, in an attempt to make the long voyage in the Ark more comfortable for both human and animal inmates, covered the floors with wool from the backs of his sheep. When the waters abated and the animals left the Ark, Noah found to his surprise that they had trampled, fulled and compacted the loose wool, making a felted carpet.

2iii

Ciba Review, (1958) No. 129,2

Though the author of the article from which this item of trade lore has been taken does not make the point, urine would have played its part in this transformation. And indeed urine, like the other alkalis used in various times and places as cleansing agents, has always played a significant role in the fuller's art. It was certainly still in use in an English cloth factory in 1935.

C.O. Clark, Ancient and modern in milling and scouring, *J. Soc. Dyers and Colorists* (1950), *66*, 187–190.

Cyprus

2iv

IN 1969 an expedition from the Universities of Glasgow and Birmingham under Dr E. G. Peltenburg and Mr P. S. Gelling started excavations at the Neolithic settlement of Vrysi due north of Ayios Epiktitos(1) on the north coast, east of Kyrenia. From 1972 Dr Peltenburg was in sole charge. Many exceptionally well-preserved houses were discovered on the flat top and on the seaward slopes of a headland on the east side of a sandy bay.

Archaeology in Cyprus by K. Nicolaou, Archaeological Reports for 1975–76, No. 22, Soc. Promotion of Hellenic Studies and the British School at Athens (1976), p. 36.

(1) a village now called Çatalköy.

Entrance to the houses was through doors just over one meter high and barely half a meter wide; and communication between the houses was along narrow winding streets. The stone walls were covered with a thick grey-white plaster. Fireplaces were among the main features of the house-floors.

In one house was found in a workshop for making bone beads the flints which were used for carving them. Elsewhere is evidence for the making of stone implements and axes, and for other domestic activities. Much of the pottery here is unique in Cyprus. Spindle whorls are plentiful as are bone needles of all sizes from awls to the finest needles with tiny eyes, suitable, it is thought, for sewing linen clothes. No loom weights were found because at such a very early date there was probably no division of the warp and the weave was darned in by hand. But what interests us most is that the fulling of woollen cloth was undertaken at this site.

One house was built right on top of a natural exposure of greenish gray clay, proved by R. C. Mackenzie to be rich in montmorillonite, and lumps of this fuller's earth were found in seven of the houses, not only loose upon the floor but in vessels beside a plaster basin. One can picture the moist fuller's earth lumps being dried near a hearth and, when dry and ready for use, being kept handy beside the fulling basin in a pot where it could not be inadvertently wetted.

2iv

The floors where lumps of fuller's earth were found have been carbon-dated at about 5000 BC and stratigraphical measurements are consistent with this dating. This is not much more than a thousand years since sheep, probably dun coloured and quite small, were fully domesticated. In other villages in Cyprus the only domesticated food animal was the fallow-deer at that time. Only at Vrysi are found many bones of both sheep and fallow deer.

2v

Crete

Minoans and Mycenaeans – Greece's Brilliant Bronze Age, Joseph Judge, *Nat. Geogr. Mag.* (Feb–1978), *153* (2), 142–185.

For the history of weaving see *A History of Technology*, ed. C. Singer *et al.*, Oxford: Clarendon Press, (1954), *1*, 425 *et seq.*

IN 1967 and 1968, Peter Warren excavated a Bronze Age site occupied between 2600 BC and 2000 BC at Myrtos in southeastern Crete. The people lived in a single stone and mud-brick building of more than ninety small living and working cells, tiny open spaces and passages, a veritable hive of humans. As the excavation proceeded and potter's wheels and loom weights accumulated, it became apparent that the Myrtos settlement was *a community of cottage industries. Pottery was thrown on the slow, hand-turned wheels and cloth was woven, dyed and fulled, which implies trade, the gathering of surplus wealth, and the rude beginnings of the magnificent civilisation that was to come.* If the weavers of Vrysi used a primitive ground-loom, the warp-weighted loom was used at Myrtos.

MESOPOTAMIA

3

"Open sesame!"

Thousand and One Nights: Ali Baba and the Forty Thieves

From sesame seed to saponification

3i

ALONG the Tigris and the Euphrates there is no fuller's earth of high quality. It is possible that fleeces and then cloth were fulled in the river mud in the earliest times, but in recorded history we see that other detergent materials had come into use. Indeed the bleacher or fuller took his name from the soapwort plant. This was called *azalag* from *mê šelag*, a solution of soapwort. He also used wood ashes (*nimur, tumru*) or soda when washing in a large tub (*gakkul, namzîtu*) in which he stirred the contents with a staff or pole, since mineral alkali would have damaged the skin of the feet. Finally he used a special bat (*mazûru*) to beat the cloth.

R.J. Forbes, *Studies in Ancient Technology*, Leiden: E.J. Brill (1956) *4*.

Mesopotamia and northern Syria were ancient centres of the woollen industry, unlike Egypt which produced linen of excellent quality; and there was no religious ban on the wearing of woollens. The city-state civilisation of the Sumerians was the oldest in Mesopotamia. They were responsible for the first syllabic method of writing, which emerged some time before 3000 BC, though I suspect numeracy preceded literacy to permit the buying and selling of goods. Anyway, by the time of its zenith, between 2100 and 1950 BC, the language was of course centuries old. By that time the fuller's craft was firmly established and was already highly developed.

In Mesopotamia different norms of taste prevailed from those in Egypt. In art and life, the Egyptian loved above all clarity of form, long retaining the simple waist-cloth which did not obscure the line of the body, and, when he did cover himself to a greater extent, he preferred for the same reason linen materials of fine texture. The Babylonian was a lover of pomp and luxury and in Mesopotamia the loin-cloth began not later than 2800 BC to be discarded in favour of clothing of heavy stuffs, in time richly patterned and elaborately fringed with looped woollen threads.

In Akkadian, fulling was called either *mazarū* or *kabāsu*. The former is named from the fulling-billet *mazûru* and the latter in its primary meaning is to trample; these words may represent fulling by hand and feet respectively.

About 2350 BC the Semitic inhabitants of the more northerly parts of Mesopotamia reduced the cities of Sumeria and introduced the Akkadian language, which became the official medium of the Babylonian empire.

Recipes for the boiling of oils with alkali occur in laundry accounts, the resultant soap was not separated but used directly for washing, scouring or fulling. The recipes prescribing olive oil and potash in various proportions are found in accounts dating as far back as Gudea – about 2100 BC – in Sumerian Lagash.

In the lexicons the Assyrians compiled around the 7th century BC for the instruction of their scribes in the old Sumerian language their own had for general purposes superseded, the earliest fulling vocabulary has been preserved. They note *azalag*, *zălag* and *mudru*, all meaning *washer* or *fuller*, and *dim-tur*, a *fulling-billet* or *beetle*. The azalag was apparently a slave attached to the temple whose charge it was to clean the priestly robes, a task he performed with detergent and the beetle.

Hammurabi ~1728 – ~1686 BC.

In Nineveh the launderers and fullers lived in a special quarter of the town.

Hammurabi, founder of the Babylonian Empire, conquered the country of Sumer and therefore of the Akkadians, from his power bases of Assur and Nineveh. About 1750 BC fulling had become a technique with equipment of its own and a distinct and specialised full-time trade. It had become a factory occupation with statutorily regulated wages. In the fulling trade the recognised grades were apprentice, guild craftsman and master-fuller.

The famous Hammurabi Code, incised on a stele of diorite, discovered in 1902 and now in the Louvre, is not wholly readable. The wage records have been effaced, and the only rate now legible is that paid to joiners. This was four grains of silver a day, equivalent to one sixth of a gram. It cost from six to fifteen shekels to buy a talent (about sixty pounds) of wool; a robe might cost anything from one third to five sixths of a shekel. Assuming that a fuller received similar wages to those of a joiner, it would cost him from three to nine weeks' wages to buy a cloak, and take from one to three years, working continuously seven days a week, to accumulate enough money to buy a talent of wool. Clearly few journeymen in the Babylonia of this period could hope to set up as masters – unless perhaps by marrying an employer's daughter, something not easily achieved in oriental society, as Jacob found out. Inscriptions of a considerably later period (6th century BC) record systematic apprenticeship regulations, that must have provided at least some sort of ladder to be climbed. Fullers had to serve two years (as against a requirement of five years for weavers – due probably to the great variety and richness of the textiles Babylonia was by now famous for – and only fifteen months, plus nine months' probation, for a baker).

Genesis, 29, 16.

Meissner, *Babylonien und Assyrien*, (1920–1925), translated from the German.

In effect the 282 clauses of the Hammurabi Code were intended to regulate every sphere of human activity. They describe an organisation on a national basis, with employers' guilds, each with its guild-steward entrusted to negotiate with the state tax-gatherers, and production in what were really factories; here the workers – fullers, spinners, weavers, dyers – who were both freemen and slaves, received, except of course the slaves, wages that were regulated by statute. From cuneiform inscriptions from Asia Minor it is clear also that before 2500 BC. Assur, later the centre of the great Assyrian empire, maintained trading centres there. These were virtually independent colonies, administering their own affairs and holding their own courts of justice and arbitration. The courts fixed rates of interest, financed business undertakings, and generally acted as the bank of the colony. The whole organisation bears a striking resemblance to that of the foreign stations of the Hanseatic League in the Middle Ages, even to the point that the principal source of income was here likewise the wool trade.

None of the written or pictorial records of the older civilisations in Mesopotamia illustrate the details of the fuller's craft. A piecing-together of archaeological evidence, however, enabled Meissner to present a fairly comprehensive picture of Assyrian fulling operations in about the 7th century BC.

loc. cit.

> *When the woollen fabric has been woven, the walker (qāsiru) receives it and tramples it thoroughly in the fulling-pit, so that the threads become felted together. He then beats it with a club and raises the nap with teasels. This club is the chief tool of the fuller (as̆laku), who remorselessly beats soiled linen and woollen clothes with it, after the main part of the dirt has been removed in the washing trough by means of soap, or a mixture of potash, soda lye and alum. The hot sun of Babylon then does the rest, bleaching the fabrics beautifully white and clean.*

Raises the nap... makes the surface soft and woolly with the dried, prickly heads of the Dipsacus plant (Fig. 17).

According to Forbes *some kind of soap was manufactured, certainly as early as the UR III period, when the Sumerians boiled oil and alkali together. Various recipes mention the use of palm oil, castor oil, potash, soda, resin and salt to compound this soap. Still soap as such was not separated from the mixture and we have no special term for soap. In washing wool the use of beech-ash is strongly recommended and textiles in general are known to have been washed with "soap", soda, vegetable alkali and alum. In some cases the lye is obtained by burning rue...* In fact, many detergents were in use.

loc. cit.

c. 2250 BC.

More clearly put, the Sumerians discovered *saponification*, i.e. the reaction between alkali and oil or fat.

Martin Levey of Temple University, Philadelphia, Pa., describes the various oil processing methods used in Mesopotamia. Filtered oil was considered clean enough for edible purposes since it was used in the preparation of food dishes. Oil of first pressing and oil of the best quality were distinguished. Sesame oil is useful in soapmaking but preferably used as an edible oil. *An account of the first millenium BC tells us that ... 12 qa of sodium carbonate (in the form of aḫulla ash), 6 qa of cypress, 6 qa of sesame (probably the seeds) were for the washing of the stones for the servant girls.* [I should understand this passage if it read "on the stones" – where clothes were washed]. A "qa" was a dry measure.

Technology of Oils, Fats and Waxes in Ancient Mesopotamia, M. Levey, *Centaurus*, (1957), *5* (2) 151–163.

quoting R.P Dougherty, Archives from Erech, time of Nebuchad-nezzar and Nobonidus, Newhaven, (1923), 212: 1–3.

Another text implies that oil was used in a laundry and that it was heated for washing purposes, probably with soda, to make soap. This was on one of the thousands of cuneiform inscriptions found in Nippur in north Sumeria.

and quoting A.T. Clay, Documents from the Temple Archives of Nippur, *Phila.*, (1906), 148:50.

3i

V. Scheil, *Revue d'Assyriologie*, (1921), *18*, 9.

Chemical, Culinary and Cosmetic Arts, R. J. Forbes, (1954) in Singer *et al.*'s *A History of Technology*, *1*, 261.

Glucosidal compounds of steroid structure.

στρουθίον

Medicinal soap was also made thus: *fat from a sheep's kidney, soda ash, sweet smelling plant ḪAL . . . thou shalt cook. Thou shalt apply to his seat, he shall be cured.*

Ashes from two common species of glasswort (*Chenopodiaceae*), rich in soda, are alluded to in ancient Mesopotamian texts as the alkaline component in many glass-recipes, writes Forbes (1954). In some cases there is added to their name in the text the adjective *ḳalâti*, meaning burnt, which is the ultimate root of our term *alkali*; the immediate derivation being from the Arabic al-qālī, the plant ash.

It was not always the ashes of plants which were used by fullers but also the saponaceous juices of plants, containing saponins; the plants particularly used were soapwort (*Saponaria officinalis* L.) and *Gypsophila struthium*. L. Pliny speaks of *radicula*, which he says furnishes "a juice extensively used in washing", and describes as being found in best quality in Asia and Syria and "beyond the Euphrates", and as having been called by the Greeks *struthion*.

3ii

EGYPT

Richard A. Caminos, *Late Egyptian Miscellanies*, London: Oxford University Press, (1954).

Perchance Amun (10, 4) will come in peace, the sweet breeze before him. 'May he cause me to become a vulture's wing like a (fully) manned mst'i-boat: so say the herdsmen, in the field, the fullers on the river bank, and the Medjay that are coming out from the district, and the gazelles upon the desert.

Papyrus Anastasi, IV, 10,4 to 10,5

Papyrus Anastasi, IV, 10, 5.

Pap. D'Orbiney, 10.9; 11.1.

THIS is from a story which has come down to us in four papyri of the twelfth Egyptian dynasty (c. 2000–1780 BC). The phrase we are interested in is

the washermen are on the river bank and again, in the same story occurs the phrase:

the washermen of Pharoah are on the river bank. (*Fuller* was here represented by the hieroglyphic group , or, in transliteration, rḫty).

That this was customarily their work-place comes out very clearly in this *Tale of Two Brothers*:

A. Erman, *The literature of the Ancient Egyptians*, trans. A. M. Blackman, London: Methuen, (1927), 150.

And many days after this Bata went to hunt, according to his daily wont. Then went the maiden forth to walk about under the cedar which was beside her house. And the sea saw her and swept up behind her, and she started to run from before it, and went into the house. But the sea called to the cedar, saying: "Catch hold of her for me". And the cedar took away a lock of her hair, and the sea brought it to Egypt, and laid it in the place in which were the fullers of Pharaoh. . . . And the smell of the lock of hair got into the garments of Pharaoh, and One [*Pharaoh*] *contended with them daily, and they wist not what they should do. And the chief fuller of Pharaoh went to the river-bank and was sore vexed because of these daily contentions. And he stood still upon the sand, and took his stand over against the lock of hair that was in the water. And he caused someone to go down* [into the water], *and it was brought unto him; and there was found an exceedingly sweet smell. And he took it to Pharaoh.*

As Forbes says *the common Egyptian detergents were natron, fuller's earth, and perhaps pounded lupins, the last of which is still used.* Elsewhere he includes potash, soapwort (*Saponaria officinalis*), asphodel and other detergent plants. By far the commonest detergent was *natron* (sodium carbonate), which came from three sources, (1) the Wadi Natrun in the western Desert, (2) near the ancient merchant port of Naukratis in the Delta, and (3) a group of five localities near El-Kab in upper Egypt. The natron, essentially sodium carbonate, forms as crusts and layers especially where the desert is annually flooded. The natron from all these sources was an important state monopoly in Ptolemaic times (from ca. 320 BC.)

In Singer *1*, 261.

Forbes, *op. cit. 4*.

From Strabo, XVII. 1, 23, cap. 803, and Pliny, *Nat. Hist.*, 31 , 109–111.

Forbes conjectures that natron, perhaps mixed with clay, was used as soap. A "black cotton soil", if freed from gritty impurities, would have served this purpose, since the main clay mineral is montmorillonite, although a modern survey of *Ancient Egyptian Materials and Industries*, by Dr. A. Lucas, Chemist in the Department of Antiquities in Egypt, and presumably in a position to make detailed investigations, makes no mention of fuller's earth. Nor is any occurrence of it recorded in Hume's *Geology of Egypt*.

In Singer *1*, 260.

London: Edward Arnold, 2nd ed., (1934).

Nevertheless, Nile mud was rich enough in montmorillonite to serve not only as a fuller's earth but also as a hair detergent. It was often fashionable in ancient Egypt to wear the hair long, often in a plait down the back, and to oil it heavily – hence the perfume of the lock of hair in the story quoted.

Early accounts suggested that Nile mud is mainly illitic but now it is known that illite, the mica-like clay mineral, has been developed only at depth. The clay which comes from the White Nile and Blue Nile is essentially montmorillonite.

H. Handi & R. Iberg, *Z. PflErnährung-Düng.*, (1954), *67*, 193–197.

A. H. Weir, E. C. Ormerod, and I. M. I. El Mansey, *Clay Minerals* (1975), *10*, 369.

"Soap" was something the Mesopotamians were alone in using in pre-classical antiquity – indeed in pre-medieval times – yet notwithstanding the cleanliness of the ancient Egyptians, and the frequent ritual ablutions prescribed for their priests, there is no sign that soap was known to them. An Old Kingdom picture shows a square object lying by bowl, jug and towel which has often been taken to be a cake of soap, but is more probably one of natron or natron mixed with clay. Certain medical recipes indicate the boiling of oils and fats with large particles of alkaline substances, but the soap thus formed is never mentioned as such, or so far as is known used for cleansing purposes, and hieroglyph, demotic and Coptic words taken to mean "soap" are all doubtful. There is a similar absence of solid soap from the record in ancient Palestine.

Grace M. Crowfoot in Singer, (1954), *1*, 432.

A horizontal loom is rather well depicted on the inside of a predynastic dish from Badari about 4400 BC. There were large flocks of sheep from the earliest times in the Nile Valley, so it is not improbable that cloth was made of wool as well as of linen. Eventually woollen clothing was condemned by the priests as unclean, whereas textiles from flax were declared by them to be the first thing the gods created in this world before moving in to live in it. Flax is smelly when being retted but at least it requires no scouring and felting.

The hot climate, the Egyptian custom of anointing the hair, and introduction of the fashion of a greater covering than the loin-cloth, all contributed to the importance of the launderer, and fulling meant in early Egypt almost exclusively laundering.

How much of it was ever done domestically, and if so for how long, it is impossible to say. Probably by the time of the MSS we have quoted washing was for the most part "put out" to people who made of it their trade. There is a poem by an Egyptian lover that runs:

W. Max Müller, *Die Liebespoesie der Alten Ägypter*, Leipzig: J. C. Heinrich, (1899), 43.

Would that I might be my fair one's launderer, if but for a single month! Then could I wash the sweet oil from her headcloth – A labour of love would it be: that would I do without pay.

George Steindorff and Keith C. Seele, *When Egypt ruled the East*, Chicago: Univ. Chicago Press, (1942), 125.

It was an even stronger profession of affection if, as seems to have been sometimes the case, the fuller's occupation was held generally to be a contemptible or even a noisome one.

In the tale from which the first two quotations of this chapter were taken occurs the contemptuous dismissal:

The story of the eloquent peasant, lines 168-9.

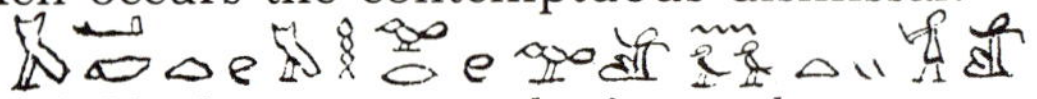

Behold, thou art a wretch of a washerman

This was not, however, invariably the case. A stele again from the 12th-dynasty commemorates the Lady Khu and her two husbands, one of whom had been a fuller. Beside the inscriptions (prayers for funeral offerings at various feasts) are pictures of the lady and her husbands and children bringing gifts – Sa-Hathor, the fuller, has the traditional fuller's bundle of clothes on head and cloth in hand. Here the fuller is a member of the upper classes, proud of what must be considered a dignified occupation.

Fig. 3 Sa-Hathor, the fuller. Hieroglyphic texts from Egyptian stelae, etc., in the British Museum, (1912), Plate 16.

Possibly Sa-Hathor was what we should term a "master-fuller", with a number of employees. Little is known of the craft organisation in the early period in Egypt. By the fact that fulling was primarily a launderer's job it may have tended to be less "regulated" than documentary evidence shows it to have been in Babylonia. Towards the end of the pre-Christian era, however, in the Ptolemaic period, fulleries, except in the less urban districts, were owned and controlled by the State.

Of actual fulling processes there is no description in the literature – for the Egyptians, as for others, laundering at least was so much a matter of course that occasion for detailed account scarcely arose. A 12th dynasty painting however shows fullers at work (Fig. 4).

Fig. 4. Fullers using sloping washing boards 12th century BC.

And in the tomb of Khnemhetep at Beni Hasan, of the same period, murals show a series of fulling operations (Fig. 5).

Fig. 5. Men wringing out wet cloth; men at work in a fulling trough; a fulling beetle; wet material being handed out of the trough; and a fuller tying up a bundle of clothes such as Sa-Hathor carries on the stele – all these operations are common to laundering and cloth-making.

The Papyrus Ebers, the greatest Egyptian medical document, refers to a waste product of the fullers.

In one preparation, an ingredient, 

translated as fuller's oil (or grease of fullers), seems to have been what we now call lanoline. The passage reads *another* [*medicament*] *to soften the members: fragrant myrrh|sawdust of pine|powder of green corn|fuller's oil|are boiled together and* [*it*] *is bandaged therewith for four days*. The "lanoline" acted as an emollient in this plaster. At that time it is possible that the wool grease was yielded by urine-scouring because this technique forms a much weaker emulsion than when soap or fuller's earth are used. The wool wax separates more easily, so it can be skimmed, washed and sun-bleached; mixed with water the product would have been lanoline.

W. Smith *Dictionary of the Bible*, (1863).

Wilkinson, *Manners and customs of the Ancient Egyptians*, 3rd ed. (1847), *3*, 162.

P.E. Newberry, *Beni Hasan*, London: Kegan Paul, (1893) pl. 29. Tomb No. 3, main chamber, west wall, left hand side, top row.

For photographs, see Luise Klebs, Die Reliefs und Malereien des mittleren Reiches, *Abhandlungen der Heidelberger Akademie der Wissenschaften*, Heidelberg: Caxe Winters Universitäts-Buchhandlung, (1922), *122*, 123.

Papyrus Ebers, translated by B. Ebbell, Copenhagen: Levin and Munksgaard, (1937), 99 The world's oldest *book*, written 1553–1550 BC from much older source material.

Wool Wax, Chemistry and Technology, E. Vernon Truter, London, Cleaver-Hume Press (1956), 110.

3ii

Pap. Ebers, ciii, p. 119.

In another preparation

"dirty water of washermen" or "washermen's slops" is specified; – which sounds like suint; also

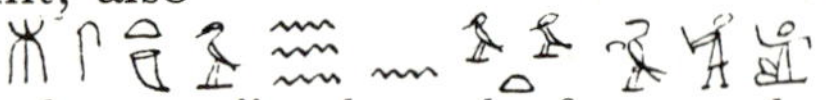

"refuse of fullers or washermen", where the first word means "refuse", "lees", "dross" or similar rather than solid waste, suitable as an emollient. *If he is ill in his nape and his eyes are dim sighted... what is to be done against it: myrtle, washerman's slops, pignon, fruit of š3mś, mixed with honey, are applied to his nape and bandaged there for four days.* This section is said to have been written as long ago as the 1st Dynasty in the time of Usaphais, the Justified.

pignon, the edible nut of *Pinus pinea*, "pine nuts".

3100–2890 BC.

Natron, as stated, is a naturally occurring form of impure sodium carbonate. The word is a 17th century European variant of the Arabic natrūn, and from it came *natrium* and hence our symbol for sodium, Na.

It has also been suggested that 'natron' is derived from the Hebrew *natara*, to bubble, with reference to the effervescence caused when vinegar is added to soda.

1567–1320 BC.

2613–2494 BC.

Natron has been found in vases dating from the 18th dynasty, and Lucas considers it probable that it was used for embalming royalty in the time of the 4th. Its use largely eclipsed that of common salt, even as a preservative, and it was used in the manufacture of glass, glazes and the blue and green frits used in pigments, and for bleaching linen. The importance of the place it held as a cleansing agent is clear from the fact that the Egyptian term for the crude product came to mean "purify oneself", and the name given to it in its purified form "nṯrj" (Greek *nitron* or *litron*, Latin *nitrum*) bears a close relation to the words "*ntr*" ("god", viz. "the pure one") and "*sntr*" ("incense"). Nitrum was burned with incense to the gods. So intimately are the concepts purification and natron linked, indeed, that not only do names for qualities of natron for purification purposes come to mean "purify with natron or incense", but the reverse happened, in the case of a term meaning originally "to purify" but later used for "natron". The believer would purify his mouth either by chewing natron or by rinsing it with a solution of natron often referred to as natron lye.

And Akkadian "nitiru".

R. J. Forbes, *Studies in ancient technology*, Leiden: E. J. Brill, (1956), *3*.

Certainly with natron so abundantly and conveniently available – it could be obtained in either granular or lump form – there was little incentive to look further for detergents.

A relaxation is apparent in the Egyptian attitude to the wearing of wool by the 5th century BC, though as Herodotus noted *nothing of wool is brought into temples, or buried with them; that is forbidden.* A loose woollen mantle at least was customarily worn over the tunic – and with this would have come of course increase in felting. But natron, together probably with urine, doubtless remained the main cloth-fulling agent. The main use of earth likely in Egypt in fulling or laundering operations may have proceeded from possible import by the Greeks, who

Herodotus, 2, 81.

were settling in the Delta from the 7th century BC, of their cimolian fuller's earth.

From the Egyptian word for fuller or washerman, rḫty, is derived the Coptic ⲣⲁϩⲧ̄, ⲣⲱϩⲉ etc., meaning *wash*. The word occurs of washing clothes several times in the Old Testament.

e.g. Exodus 19. 10, Leviticus 11.25.

In the Coptic Medical Papyrus of Cairo (line 274) an ointment for wounds consists of hog's grease and ⲟⲙⲉ ⲛ̄ⲣⲁϩ̄ⲧ, which literally means *washing mud* or *clay*, – almost certainly fuller's earth.

In the well-known passage in the Gospel of St Mark, where the Greek has γναφεύς, the Coptic has ⲣⲁϩ̄ⲧⲛ̄ⲧⲱϭ̄ⲥ, literally "tread washer" from the verb ⲧⲱϭ̄ⲥ "to tread."

Mark 9,3
The stroke stands for E, commonly omitted in writing.

There is another Coptic word for fuller's earth, ⲁⲗⲓϫⲓ, which is the same as the Arabic طفل. It is later in date than the classical authors. It occurs in the Scalae or glossaries of Coptic words with their Arabic equivalents which were drawn up when Arabic was becoming the spoken language of the country.

Dyeing

FULLER'S earth was also used by dyers. *Wool was often first dyed blue with indigo and then dyed again with some other dye like madder; the result yielded a purplish colour.* A recipe for this re-dyeing with madder can be found in the earliest alchemical papyrus. Ashes are scattered over the indigo-dyed wool, and then trodden with the feet. The wool is then rinsed in water and clay (fuller's earth) and pickled in alum. Finally the madder is dissolved, together with some flour of beans (to make the water softer), and the treated wool dyed for the second time. Fuller's earth has been used in the present century for preventing indigo-dyed serges from bleeding.

R. J. Forbes in *A History of Technology*, Oxford: Clarendon Press, (1954), *1*, 249.

This 3rd c. AD papyrus is quoting records of the 3rd c. BC and probably much earlier.

Cosmetics

FULLER'S earth is not particularly good as a face-powder except in combination with other ingredients – lacking the whiteness of kaolin and, where whiteness is not sought after, having in any case little covering power and too drying an effect on all but very oily skins. An Egyptian beauty of 1500 BC, however, had, from the inscriptions on a carved powder box from the temple of Queen Hatshepant, four different face-powders, one for daily use, one for hot weather, one for winter, and one for spring – fuller's earth as an ingredient might have had its advantages for, say, the hot season.

3iii

Switzerland and Scandinavia

The first woollen garments to have survived

SINCE wool is more perishable than linen, prehistoric samples of woven woollen cloth have rarely been preserved. Thus in the third millenium BC the Swiss lake-dwellers were producing linen fabrics of quite complex weave and colour patterning. The first signs of woollen fabrics in Europe come with the European Bronze Age – indeed whole articles of woollen clothing have survived in Scandinavia. No doubt woollen cloth was made much earlier in many northern areas.

C. Singer *et al*, *A History of Technology*, (1954) *1* 442.

At Treenhøi, near Ribe, in Jutland, the entire dress of a Bronze Age chieftain has been discovered. The body was wrapped in a coarse woollen cloak, with a plush-like inner surface of fleecy fibres. Shawls, about five feet by four, and with long fringes, covered the feet and supported the head. Other garments comprised a woollen shirt, with a hole in the top for the neck, fastened at the waist by a long woollen band; a thick, hemispherical woollen cap about six inches high; and a pair of small woollen garments that probably served as leggings.

id, plate 17A

Thomas Froncek, *The Northmen*, (1974) Time Inc., p. 104.

At Egtved, also in Denmark, there have come to light from this period a girl's woollen jacket or jumper, and a woollen skirt, found together. The jacket, like the chieftain's shirt from Jutland, was made from a strip of woven material with a minimum of cutting. The skirt is of a type found nowhere else in the world, and at no other time in Scandinavia except during the warm, sunny centuries of the Bronze Age. This provocative miniskirt consists of a continuous thread of wool passed up and down in large loops over a band at the top – the effect is curiously like that of a grass skirt – with the loops tied at the bottom into metal rings or tubes and lashed by a row of twining. The garment was worn low on the hips and ended at mid thigh. Between the top of the skirt and the bottom of the tunic the mid-riff was bare, Indian fashion. The date of these garments was 1500 BC.

The wool at this time came from sheep of a primitive race, the wool containing long, coarse hairs. The cloth is quite thick in texture and has therefore been fulled.

This type of pile fabric is also found, with pile tufts run in with needles after weaving. In this case there was less need for thickening by fulling.

Felt hats in the Scandinavian Bronze Age

FELT is made by beating, rolling and pressing animal hair or flocks of wool into a compact mass of even consistency. The art is older than weaving and came into Europe from the Slavs, who in turn had it from Scythian-Siberian-Turkish people, in the late Bronze Age. Felt is essential to the nomads of northern and central Asia, in Europe it was often used for hats, but *Caesar's soldiers made themselves tunics of felt or raw hide and in some cases quilted jackets, as a protection against Pompey's archers.*

R.J. Forbes *Studies in Ancient Technology*, **4**, (1956), E.J. Brill: Leyden.

Pliny confirms that *self-felted fleeces make clothing, and also if vinegar is added withstand even steel*, and even fire. It is known now that the felting action is enhanced by acid solutions.

Caesar, *Bellum civile*, III, 44; Pliny, *Nat. Hist.* 8, 1920.

The classical world clearly derived its knowledge of felting from the nomads of the northern steppes.

Forbes, *loc. cit.*

4 PALESTINE

And hise clothis weren maad ful schyning and whight as snowe, whiche maner whighte clothis a fuller may not make on erthe
Wyclif, Marc 9.3 (Later version, c. 1396).

And his raiment became shining, exceeding white as snow; so as no fuller on earth can white them.
Authorised Version, (1611), Mark 9.3.

C. Singer, *1*, 432–3.

A. Mallon, R. Koeppel, and R. Neuville, 'Teleilāt Ghassūl', Scripta Pontificii Instituti Biblici, Rome, (1934).

G. Loud, 'Megiddo', Vol. 2, Univ. of Chicago, Orient. Inst.; Publ. 62, Chicago (1948).

Ghassūl, 19km ESE of Jericho, in Peraea, at the NW corner of the Dead Sea.

G.T. Faust, J.C. Hathaway and G. Millot, *Amer. Mineral.*, (1959), *44*, 342–370.

V.J. Linhart, *Scotsman*, (1982).

Proverbs, 25, 20

i.e. sodium carbonate

BEFORE we reflect on these moving words on the Transfiguration, we shall try to trace the history of fulling in Palestine, beginning with some archaeological finds at Ghassūl. Several fragments of plant-fibre textile were found at Teleilāt Ghassūl; they were carbonised and unidentifiable. On this site dating from ~3000 BC were some interesting weaving tools, including a needle-shuttle, a possible warp-spacer in bone, and a number of rather clumsy stone and pottery spindle-whorls. Needle-shuttles of this period were also found in an early cave at Megiddo.

Barrois states in his *Manuel d'Archéologie Biblique* that the ancient Palestinian fullers used a mass of fine earth and ashes (*amas de terre fine et de cendres*) as found in the excavations directed by the Papal Biblical Institute at Teleilāt Ghassūl, and there are recent reports of clay being discovered here. The name of the place is a literal translation of the Arabic, meaning the "mounds of ghassūl", and ghassūl derives from the verb meaning "to wash or cleanse". Indeed the word is used by Arabs in North Africa to denote fuller's earth proper, that is smectitic clays, to be found there. The ghassoul from the Djebel Ghassoul mine in Ksabi province, Morocco, is the rare variety called hectorite. In shops in Fez "mineral shampoo" can still be bought and has probably been on sale for centuries.

As "ghassūl" is also used in Iran to denote various "washing herbs" yielding barilla-type ashes rich in sodium carbonate, we cannot say what detergents were used in these early times in Palestine, though natron was familiar in Biblical times. The phrase *as vinegar upon nitre, so is he that singeth songs to a heavy heart* makes little sense since vinegar does not noticeably affect nitre (potassium nitrate, or saltpetre). It was not until Boyle in 1680 obtained a sample of "*Egyptian* nitre" and found by experiment that it effervesced readily with acids that the misunderstanding was made clear.

In Jeremiah the Authorised Version reads *For though thou wash thee with nitre, and take thee much soap, yet thine iniquity is marked before me, saith the Lord GOD*. Here again nitre should read natron and soap was probably one of the plant ashes rich in soda, such as borith, obtained from the soda plant, *Salsola kali*. Here and in other Near Eastern countries early use was probably made of another species *Salsola soda*, found in Syria and the Levant, ashes of which were later imported into Europe for soap-making as "polverine" or *rochette* or, when obtained from Spain, *barilla*. *Salsola* ash, unlike other plant or wood ash which is potassic, contains sodium compounds and is richer in alkali.

Jeremiah, 2, 22

See 'The Trade in Ashes and Kelp, Ch. III in *The Chemical Revolution* A. and N.L. Clow, London: Batchworth Press, (1952), 65 et seq.

Even today no substantial deposits of calcium montmorillonite have been discovered in Israel, this commodity being imported from Cyprus and elsewhere. Apart from the mineral and vegetable alkalis, stale urine and saponaceous plants were used.

The complex provisions of the Mosaic law for ritual cleansing of what was regarded as "unclean" brought the fuller into closer contact with the priesthood even more than in Egypt. The purification wrought by fullers provided an analogy with the purification of the spirit:

> *But we are all as an unclean thing, and all our righteousnesses are as filthy rags.*

Isaiah, 64, 6

> *Now Joshua was clothed with filthy garments and stood before the angel. And He answered and spake unto those that stood before him saying, Take away the filthy garments from him. And unto him He said, Behold I have caused thine iniquity to pass from thee and I will clothe thee with change of raiment.*

Zechariah 3, 3

Of the capable wife it was written *she keeps herself busy making wool and linen cloth*.

Proverbs 31, 13 (Good News Bible)

The "Philosopher" who wrote Ecclesiastes advised people to work hard and to enjoy the gifts of God as much and as long as they could, – after all, *a living dog is better than a dead lion*. Later he writes *Go thy way, eat thy bread with joy, and drink the wine with a merry heart; for God accepteth thy works. Let thy garments be always white; and let thy head lack no ointment*.

Ecclesiastes, 9, 4

Ecclesiastes 9, 7, 8

The symbolism of whiteness runs throughout the Old Testament. In Daniel's vision we read *I beheld till the thrones were cast down, and the Ancient of days did sit, whose garment was white as snow, and the hair of his head was like pure wool*. And in the Revelation of St John: *I beheld, and lo a great multitude, which no man could number, of all nations, and kindreds, and peoples, and tongues, stood before the throne, and before the Lamb, clothed with white robes, and palms in their hands*.

Daniel 7,9

(one who had been living for ever)

This chapter begins with the description of the climax of the Transfiguration:

οἷα γναφεύς ἐπὶ τῆς γῆς οὐ δύναται λευκᾶναι

Mark 9, 3

In recent translations, the fuller himself is not referred to – *whiter than anyone in the world could wash them* conceals the craftsman.

Good News Bible Collins/Fontana 1976

There are in the Bible other references to the scouring agents used;

4

Malachi 3,2

מכבס = fuller

בֹּרִית = herba borith

John Eachard, . . . *the contempt of the clergy*. . . (1670), 56

שְׂדֵה כֹּבֵס
ἀγρός τοῦ γναφέως
ager fullonis

highway = an embanked road

The same story is told, with identical phrasing as to site, in Isaiah 36, 2.

Note: Josephus, B.J., 5. 4 §2, mentions a fuller's monument near the NE corner of the third wall.

II Kings 18, 27

– such as in Malachi: *But who shall abide the day of his coming? and who shall stand when he appeareth? for he is like a refiner's fire and like fuller's soap.* This was not soap, but borith, the ashes of the soda plant. It is not the mistranslation which engages our attention but the splendid metaphor, one of thousands which grace the books of the Bible. How much more dignified they are than John Eachard's *The blots of sin will be easily taken out by the soap of sorrow and the fuller's earth of contrition* (see also Ch. 14)

Whatever scouring agent was used, and I suspect that stale urine was one of them, the trade was noisomely smelly and the occupation of fullers, despite the excellent end-products, came to be looked down on. Fullers had therefore to ply their craft without the city walls.

The exact siting of the fulleries of Jerusalem has formed a point of departure for some intriguing speculation. There are several biblical references to a "fuller's field" which, if it is the same one in each case, would appear to have lain outside the walls of Jerusalem, though within earshot of them, near a watercourse, and on one of the main roads of access to the city. In Isaiah VII, 3: *Then said the Lord unto Isaiah, Go forth now to meet Ahaz. . . . at the end of the conduit of the upper pool in the highway of the fuller's field.* And when Eliakim was sent out by Hezekiah (II Kings XVIII, 17) to meet Rabshakeh at the head of an Assyrian host, the encounter took place *by the conduit of the upper pool, which is in the highway of the fuller's field*, and Eliakim in the course of it begged Rabshakeh not to conduct the conversation in the Jewish tongue since he could be heard by citizens on the wall. It is the direction in which the pool lay that is controversial. The only road by which a large body of troops could have approached the city was from the North – the route of the army of another Assyrian in Isaiah X – so that it might be taken to be in that direction. Early Hebrew fullers are also said, however, to have walked the cloth in the *En-rogel*, or fuller's fountain (a spacious trough in which men worked side by side), which lay south-east of the city, and it might well be that Rabshakeh and one or two immediate associates had left the army and advanced along the east side of Mount Moriah to En-rogel, to a convenient place under the temple walls for colloquy. If smell was an important factor the most likely site would be to leeward of the city – and in view of the necessity to avoid contaminating the water-supply it was certainly downstream.

Proof of the use of urine has been professedly adduced with the taunting words of Rabshakeh to Hezekiah's emissaries at the meeting near the fuller's field already cited. He refused to speak Syrian which the men on the walls might not understand: *Hath my master sent me to thy master, and to thee, to speak these words? Hath he not sent me to the men which sit on the wall, that they may eat their own dung and drink their own piss with you*? The men may have been sitting near a drain leading through the wall to the fullery: equally the taunt would seem in any case otherwise a natural one as addressed to ambassadors of a beleaguered city. Judgment would seem to rest with the Jerusalem sewage system.

Cimolian earth may well have been imported in Roman times; and certainly it is referred to in the Babylonian Talmud. This is a continuation of Jewish teachings after the Old Testament, dating mainly from the period 10–220 AD but edited in large measure by Rav Ashi (352–427 AD).

The prescribed quantity for wood is as much as suffices to cook an (easily boiled) egg; and the different spices are counted together; nut shells, pomegranate peel, isatis, and cochineal, as much as suffices to dye the edge of a small piece of cloth; lime, alkali, Cimolia chalk, vegetable soap, as much as suffices to wash the edge of a small piece of cloth. R. Jehuda says as much as will suffice to remove a blood stain. We have here some kind of quantification applied to raw materials and an early precursor of the launderometer of today.

Babylonian Talmud, Volume 1, Sabbath 9, 5, translated by M. Rodkinson and Isaac M. Wise, New York Book Co., (1896).

Mark 9.3 may also have inspired Paul Claudel to write (1935) *My habit of white that all my brethren from Paris and from Rome did defile with such defilement, nor fuller's earth nor soda can effect its cleansing.* So spoke Brother Dominic, a Dominican Priest, referring to the conduct of the trial of Joan of Arc.

Joan of Arc at the stake, Honegger's Dramatic Oratorio, Royal Albert Hall, 6 Dec. 1949. Play by Claudel written (1935), translated by Dennis Arundell.

5

CLASSICAL GREECE

R. Claiborne and R.D. Biggs, *The Birth of Writing* (1974), Time Inc. pp 22, 135.

THE Cretans, who began writing at some unknown date, were by the 17th century BC employing two scripts, called Linear A and Linear B. Michael Ventris, a linguistically gifted British architect, found that Linear B was an archaic form of Greek, while Linear A has not been translated. Many of the Linear B inscriptions were records of business transactions and palace inventories. Among them were numerous wool weight tallies.

In Classical times inventions were assigned to gods, heroes, mortals and the Cyclops, indiscriminately. Pliny was often non-committal, as in the first establishment of laws: *she (Ceres) was the first. . .; though, according to some, it was Rhadamanthus*, but with more honesty he confesses to defeat over the origin of alphabetic script. However he is more precise about the wool industry:

Pliny, *Natural History*, 7, 57.

Aegyptii textilia, inficere lanas Sardibus Lydi, fusos in lanificio Closter filius Arachnae, linum et retia Arachne, fulloniam artem Nicias Megarensis, ...

(The Egyptians invented weaving; the Lydians of Sardis the art of dyeing wool. Closter, the son of Arachne, invented the spindle for spinning wool; Arachne herself linen cloths and nets; Nicias of Megara, the art of fulling cloth. . . .)

Μέγαρα

Who Nicias was, or how Pliny heard of him, is unknown. It was a natural enough thing, however, for a Roman to think first of fulling as a Greek art – by Pliny's day most of the fullers in Italy were of Greek origin. And certainly it was the Greek initiative that brought effective cloth-production techniques into use in Europe.

It is a matter of general agreement that in the main the chemical technology of the classical world is little more than a perpetuation of the usage of the older civilisations from which it was derived – neither the Greeks, who first became conversant with it, nor the Romans, who learned it from them and later made use of their records, contributed much that was new. However, if the art of dyeing wool was made in Lydia, in Asia Minor, the Greeks attributed the art of fulling to Megara on the mainland. Classical authors do not refer at all to fuller's earth in Cyprus, but varieties of fuller's earth were known in Thessaly and at Bubon in Lycia, in Asia Minor, East of Rhodes.

Pliny. *Nat. Hist.* 35. 57.

Two factors, however, combined to make the Greek contribution to the fulling and finishing processes a more positive one. Interest was concentrated to a much greater degree than in any of the earlier civilisations on manufacture of cloth from wool – Egypt had won renown for fabrics of linen, India, of cotton, China, of silk, and Mesopotamia for wool textiles among others: it was the Greeks, and after them the Romans, for climatic reasons so much more dependent on wool as their native raw material, who were to raise production of woollen textiles to comparable standards. And one of the most important means which enabled them to do this was the fuller's earth discovered in Greece, which proved to be superior to any fulling agent elsewhere and hitherto in use.

Discovery of this Kimolian earth – so-called after the Aegean island of its origin, north east of Melos – led to the change from alkalis to earth as the main scouring and fulling agent, and was almost certainly responsible for a major innovation in the processes of cloth production. Kimolian earth became the best known detergent of antiquity, from one end of the Mediterranean to the other. Its penetration of the markets over so wide an area led to the search for local substitutes and smoothed the way for the cloth industry of a large sector of Europe as a whole throughout the Middle Ages and beyond. How far back in time traffic in Kimolian earth went is unknown, but Homer (c. 700 BC), recording events of ca. 1250 BC, describes only the use of hot and cold water and specially-made fulling troughs.

Homer, *Iliad*, xxii, 145–156

οἱ δὲ παρὰ σκοπιὴν καὶ ἐρινεὸν ἠνεμόεντα
τείχεος αἰὲν ὑπὲκ κατ' ἀμαξιτὸν ἐσσεύοντο,
κρουνὼ δ' ἵκανον καλλιρρόω· ἔνθα δὲ πηγαὶ
δοιαὶ ἀναΐσσουσι Σκαμάνδρου δινήεντος.
ἡ μὲν γὰρ θ' ὕδατι λιαρῷ ῥέει, ἀμφὶ δὲ καπνὸς
γίγνεται ἐξ αὐτῆς ὡς εἰ πυρὸς αἰθομένοιο·
ἡ δ' ἑτέρη θέρεϊ προρέει ἐϊκυῖα χαλάζῃ,
ἢ χιόνι ψυχρῇ, ἢ ἐξ ὕδατος κρυστάλλῳ.
ἔνθα δ' ἐπ' αὐτάων πλυνοὶ εὐρέες ἐγγὺς ἔασι
καλοί, λαΐνεοι, ὅθι εἵματα σιγαλόεντα
πλύνεσκον Τρώων ἄλοχοι καλαί τε θύγατρες
το πρὶν ἐπ' εἰρήνης, πρὶν ἐλθεῖν υἷας Ἀχαιῶν.

Then they [*Hector, with Achilles in pursuit*] *rushed by the look-out place and the wind-waving fig-tree, always under the wall along the public way, and reached the two fair-flowing springs, which rise to feed deep eddying Skamandros. The one, indeed, flows with warm water, steam arising from it as of burning fire; whilst the other flows forth even in summer time, like unto hail, or cold snow, or ice from water. There, beside them, are wide, handsome stone basins, where the wives and fair daughters of the Trojans used to wash their splendid garments, in the old days of peace, before the sons of the Achaeians came.*

Translation based on that of T.A. Buckley, Bohn's edition, (1857), p. 406.

Homer refers elsewhere to spinning and weaving, but the description of the washing-troughs suggests that at least some of the clothes were woollen.

πλυνοὶ

Long after the Homeric Dark Age, when the city-states were emerging, the Greek cloth industry was in full swing. Achaeans colonised Sybaris in the south of Italy and exploited the sheep-rearing ability of the natives and for over two centuries amassed such wealth that their life of luxury became proverbial; but when in 510 BC their neighbours turned on them and massacred them, the mourning-clothes donned back in Ionia by the Milesians were not only for their Greek kinsmen but for the loss of a prosperous industry which depended in large measure on supplies of wool from Sybaris.

The more successful settlement at Marseilles is described in ch. 7i,

Naturally, urine and vegetable alkalis would have been available; and various techniques of scouring and fulling wool may have been practised at one time, especially since fullers in the Greek cities tended to be foreigners. The Athenians in particular took pride in the welcome given by their city to strangers and were especially tolerant in their treatment of μέτοικοι, or resident aliens. These people were attracted in large numbers to the growing woollen industry, the women and girls finding employment in spinning, weaving, sewing and embroidery, which were regarded as superior occupations, since the arts of Athene were still cultivated in the home by the most aristocratic ladies, and the men taking readily to the various processes of fulling, which was a craft reserved, for a time at least, for free residents alone. Those among the latter who had learned their trade in eastern countries would probably exert to some extent a conservative influence, preferring the fulling agents they had been used to.

metics

It would seem that white breeds of sheep had been firmly established already in Homeric times. Homer uses the word ἀργεννός, white, almost always of sheep, as in the phrase ἐπ… ἀργεννῇς ὀΐεσσι, amid their snowy sheep. Woollen clothes were always white except when they were dyed black for mourning.

Iliad vi, 424

Greek dress

A FEW words may be given here about Greek dress. Wool was the most common fibre in most ancient times and may have remained so among the lower orders of society in classical times. For the privileged classes the peplos or outer garment was made of wool, but linen underclothes were worn in Athens, often of a fine, crinkly and clinging textile. In Sparta the peplos was the only garment worn by young women; in their Dorian fashion the cloth, folded at one end to form a bib, was sewn together or pinned above the right shoulder, so that the legs and back were revealed on one side. The style is well shown in the drawings of a *bacchante with crotals, dancing, showing the opening at the side of the garment*, but the Spartan girls were allowed no embroidery. The lack of protection inured the wearers to cold, – part of the military customs of the Dorian upper class.

Fig. 5 πέπλος

small, spherical bells

Ancient Greek Female Costume, J. Moyr Smith, London: Sampson Low, *et al.*, (1882), pl.19.

The Ionic chiton, used in Athens differed from the peplos in several ways; it was often of linen, but always sewn up along the right side and

Fig. 6 χιτών

Fig. 6. Peplos.
A plain variety of this garment was used in Sparta.

Fig. 7. Chiton.
As worn in Athens.

J. Moyr Smith, *Ancient Greek Female Costume*. London Sampson, Low, Marston, Seale & Rivington (1882), plates 18 and 19.

5 *Id.*, pl. 18
διπλοΐδιον
from διπλοΐς
a double cloak

hanging from both shoulders, as in the picture of *a girl with chiton made of one large piece of cloth*; it is embroided on the upper or diploidion part looking like a bib.

Fig. 8i How the Doric peplos was made from a large oblong of woollen cloth

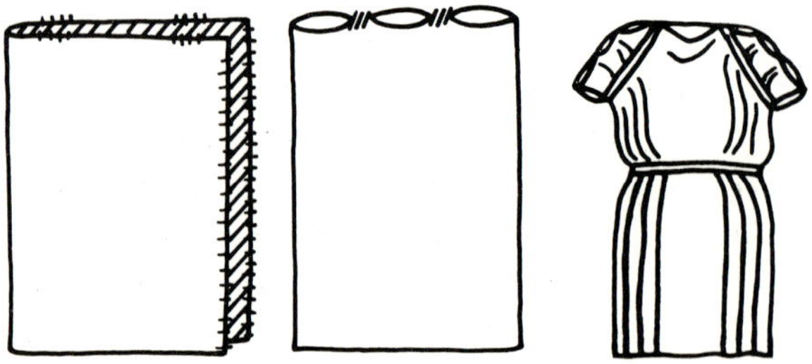

Fig. 8ii How the Ionic chiton was made, often of a crinkly sort of linen. After Charles Seltman (1956)

Pan Books: London (1956), pp. 82 and 83

ἱμάτιον

In the development of the Scottish Highlanders' kilt, the folds became much deeper.

The method of making up the peplos and the chiton is shown in Fig. 8 from Charles Seltman's "Women in Antiquity" (1956). Over both Doric and Ionic garb there was often worn a *himation*, usually a rectangular piece of cloth, in the form of a shawl, stole or a cloak, according to its shape and the manner in which it was worn. Sometimes it was worn alone (ἀχίτων). Both the peplos and the chiton fell in attractive folds which were often maintained by small weights sewn into the hem.

Wool processing

IN CLASSICAL times, certainly, the Greeks would have been long conversant with the wool-processing techniques of the Middle Eastern world and Egypt. The 7th century saw the establishment of what became a flowering Greek settlement at Naukratis on the Nile delta. From this trading town Egyptian alkali was exported to the Greek mainland, probably at a very reasonable f.o.b. price since the natron was produced nearby.

Kimolian earth had one limitation to its use; it tended to adsorb basic or other fugitive dyes. As long as most of the clothing was of naturally coloured wool or of white, Kimolian earth was the detergent of choice.

Aristophanes, *Frogs* (405 BC) 713.

In Greek theatre Aristophanes was the first to mention Kimolian earth – in The Frogs, brought out in c. 405 BC.

Κλειγένης ὁ μικρός, ὁ πονηρότατος βαλανεὺς
ὁπόσον κρατοῦσι κυκησιτέφρου
ψευδολίτρου τε κονίας
καὶ Κιμωλίας γῆς ...

That Cleigenes shifty and small,
the wickedest bathman of all
Who are lords of the earth, which is brought
from the isle of Kimolos, and wrought
With nitre and lye into soap
Not long shall be next us, I hope.

translation by B.B Rogers.

The stock-in-trade of the bath-attendant had also been rendered *à merveille* by a Radley schoolboy as "his terraqueous scouring-ball."

Dioscorides confirms the value of this mixture as a kind of soap:

σὺν νίτρῳ ἢ γῇ Κιμωλίᾳ σμῆγμα κράτιστον

the best soap (is made) with nitron and cimolian earth.

The product was a body detergent and not a fulling agent. Sodium montmorillonite is not used for fulling because it tends to plate the fibres.

The mixture, to be more precise, was sodium montmorillonite, an impure form of sodium carbonate, and plant ash. This is the first description of the synthesis of bentonite, by double decomposition: the products of the reaction being sodium and potassium montmorillonite and calcium carbonate. Two thousand three hundred and forty five years later, Aristophanes was cited to legal counsel engaged to advise on the validity of a German patent.

Although Cleigenes was not using his mixture for fulling, the combination of ingredients suggests quite a high degree of technical skill and a long time of development before the play was written – possibly to distract men's minds from Athens' last effort in the Peloponnesian war.

As we know, Nausicaa, king's daughter that she was, washed her own and her family's clothes. Domestic washing or laundering of clothes was never entirely abandoned in Greek, or Roman, living. With improvements in standards of textile production, however, woollen cloth had increasingly to be entrusted to a specialist fuller, who with his industrial experience knew how to restore the texture and nap of the cloth as well as might be. In Periclean Athens, soiled attire went normally out to the fuller – an arrangement productive of leisure but not always of peace of mind:

Μὰ τὸν Δί', οὐ γὰρ οὐδαμῶς μοί σύμφερον
καὶ γὰρ πρότερον ἐπανθρακίδιον ἐμπλήμενος
ἀπέδωκ' ὀφείλων τῷ γναφεῖ τριώβολον.

Aristophanes, *Wasps*, 1128, (422 BC.)

'Tis not expedient – No, by Zeus, 'tis not. 'Twas but the other day I gorged on sprats, And had to pay three obols to the fuller. Thus spoke Philokleon to Bdelukleon in *The Wasps*, in a disclaimer of ambitions to the high life.

So well established were fullers and launderers in late urban Greek life that they are quite often alluded to in the literature. Theophrastus, in his *Characters*, exemplifies two of his "types" by their attitude to them:

Theophrastus, *Characters*, 22, 18.

εὔδον μένειν, ὅταν ἐκδῷ τὸ ἱμάτιον ἐκπλῦναι

When he has sent his coat to be cleaned by the fuller, he will stay at home; and the avaricious one – in what reads as an intriguing anticipation of

5

Theophrastus, *Characters* 30, 13.

modern insurance fraud:

τὸ ἱμάτιον δὲ ἐκδοῦναι δεινὸς, οὐχ' ὅς βέλτιστα
ἐργάσεται, ἀλλ' ὅταν ᾖ ἄξιος ἐγγυητὴς τοῦ γναφέως

apt to send his coat to be cleaned, not to the best workman, but wherever he can find sterling security for the fuller.

Herodotus, IV *Melpomene*, 14.

The story may be connected with the Pythagorean theory of transmigration. In fact Aristeas 'disappeared' a second time.

Arimaspians, a people that have only one eye.

Pindar, fr. 271.

Fulling both for cloth-making and for laundering as a specialised craft was regarded as having the status of antiquity. There is a story told both by Pindar and Herodotus concerning Aristeas, a man of very good family, who entered into a fuller's shop in Proconnesus and died there suddenly. The fuller closed his workshop and went to tell his relations so that burial arrangements could be made. A traveller insisted that he had spoken to Aristeas on the road to Cyzicus, but they went to the fuller's shop taking with them what was necessary for carrying the body away, only to discover that Aristeas was not to be seen either dead or alive. He reappeared seven years later and wrote the poem *Arimaspes*.

Herodotus wrote in the 5th century BC, but Pindar wrote in the 6th. At least we know that fullers' shops existed in the earlier period.

Aristophanes, *The Plutus*, 166 meaning *wealth*, a moral allegory of 408 BC.

From a passage in Aristophanes' *The Plutus*,

ὁ δὲ γναφεύει γ', ὁ δέ γε πλύνει κώδια

one is a fuller, one a sheepskin washer, it is clear that the two trades were separate.

Xenophon, *Agesilaus*, 1, 2c.

In a despatch to Agesilaus, King of Sparta, from Asia Minor, Xenophon, the great soldier-historian who led the memorable Greek retreat from Babylonia after Cyrus's disastrous death, writes of war-preparations in Ephesus:-

ἐν τῇ ἀγόρα... οἵ τε χαλκότυποι καὶ
τέκτονες καὶ σιδήρεις καὶ σκύτεις καὶ
κνάφεις πάντες πολεμικὰ ὅπλα κατασκευάζον

In the market place. . . the copper-smiths and carpenters and iron-workers and shoe-makers and fullers were all preparing weapons for war.

Among the followers of these "reserved occupations", the fullers may have been making felt for targes or for soldiers' protective armour.

Fulling in the reflective literature

Heraclitus, section 59.

ABOUT 500 BC Heraclitus, a pre-Socratic philosopher, born at Ephesus in Ionia of distinguished parents, developed a system of philosophy in which all things were one, and that this one was also the many, the "opposite tension" of opposite constituting the unity of the one. Hot and cold, good and evil, night and day and so on are the same in the sense that they are inseparable parts of one and the same thing. As a rather happy concrete example he says that:

cf. Mark Twain, writing of a street in an old German city "it was straighter than a corkscrew but not so straight as a dog's hind leg."

for the Fuller's [*screw*], *the way straight and crooked is one and the same.* The literal translation is *the straight and the crooked way of the fullers is one and the same*. The passage has meaning if a screw press is borne in mind.

Likewise Plato in a disquisition to the young Socrates about the two arts involved in all production, the contingent cause and the actual, as a representative, perhaps even a generic activity, says:

Plato, *Politician*, 282 A.

Τῶν αἰτιῶν δὴ πλυντικὴν μὲν καὶ ἀκεστικὴν
καὶ πᾶσαν τὴν περὶ ταῦτα θεραπευτικήν,
πολλῆς οὔσης τῆς κοσμετικῆς, τοὐνταῦθα
αὐτῆς μόριον εἰκὸς μάλιστα περιλαμβάνειν
ὀναμάζοντας πᾶν τῇ τέχνῃ τῇ κναφευτικῇ.

The arts, and the other preparatory arts which belong to the causal class, and form a division of the great art of adornment, may all be comprehended under what we call the fuller's art.

And it is the analogy chosen by Hippocrates with which to illustrate his conception of the art of the physician:

Hippocratic writings, *Regimen*, 1, 14. English translation by W.H.S. Jones, London: Wm. Heinemann. (1931) IV, 252–253.

Καὶ οἱ γναφέες τοῦτο διαπρήσσονται·
λακτίζουσι, κόπτουσι, ἕλκουσι, λυμαινόμενοι
ἰσχυρότερα ποιέουσι, κείροντες τὰ
ὑπερέχοντα καὶ παραπλέκοντες καλλίω
ποιέουσι· ταὐτὰ πάσκει ἄνθρωπος.

This also do the fullers. They trample, strike and pull; by maltreating they make stronger; by cutting off the threads that project, or by weaving them in, they beautify. The same happens to man.

Medical and technical literature.

THE literature that emerged with Greek civilisation, however, affords much more than such purely sociological ideas. In the writings of its natural historians is to be found a body of information of a specifically technical kind, such as is nowhere earlier available.

Items regarding the use of earths have been cited from, for instance, Egyptian medical papyri, and much older Sumerian recipe-lists, but they have for the most part been isolated "finds"; and though the Sumerians are known to have classified 180 minerals according to hardness and characteristic shape and colour, it is with the writings of Theophrastus, Dioscorides, and the Hippocratic collection, and of Pliny who quarried from these including passages no longer extant, that the material archaeological evidence of industrial-chemical usage is for the first time supplemented by systematically compiled and comprehensive written information.

The compilations of course may very easily mislead – chemistry was still far from an exact science – and much has always to be supplied by deduction. In the case of minerals, nevertheless, and hence of the earths that more particularly interest us here, duly critical use of the Greco-Roman written sources enables a fairly accurate picture to be formed of what went on.

Although Hippocrates flourished *ca.* 400 BC and was a contemporary of Aristophanes, the 'Hippocratic Collection' runs over centuries, the earliest works being of more than one author. *Epidemics* i and ii are relatively old or even "authentic".

Epid. i. 21

In *Epidemics* vol. i 'he' wrote of the death through suppuration of the serving-maid of Skymnus the fuller, and of Phanocritus, who lay sick at the house of Gnathon the fuller.

Dis. Women, ii. 114

In *Diseases of Women* the scum or refuse of fullers is used as a medicament in certain cases of uterine hemorrhage. It is interesting to note that 'lanoline' was used internally.

Epid, iv. 36; v. 59; vi. 79 & 81.

Occupational diseases of fullers are described in *Epidemics* iv, v, and vii.

In each passage quoted the word for fuller was κναφεὺς or γναφεὺς.

Theophrastus, ΠΕΡΙ ΛΙΘΩΝ *On Stones*, translated by E.C. Caley and J.F.C. Richards, Columbus, Ohio: Ohio State University (1956)

Theophrastus, *De Lapidibus*, translated by D. E. Eichholz, Oxford: Clarendon Press (1965).

B. Farrington, *Greek Science*. (1949), 17.

Written in or shortly after 315–314 BC. Theophrastus *On Stones* is the earliest known scientific work dealing with stones, industrial minerals and artificial products derived therefrom. The book, of unique importance in the history of mineralogy and chemical industry, has an account of what were then considered the major earths.

Theophrastus was born between 372 and 370 BC at Eresos in the island of Lesbos, his father being Melantas, a fuller. He studied under Plato, accompanied Aristotle on his wanderings, returned with him to Athens in 335–334 and helped him to establish the Peripatetic School in the Lyceum. Theophrastus succeeded Aristotle in 321. His background has been held to have been of advantage to him in developing an empirical approach to natural history. He was also an Ionian. In its general alignment this book, one of many, is an attempt at classification on Aristotelian principles rather than an exhaustive description. Nevertheless it is highly informative. The more scholarly translation is by Eichholz (1965).

For Theophrastus

Theophrastus, *On Stones*, ix, 62

αἱ δ'αὐτοφυεῖς καὶ ἅμα τῷ περιττῷ τὸ
χρήσιμον ἔχουσαι σκεδὸν τρεῖς εἰσιν ἢ τέτταρες
ἥ τε Μηλιὰς καὶ ἡ Κιμωλία καὶ ἡ Σαμία καὶ
ἡ Τυμφαϊκὴ τετάρτη παρὰ ταύτας ἣ γύψος.

For identification, see R.H.S. Robertson, The earths of Theophrastus, *Classical Review*, (1958), New Series, *8*, 222–223

there are, roughly speaking, three or four natural earths which are useful as well as unusual, namely the Melian, the Cimolian, the Samian, and finally besides these the Tymphaic, otherwise known as "gypsum".

"Milowite", J.N. Wilson, *Chemical Trade Journal*, (1935), *97*, 28; *Sands, Clays and Minerals*, (1935), *2*, 127–130.

Dioscorides, 5, 159 (Wellmann). Pliny, *Nat. Hist.*, 35, melinum was also used as a depilatory because of its abrasiveness, being quartz.

Kimolian earth must have been very well known, since Theophrastus says no more about it. Melian earth, used in painting pictures, was rough, not greasy and of a loose texture, and probably the naturally-occurring fine silica, known in this century as MILOWITE. Dioscorides adds that it makes a sound when rubbed between the fingers. The first and third of these properties were explained by Eichholz on the grounds that Samian earth was "greasy, dense and smooth," and of no value to painters. Samian earth, probably a kaolin, was extracted from a narrow sloping vein. The lower quality material was nearest the headwall and footwall; and the best quality, selectively mined, bore the trade name THE STAR, either because it was stamped with this figure or because it was highly reflecting. According to Dioscorides it clings closely to the tongue; this and its use in plasters and salves also suggest that the clay was kaolin. Deposits of kaolin, no doubt from a more accessible site, were worked in ancient times in the island for making ceramic ware. The STAR grade of earth, according to Theophrastus, was used mainly, if not entirely, for treating cloaks, presumably because its white crystallites gave the cloth a clean and white appearance as in pipe-clayed webbing or china clay-coated art paper.

id., 63 ἀστήρ

Himatia

Also for treating cloaks was *the Tymphaic earth, which is called "gypsum" by the people of Thessaly and the neighbourhood. Gypsum is most plentiful and most easily discerned in Cyprus because only a little soil need be removed when it is dug there.* Theophrastus proceeds to describe the use of plaster of paris in grouting and mentions its heat of wetting (with some exaggeration). The fuel used for making plaster of paris was ox-dung; this was said to burn hot, but the steam given off may have been of advantage in controlling the dehydration.

id., 64 γύψος

id., 65

For a discussion of grouting see Chapter 65

Pliny confirms that the Tymphaean earth, like the Umbrian, was used for imparting lustre to cloth. Even in the present century gypsum has been recommended for cleaning woollen articles by being dusted on to them in a special machine and then removed – there is no reason why a similar procedure should not have been devised over 2000 years ago.

Dioscorides states that Kimolian deposits yielded two types of earth, one white and one reddish (ἐμπόρφυρος), and, whether or not following him, so do Pliny and Oribasius. Ovid also refers to "cretosaque rara Cimoli". Both red and white varieties were also used in medicine: mixed with vinegar they served to dissipate swellings and check discharges.

Dioscorides, 5, 176 (Wellmann 5, 156)

Ovid, *Metamorphoses*, vii, 463 (and the chalky fields of Cimolos).

Both white and reddish fuller's earth can be seen to this day in the huge quarry which supplied the most popular detergent of antiquity; and these colours are mentioned in pharmaceutical works for the next 2000 years. However, Encelius (1551), quoted Dioscorides, Pliny and others but added *Celsus cimoliam terram laus dat, quae set subcaerulei colores – Celsus praises cimolian earth with a bluish tinge* – an excellent observation since the white and the reddish occur on the weathered surface of the deposit.

Christopherus Encelius, *De re Metallica*, Frankfurt (1557) [earlier edition (1551)].

Kimolian earth is now known to be calcium montmorillonite, but Caley and Richards unfortunately say that it is a particular variety of sepiolite, a hydrated magnesium silicate – an error which has been copied elsewhere. There should be no doubt, since Cimolian earth is

op. cit., 209

See also *The Growth of the Pharmacopoeia*, Chapter 17 ii

still quarried in Kimolos as well as in the larger, neighbouring island of Melos.

Pliny, *Nat. Hist.*, 35, 194
Dioscorides, 5, 174
(Wellmann, 5, 155),
Oribasius, Bk 13.

The earths mentioned by Theophrastus are not of course the only ones in use in ancient Greece. The island of Chios, a traditional birthplace of Homer, yielded one similar to the Samian. It was used as a substitute for soda in the bath, and for beautifying the skin – though whether as a face powder or a cleansing agent we are not told. From Lemnos (now Stalimene) came three varieties: one very similar to the Kimolian, and used likewise for fulling; another a red clay, Miltos, used in painting; and the third renowned from earliest times for its healing properties.

Pliny, *Nat. Hist.*, *37*, 173

Bohn's Edition (1857) VI, p.453, note 80

Dioscorides, *m.m* V. 161

μωρός, thick,
earth ἡ χθών

cf. moron, thick or stupid.

Pliny also lists among his mineral species: *morochthos, colore porracea, lacte sudat.* "Morochthos" is leek-green in colour and exudes milky moisture. This sounds very unlikely: indeed Bostock and Riley admit that the reading here is very doubtful. It is interesting that Dioscorides in his *Materia Medica* says of the milk-white rock morochthos that it is used both by linen weavers for cleansing their fabric and medically to staunch blood and empty ulcers. The name itself means sluggish earth. We do not know if the earth made a viscous mixture with water or whether it got its name from its styptic property causing blood to thicken. Could this have been a fuller's earth?

Athenaeus, of Naukratis in Egypt, fl. end of 2nd & beginning of 3rd century AD.

Needless to say fuller's earth was not the sole Greek textile detergent; there was always stale urine. As the entertaining Athenaeus says in the Deipnosophists

τὰ γοῦν ἱμάτια τούτῳ χρώμενοι
ῥύμματι πλύνουσιν οἱ γναφεῖς

at any rate the fullers cleanse garments by the use of it as a washing agent.

Lemnian earth 6i

Fig. 9. The annual sealed earth ceremony at Lemnos. Parke–Davis & Co. Ltd. (1952); from an oil painting by Robert A. Thom; by courtesy of Warner–Lambert (UK).

6i

THIS medicinal earth has had an interesting social history. Prized by the ancients as a sovereign cure for festering wounds and snake-bites, and believed by modern Turks to possess the power of making poison drunk from a vessel made of it harmless, it has from time immemorial had supernatural associations, and, presumably originally in virtue of the latter, been dug only in a once-yearly ceremony.

Jessie M. Sweet, Sir Hans Sloane: Life and mineral collection, *Natural History Magazine*, (1935), *5*, No. 36 145–164

In classical antiquity, the story ran that Hephaistos (Vulcan), thrown out of Olympus by his father Zeus, had fallen on Lemnos, and lying there with a broken limb had had this healed for him by the Lemnian earth, dug now from the hill on which he had lain. The annual digging took place on the 6th May, in a ritual that doubtless antedates the myth and was presided over by a priestess of Artemis. The single waggon-load of earth taken was rolled into small cylinders, each of which was then stamped with the likeness of the goddess – becoming known in consequence as *terra sigillata*, and constituting one of the early examples in history of a trade-marked therapeutic agent. The cylinders – or troches – were distributed over most of the western world.

Not to be confused with Samian ware also called *terra sigillata*.

See Robert Meldau, Zeichen, Waren-zeichen, Marcken, (1967), Bad Homburg: Verlag Gehlen.

According to Dioscorides the seal bore the likeness not of the goddess Artemis but of a goat. Galen the physician (c. 130–200 AD) landed at Hephaistias in Lemnos on the very day when the priestess of Diana (Artemis) led the ceremony of the digging of the earth by spreading an offering of wheat and barley. The reddish, somewhat greasy earth was found behind certain stones. One waggon load was taken from the excavation on the hill called Mosychlos and taken to the town where it was dispersed in water and the grit allowed to settle. Only the mud

Dioscorides, middle of first century AD.

Jessie M. Sweet, *op. cit.*

Fig. 10. Terra sigillata (Lemnian earth) in the process of being sealed.

Reproduced from "Ortus Sanitatis"; c. 1500, by Jessie M. Sweet (1935).

resting on the sandy sediment was used, the top having been poured off. The mud was dried in the sun until it had the consistency of soft wax when it was made into troches and stamped with the sacred seal of Diana. These were put into the shade to dry until quite hard. Galen managed to get 5000 troches and found them of great efficacy in curing bites and stings of venomous animals and against poisons.

The next account of visits to the Lemnian earth mines was written in the 16th century, when it became fashionable to give presents of Terra Sigillata *to foreign ambassadors. Between 1562 and 1572 Stephano Albacario, a Spanish physician, was sent by Ogier Ghiselin de Busbecq, Imperial Ambassador at Constantinople, to report on the Lemnian earth.* The deposit he was shown near the village of Rapanidi was clearly different from what was seen by Galen. It was still dug with great ceremony, and also on one day only in the year, but now on August 6th, the Feast of Christ the Saviour. The Greeks were allowed to do the digging, but the Turkish Governor and Overlords made sure that none of the troches could be bought on the spot.

Jessie M. Sweet, *op. cit.*, p. 149, quoting *Les Commentaires de M. P.A. Matthiole . . . sur les six livres de P. Dioscoride. . . de la matière médicinale. . .* Lyon, (1572)

Some said it was dug on the Ascension of the Blessed Virgin

Pierre Belon (1555) had made a collection of the troches in Constantinople, but was unable to see the digging when he visited the island. He illustrates a number of Turkish seals said to be marked "Tin imachton", or sealed earth. When Professor D. Sinor, of Bloomington, Indiana, said that the inscription was not Turkish, it occurred to me that it was Greek in Turkish letters – "tini ma chthon" meaning "for you mother earth". Certainly by the 18th century there was a great

P. Belon, *Les observations de plusieurs singularitez & choses mémorables, trouvées en Grèce*, Paris (1555).

τινί μᾶ χθών

for you mother earth (strictly, for someone)

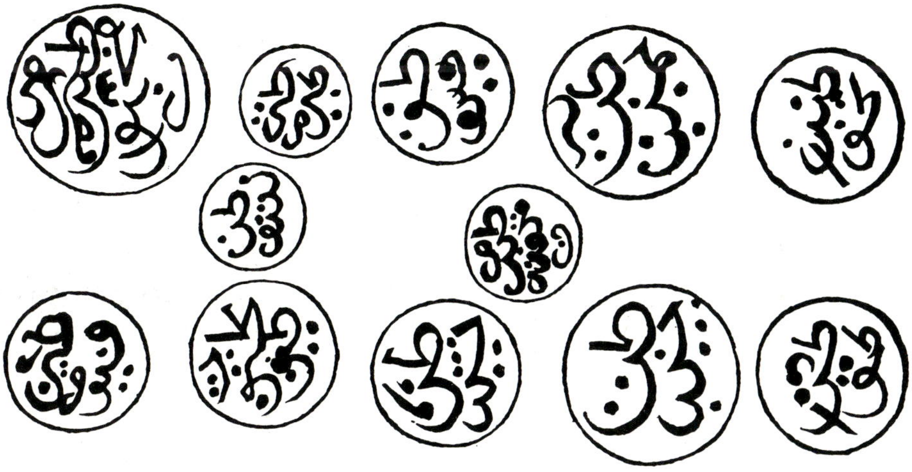

Fig. 11. "Turkish" seals for Terra Sigillata in use in Lemnos during the sixteenth century.

Reproduced from P. Belon (1555), by Jessie M. Sweet (1935)

Sweet, p.150, Fig. 5.

6i

Sweet, *op.cit.* p. 154 for photograph.
Henry Teonge, *Diary* 1675–79, (1825).

See also D.A.K. Black, A revaluation of Terra Sigillata, *Lancet* (27/10/56), p. 742 and letter by R.H.S. Robertson, *id.*, (1/12/1956)

variation in design. Twelve of the specimens in the Sloane Collection have a crescent moon, pierced by three crossed arrows and surrounded by the words Terra Sigillata.

Henry Teonge, "old cavalier" and Warwickshire clergyman, turned, in his fifties, ship's chaplain, visited the island in 1675, and in a curiously animistic diary entry, said that the clay came from *one peculiar place in the earth, where it doth come up in one night's time, like that which dew-worms throw up*. His understanding of the industry was sounder when he recorded that *the Grand Seigneur is presented first with a great share, and, after that, some others of the great ones, and after, some for other countries*.

In modern Stalimene, early in the 20th century, the digging proceeded from sunrise to sunset; and the ceremonial duties were shared by Turkish *hodja* and Roman and Greek Orthodox Christian priests, accompanied by the local magistrates. One can safely say that the controlled supply of terra sigillata represents the most successfully and consistently maintained restrictive practice in history.

6ii

Wool-fat: "Oesypus" or "Lanoline"

Dioscorides, chap. 84, quoted in Greek in *J. Soc. Chem. Ind.*, (30.4.1890) 356- 358, by H.W. Langbeck.

Soapwort = *Saponaria officinalis*.

WE have seen that wool-fat was used by the Egyptians in pharmacy. Dioscorides, however, gives a good description of how it was prepared.

> *Wool-fat is called the grease out of the filth of new shorn wool. You will prepare it in this manner: Having taken sweet, newly shorn wool, not treated with soap wort, wash it in hot water, at the same time squeezing out all the filth; having put this in a wide-mouthed vessel, pour water over it, and agitating it violently until froth is formed, pour it out again of the vessel; or agitate it vigorously with a stick until much scum and filth collects; thereupon sprinkle sea water over it, and when the fat swimming on the top has set put it into another earthenware vessel; pour more water into the tub, stir it up again, sprinkle the froth with sea water and take it out again, and do so until after the fat has been exhausted,* [*when*] *no more froth gathers. Smooth the collected wool-fat with the hand, and if there is any filth remove it forthwith, squeezing out the first water and pouring on fresh and agitating it with the hand until, applied to the tongue, it does not bite but has an astringent taste and looks brilliant and pure and white. In this condition store it away in an earthenware vessel. But all must be made in hot sunshine. Some treat the fat by washing it in cold water, rubbing it with the hands as women* [*do with*] *wax, thus rendering it whiter. Others, after washing the wool and squeezing out the grease, boil it together with water in a cauldron over a slow fire, and having gathered the fat which swims on the top they wash it with water, as has been said, and straining it into an earthenware dish containing hot water, and covering it with a linen cloth*

they place it into the sun until it becomes sufficiently thick and white. Some pour off the first water after a lapse of two days, and pour on fresh. Better is that which has not been treated with soap wort and is smooth, smelling of the oily wool, washed with cold water in a shallow dish, and having been bleached and containing in itself no hard or solid [*particles*], *as for instance that which is adulterated with wax or tallow. It possesses a power to close up wounds; to soften, especially with melilot and butter.*

Langbeck pointed out that Nicolas Culpeper had published 'The Manner of Preparing Oesypus' in 1653; and there was another recipe, also based on Dioscorides, published in 1680, but it was not until 1882 that Drs. Braun and Liebrich took out a patent for the purification of wool fat, which, mixed with water, they called "Lanoline". This was dirty yellow and still contained free fatty acids from the soap. Langbeck brought the story of trying to purify wool-fat to 1886, when he took out a patent with his assistant, Mr Ritsert. A good modern discussion of wool wax is given by E. Vernon Truter (1956). Various processes were adopted in his time:

Pharmacopoeia Londinensis, (1653)

Ger. Patent. (7 May 1886)

London: Cleaver-Hume Press, (1956), 368 pp.

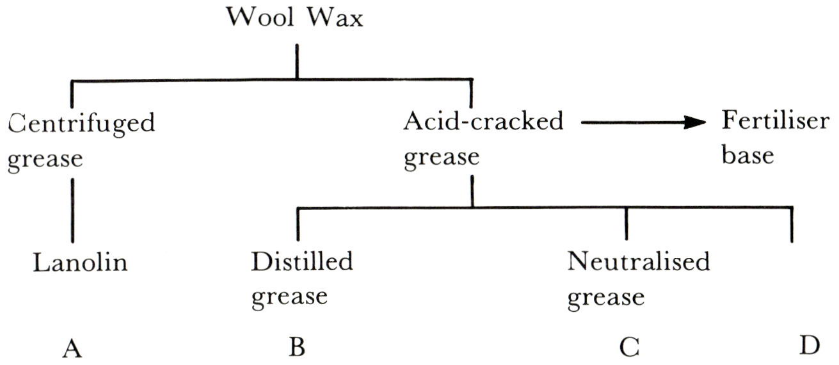

The four products A–D have their own range of uses.

7i

ROME

The Mechanics of the Greeks continued without a break into Roman use.
W. M. FLINDERS PETRIE, The Revolutions of Civilisation

Harper & Brothers, London & New York, (1911), p. 99

THE Greco-Roman phase of recurrent civilisations was a single phenomenon. The successive cultural peaks according to Petrie were: BC 450 in Sculpture, BC 350 in Painting, ca. BC 200 in Literature, ca. BC 0 in Mechanics, ca. AD 150 in Science and AD 200 in Wealth. Of these only Literature was probably bimodal, the peaks as defined by Petrie occurring earlier in Greece than in Rome.

In the Hellenistic period, after the death of Alexander the Great in 323 BC, the Greek language spread widely from its original bases and Greek-speaking craftsmen settled in Rome. Thus Pliny quotes from many Greek authors and observed the work of Greeks in their workshops. As Gibbon says *If a man were called to fix the period in the history of the world during which the human race was most happy and prosperous, he would without hesitation, name that which elapsed from the death of Domitian (96 AD) to the accession of Commodus (180 AD)* – only 84 years. After that, Rome suffered the Decline and Fall so ably chronicled by Gibbon.

Gibbon's *The Decline and Fall of the Roman Empire*, D M Low, Chatto & Windus (1960), reprint (1981), p. 1.

In this roughly drawn setting the history of fulling may now be traced.

Tradition has it that there was a fuller's guild in Rome as early as the 8th century BC – one of nine collegia established, according to Plutarch, in the reign of Numa. A king of this name did presumably once rule in Rome, but he and the institutions of his days are as legendary as the wolf's milk on which his immediate predecessor Romulus is said to have been suckled. Some kind of early organisation there must have been among fullers, however, since they had usually to share water supplies. Certainly their *collegium* at Pompeii had several centuries' history behind it. And in 220 BC the importance of fulling in the life of the community had reached proportions such that the censors C. Flaminius and L. Aemilius legislated for ensurance of standards. (As Pliny put it: *nothing was beneath the notice of our ancestors.*)

Pliny, *Nat. Hist.*, 35, 197

The text of the *Lex Metilia de Fullonibus* has not survived, and Pliny, to whom we owe our knowledge of it, unfortunately only records parts relating to laundering. The latter do not seem to have eliminated the disadvantages of having a garment laundered. There was still loss in quality, references to which in the first century AD echo those of the

Greeks. In Petronius's *Satyricon* visitors, finding a slave about to be flogged because, the steward explains, he has lost the latter's clothes in the bath, receive in reply to a plea for mercy the haughty concession: *It is not the loss I mind so much as the fellow's carelessness. He lost my dinner dress. . . It was Tyrian dye, of course, but it had been washed once already. Well, well, I make you a present of the fellow.* Martial in one of the *Epigrams* sneers at a self-styled model of generosity: *"I have given my friend", say you, "5,000 sesterces and a toga (o bounty!) not more than three or four times scoured." Magnificent gift!* But the Lex Metilia afforded the client protection in other directions. The fullers were made responsible in law for all property entrusted to them, and could be sued if they failed to return garments or returned the wrong ones.

Petronius Arbiter, fl. ca. 60 AD, *Satyricon*, 30.

Martial, *Epigrams*, x, 11

As had happened in earlier civilisations, enterprising workers in the cloth manufacturing processes or in the distribution and collection of wool for spinning, had found they could set up and run what were virtually factory establishments. By the 3rd century BC in Rome, in addition to innumerable small fullonicae, there were several such large ones, employing many workers. Fullers owning such factories often became cloth merchants as well, buying the "raw" pieces from domestic weavers and marketing them, when processed, on their own account. Thence it would be a short step to supplying the weavers with wool, on terms relegating them to the position of paid employees. There is no definite evidence of fullers acquiring such a position of control of cloth-production, but as theirs was the only process other than dyeing that could not be performed domestically and automatically involved the master at a centre of operations, they shared with the farmers-out of raw wool the position of potential dominance, and the positions of importance some of them occupied in the municipalities points to it.

Up and down the peninsula little market towns that had thriven on the traffic in wool and yarn had become manufacturing centres, using local wool to make in most cases the cheaper kinds of cloth. Other centres, like Pompeii, had wool sent in, raw or as yarn or unfinished webs, from all parts of the country, to be made into a finer quality of cloth. Increase in wealth and the growing needs of city-dwellers and the army meant expansion of demand in both the mass and luxury markets thus catered for.

As I have said, Greek techniques of fulling may have been introduced to Sybaris in 720 BC, because the colonists' remarkable prosperity was founded upon wool produced by the subjugated indigenous people in this part of the "instep" of Southern Italy. Kimolian earth could have been imported for this purpose, but perhaps not for long, because the pleasure-loving Sybarites could make more profit by exporting raw wool to Ionia and mainland Greece.

chapter 5

Sybaris was razed to the ground by the neighbouring Crotoniats, but the more prudent and hard-headed Phocaeans, who took over the

Phoenician settlement of Massalia about 600 BC, were successful in establishing an industrial port. No doubt Kimolian earth would have been shipped thither, until the Massaliots, trading deep into Liguria, may well have bought fuller's earth from Apt in Vaucluse, since there are signs that the earth was worked there before the Roman era.

Plautus, 254–184 BC, *Aulularia*, 3, 5, 34.

In Rome itself cloth-working had long ceased to be a household task. The comic dramatist Plautus, who died in 184 BC, was writing of all the tradesmen who were standing in a crowd, demanding to be paid. *Sed hoc etiam pulchrum est praequam (ubi sumptus patient) stat fullo, phrygio, aurifex, lanarius, caupones, patagiarii, indusiarii, flammarii, violarii, carinarii,... But even this is pleasant compared with when the fuller, embroiderer, goldsmith, woolworker, peddlers, border-makers for garments, makers of women's undergarments, makers of bridal veils, dyers in violet and yellow (?), stand and demand payment.* Other cloth-workers in this famous list, which illustrates so well the commercial activity about 200 BC, were wholesale linendrapers, bodice-makers, makers of kirtles; weavers, lacemakers and dyers in saffron colours – though I suspect that some translated words are speculative.

The meaning of carinarius is uncertain, the nearest word being carina, a ship's keel.

In the country the custom whereby a household or estate produced its own cloth persisted alongside increasing specialisation. From being "aristocratic" occupations, however, spinning and weaving became by the early Empire tasks the great ladies entrusted to their slaves, and fulling, always a task for the latter, was for economic reasons as early as the first century BC ceasing to be done on the estate at all: Varro (116–27 BC) writes in his treatise *On Agriculture*:

Marcus Terentius Varro, 116–27 BC *On agriculture*, 1, 16

> *...farmers in such circumstances prefer to have in their neighbourhood men whose services they can call upon under a yearly contract – physicians, fullers, and other artisans – rather than to have such men of their own on the farm; for sometimes the death of one artisan wipes out the profit of a farm.*

In the opening years of the Christian era recourse to the outside specialist craftsman had almost completely prevailed. Vitruvius in the introduction to Book VI of the Ten Books of Architecture, commenting on the need for a do-it-yourself manual of the building arts, declares:

Marcus Vitruvius Pollio 27 BC–47 AD 6, Introduction 6, 7.

> *Nobody, therefore, attempts to practise any other art in his own home – as, for instance, the shoemaker's, or the fuller's, or any other easier kinds – but only architecture, and this is because the professionals do not possess the genuine art but term themselves architects falsely.*

As home-laundering had almost disappeared so also had "homespun". The rich had found it more agreeable to make their choice from an ever-increasing range of commercially-produced fabrics, and the stage was reached when the cheaper ranges of these provided them with clothing for the slaves of their household also at less cost than the domestically-produced article. Homespun is too good for the ordinary slaves on a great estate, says Columella. It may be used with advantage

Lucius Iunius Moderatus Columella, *fl.* mid 1st c.

to the owner's purse for stewards and overseers; for the rest it is cheaper to buy. Augustus in his attempt to foster a return to what he considered the proper domestic virtues by wearing only clothes spun and woven at home by his wife, sister and daughter was seeking to stem a current already too strong to counter.

7i

Augustus, fl. 24 BC–14 AD.

Pompeii

7ii

EXCAVATIONS at Pompeii have revealed better than any literary account the art of the fuller. Pompeii was buried in the ash of Mount Vesuvius in 79 AD. Gauis Plinius Secundus, Pliny the Elder died of sulphur dioxide poisoning when he went to investigate the type of violent volcanic eruption which bears his name, and to see if he could help in any way.

Born 23 AD.
Died 26.8.79 AD.

Plinian eruptions

Pompeii at the time of its overwhelming was one of many cloth-manufacturing towns in Italy, and prominent as a centre of production of finer quality materials. The Italians had always clothed themselves in wool from the sheep that flourished all over the peninsula, but particularly in the south, and early discriminated as to the suitability for different purposes between the types produced in different localities: that of Canusium being recognised as softer that that of Tarentum, for instance, but not so glossy. The beginnings of industrialisation date back likewise very early: with *lanariae* set up by town authorities for sale of the wool of the district to incoming as well as local buyers (and perhaps for the preparation of it), and *lanipendiae* congregating in the market towns to organise putting-out of the wool for spinning.

See *Cambridge Economic History*, II, 357–358.

In the latter there was towards the end of the republic a certain competition to be faced from linen, and in the days of Augustus also from silk, both of which became fashionable among the wealthy and patrician. The authorities, whether out of genuine austerity of sentiment (the women of the wealthy classes, in revolt after generations of such austerity, had taken to wearing transparent garments woven from the low-denier threads spun by the silkworms of Cos) or of a protectionist attitude to the wool industry, for a long time frowned upon such extravagance. From time to time the wearing of pure silk was forbidden, though mixtures were allowed, and Augustus to set an example wore only wool. It was another of that emperor's losing battles, for in only two centuries, the depraved and effeminate Antoninus Elagabulus, prided himself on wearing robes wholly of silk, and refused to touch linen that had once been laundered, declaring it fit only for beggars. Linen and silk, even in fashion and approved, however, were alike privileges of the rich: the vast majority of the Italian population still wore wool.

27–14 BC
Lampridius, *Scriptores historiae Augustae*, translated by David Magie, London, (1924), *2*, 157

7ii

New York: The Macmillan Co., (1899).

In Pompeii the fullers were a powerful and wealthy body. One of the finest buildings in the town was theirs, "dedicated", the inscription runs, "to Eumachia, daughter of Lucius Eumachius, by the fullers." To quote August Mau, in his study *Pompeii, its Life and Art*:

This building, in which the fullers had set up, in a specially prominent place, a statue of the person who had erected it, must in some way have served the purposes of their trade. Clearly enough it was not a fullery; on the other hand, it was well adapted for a clothier's exchange, a bazaar for the sale of cloth and articles of clothing. Tables and other furniture for the convenience of dealers could be placed in the colonnade and the corridor; in the corridor, especially, goods exposed for sale in front of the open windows could be conveniently inspected by prospective buyers – not only by those in the corridor itself, but also by those looking in from the colonnade. The small doors between the corridor and the colonnade could be securely closed, and the entrance from Abbondanza Street could be easily guarded; there was only a narrow door at the end of the passage opening into the corridor, and at the street entrance was a porter's room connected by doors both with the passage and with the street. This evidence of unusual precaution suggests that possibly the side entrance, from its close connection with the corridor, was intended especially for the conveyance of goods to and from the building, in order that the front entrance might be left for the exclusive use of purchasers and dealers.

On the assumpton that the building was a cloth market, it is clear that the colonnade would naturally be open at all times, the corridor only during business hours; after business hours the corridor would be closed for the protection of the goods left there overnight. The windows may have been closed with shutters as in Oriental bazaars. Other peculiarities of arrangement are cleared up by this explanation. . . It is not possible, however, to make out what the purpose was of certain remains of masonry found on the south side of the court which have now disappeared, or of two rectangular elevations at the rear, or, finally, of a large stone in the middle of the court in which a movable iron ring is fastened. Our information is so scanty that we are unable to determine in all particulars what the requirements of a fullers' exchange might have been.

The fulling trade was by now attracting capitalists from outside. L. Caecilius Jucundus, who at the time of the Vesuvius eruption was renting from the state one of the large fulleries since excavated, was an auctioneer and banker. Marcus Vesonius Primus, who managed the workshop for him, was a man of influence in the political life of the city; the front of the fullery still bears the names of his nominees for office.

Excavations have brought to light two large fulling works and a third rather smaller one, (these can be identified with certainty – there were many smaller *fullonicae* less specifically designed scattered through the town), and from the special ground-plan and fittings of these, and from paintings in one of them and in a private house, together with the information from written sources, the kind of comprehensive picture

Fig. 12. Fresco from Pompeii showing cupids fulling cloth. Fulling on the left. This is a modern copy from the very much damaged original. The original is a continuous strip on a red background. It is reproduced here in two parts for convenience.

becomes available of European fulling that has been wanting in the case of earlier civilisations.

In the house of the freedman, Vettii, a series of very fine mural illustrations, painted probably about the middle of the first century, show Cupids engaged in a variety of activities – making and selling oil, practising as goldsmiths, gathering and pressing grapes and dealing in wine, and also carrying out the processes of the fullery (Fig. 12).

The quotations in this section are from Mau.

treading the clothes in vats, carding, inspecting the cloth to see that the work is properly done, and folding the finished garments for delivery to the owners.

The second set of illustrations reproduced are from paintings found in the largest of the fulleries, in Mercury Street (Fig. 13).

Fig. 13. A fuller treading the cloth, and others rinsing.

Drawn from photograph of mural

One wash and three rinses is still commonly programmed in our automatic washing machines. This is the normal procedure for *laundering*.

In the first picture the clothes are being washed. They are in four round treading vats, which stand in niches formed by a low wall. One of the workmen is still treading his allotment, steadying himself by resting his arms on the walls of the niche at both sides; the other three have finished treading and are standing on the bottom of their tubs, rinsing the garments before wringing them out.

"The next scene is threefold. In the foreground at the left sits a richly dressed lady, to whom a girl brings a garment that has been cleaned; that the woman is not one of those employed in the fullery is evident from her elaborate headdress, necklace and bracelets (Fig. 14A). In the background a workman dressed in a tunic is carding a large piece of cloth (Fig. 15A). Nearby another workman carries on his shoulders a bleaching frame, over which garments were spread to receive the fumes of the sulphur; he holds in

Fig. 14A (left). A well-dressed woman collects her laundered garment.

Fig. 14B (right) A young man hands a garment to a girl.

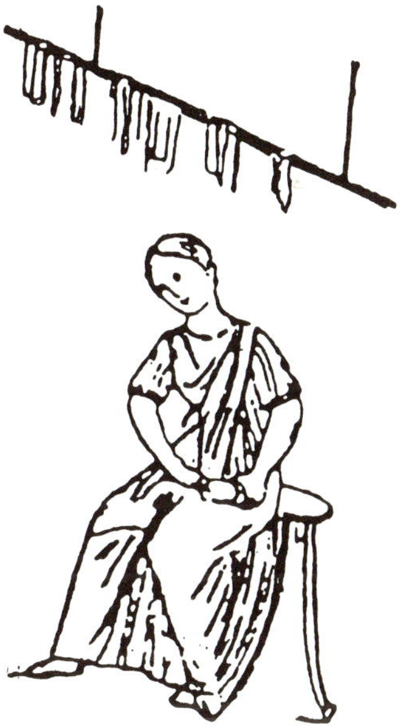

Figs. 15A and B (left and right). Carding the cloth; a woman cleaning the carding tool.

Fig. 16A (left) Carrying in the sulphur pot and bleaching frame.

Fig. 16B (right) Cloth-press operated by right- and left-handed screws.

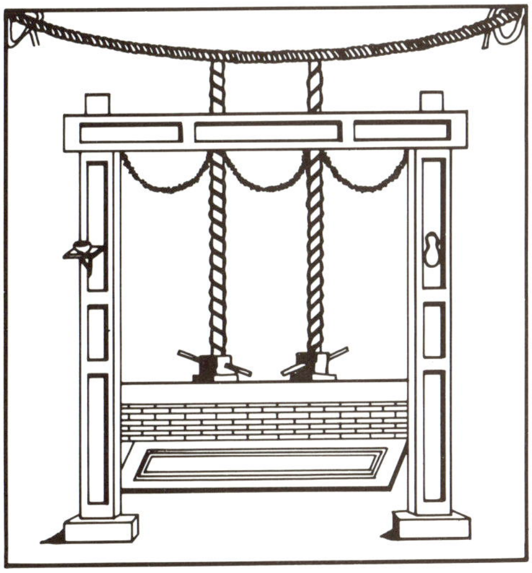

Drawings made from photograph of mural at Pompeii.

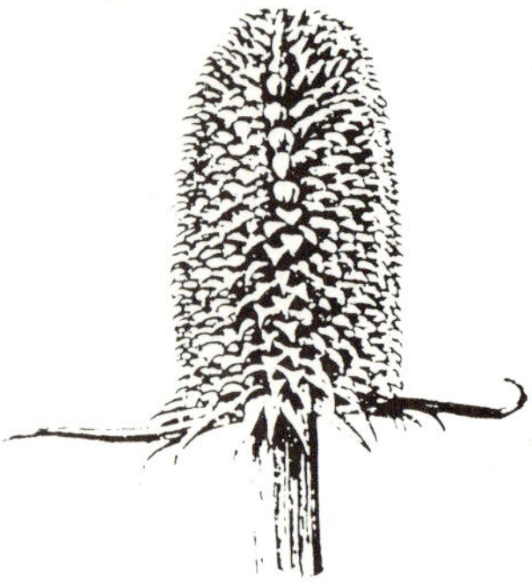

Fig. 17. *Dipsacus fullonum*, Linnaeus.
The teasel crop is still cultivated in England. See Figs. 94A-B, pp. 400–401.

his left hand the pot in which the brimstone was burned. An owl, symbol of Minerva, who was worshipped by fullers as their patron divinity, sits upon the frame; and the man underneath has on his head a wreath of leaves from the olive tree, which was sacred to the same goddess (Fig. 16A).

In the third picture a young man hands a garment to a girl (Fig. 14B); at the right a woman is cleaning a carding tool (Fig. 15B). The fourth gives an excellent representation of a fuller's press, worked by two upright screws; it is so much like our modern presses as to need no explanation (Fig. 16B). The festoons with which it is adorned are of olive leaves.

"Carding" was to give the cloth a softer finish by raising the fibre-ends to the surface – raising the nap. Pliny records that sometimes the skin of a hedgehog was used, sometimes a thistle or thorn, predecessor of the mediaeval teasel (*Dipsacus fullonum*) (Fig. 17). The thistle spines were made into the carding tool (*aena*) by mounting as in a brush.

E. P. Sandcross, Fulling by burned sulphur fumes *Amer. Dyestuffs Rep.*, **1935**, *24*, 418.

The fumigation with sulphur was to bleach the cloth. It was standard practice in Italy and was still a widely-used method early this century. E. P. Sandcross (1935) gives as the main reason for its continued practice the fact that it is considerably less expensive than the alternative (and more permanent) bleaching with hydrogen peroxide. The wicker-frame has given place however to a lead sheet roof with aluminium hooks, and sulphur stoves are now built of wood, concrete or brick.

Pliny, *Nat. Hist.*, 35, 50.

Pliny enters into some detail as to the type of sulphur then favoured for the purpose. Other authors were alive to picaresque possibilities.

Appuleius, c. 155 AD *Metamorphosis*, Book 9, trans. T. Taylor, London, **1822**, 209

Apuleius tells how

the wife of a fuller, who was a secret adulteress, embraced a young man in the house of her husband just before dinner and her husband arrived. So being disturbed in her act of connection, she hid him in a wicker coop, which being plaited from willow twigs into an erect heap, and smoked with the white fume of sulphur, was used for the purpose of whitening the cloth that was placed round it. The young man was oppressed by the strong odour of the sulphur, that living mineral, [*which*] *caused him to sneeze violently; and he was nearly suffocated.*

cochlea

The method of use of the screw press is clear from the illustration. Pressing gave the cloth smoothness and lustre. Even tightening from the two sides was facilitated by the two screws. Pliny attributed this development to the ingenuity of his own time. We have seen that the single-screw press was a Greek invention and known to Heraclitus.

c. 500 BC

Both the Mercury Street fullery and the other large one excavated, on Stabian Street, had been converted from private houses. They were not purpose-built. In the words of Tenney Frank, the big Pompeiian *fullonica with its expansive system of vats, its complex trade-processes, and its group of skilled workmen, may fairly be called a factory; but it is at once characteristic of ancient conservative methods that no fullery outgrew the relatively narrow confines of the ordinary dwelling-house.*

T. Frank, *Economic survey of ancient Rome*, (1937), Baltimore.

In the Stabian Street establishment, that managed by Marcus Vesonius Primus, part of the house was retained for domestic use – the *atrium*, or entrance-hall, and some of the rooms opening off it, which included a kitchen. The fullery occupied a shop next to the entrance and the peristyle.

In the former are the foundations of three treading vats, and on the opposite side an oblong depression in which the press was placed. The peristyle contains three large basins of masonry for soaking and rinsing the clothes. A jet of water fell into the one next the rear wall from which it ran into the other two through holes in the sides. Along the wall is a raised walk on a level with the top of the basins, into which the workmen descended by means of steps. At the ends of this walk are places for seven treading vats, five in one group, two in the other. The wall above is decorated with a long sketchy painting, in which the fullers are seen engaged in the celebration of a festival – doubtless the Quinquatrus, the feast of Minerva; the celebration is followed by a scene before a magistrate resulting from a fight engaged in by the celebrants.

See Fig. 27 used in the discussion of Roman Britain.

Conversion of the house in Mercury Street had included replacement of the columns of the large peristyle by massive pillars supporting an upper gallery for drying of clothes.

At the rear are four square basins, the two larger of which are about six feet across; the water passed from one to the other as in the basins of Primus's fullery. In the corner near the last basin are six rectangular niches for treading vats, separated by a low wall. There is a vaulted room at the right of the peristyle, with a cistern curb, a large basin of masonry, and a stone table. Here a substance was found wnich the excavators supposed to be soap, but which was doubtless fuller's earth, like that found in the establishment on Stabian Street.

Four large *fullonicae* were excavated in Pompeii in 1929. The largest belonged to L. Veranius Hypsaeus (VI, 8, 20), another to M. Vesonius Primus (VI, 14, 20, 21, 22) and another to the freedman Stephanus (1, 6, 7), whose installations are the best preserved – all examples of an industrial enterprise set up in a private house.

L. Jacono, Notizie scavi 1929, *Rend. Accad. Archeologica Napoli*, (1934). The numbers refer to districts, blocks and houses.

Fulling operations, as they are to be deduced from paintings and building remains, clearly involved treating the one piece of cloth in several different vats. Mau in his commentary on the illustrations inferred that a first insertion, for trampling, was followed by several rinses, but in the more detailed account he devotes to the processes, he speaks of a first washing (and, presumably, rinsing), followed by beating and stretching, and then by washing and drying a second time, *for cleaning and shrinking*. Beating was apparently the method of felting the cloth. Stretching was presumably done on wooden frames or "tenters", such as were used in the Middle Ages. The felted material was then carded, as shown in the mural, to raise the nap, but the last stage of shearing to make a smooth finish is not shown in these pictures. After this came the bleaching, especially of woollens, with sulphur fumes, and finally smoothing in the press – the fuller had then only to cut it into suitable lengths to produce virtually finished garments,

Scouring followed *by fulling* is here discussed.

Roman dress, traditionally austere, requiring scarcely any fitting and only the simplest sewing.

Until the end of the 2nd century BC free citizens, men and women, wore the *toga*, and those not entitled to this the scarcely less simple *pannela*, a bell-shaped cloak with an opening for the head. Worn by free men under the *toga* was a plain sleeveless shirt, the *tunica*, consisting simply of a front and back piece sewn together and belted round the waist. The dress of women of this status comprised, with the *toga*, the *fascia pectoralis* or breastband, one loose undergarment, the *subucula*, not unlike the *tunica*, and the *tunica interior*, worn inside the house, which was of equally simple cut. During the 1st century BC the women's *toga* tended to be replaced by the *stola*, a kind of long *tunica*, which was fitted with short sleeves. The *palla*, however, that completed the costume, and was worn also by those not entitled to the *stola* as combined robe and cloak, was simply a square piece of cloth worn draped in either of two ways, in one of them fastened on the shoulder by brooches. Tailors are mentioned rarely in Latin literature, except as stitching or mending clothes.

The earth ready for use and found by excavators in both the Stabian Street and the Mercury Street workshops, was lost by bombing in World War II, but a newly-found sample from "House No. 15 to Reg. I. Ins. XI" was obtained through the kindness of Dr. G. Novelli from the Superintendent of Antiquities of the Provinces of Naples and Campania.

S. De Luca, Meeting of 13 April 1878, Royal Association of Naples, Academy of Physical Science and Mathematics, vol. *17*.

Early in 1878 Professor Sebastian de Luca had reported the discovery of a large quantity of fuller's earth at Pompeii near to some washing tubs but in a separate and sheltered spot. The creamy, greasy material had been contaminated by lime mortar, ash and possibly other materials.

W. Thompson, *Rept. Brit. Assoc. Advanc. Sci.*, (1879), *49*, 321.

William Thomson about the same time took a sample of fuller's earth from a square tank in one of the Pompeian fullonicae; it was pointed out by the guide as "soap". Similar material was also found in private houses for use as a household detergent. Thomson's chemical analysis is I in Table 1. No. II is an analysis I had carried out in 1967. No. III is of bentonite from Ponza by W.H. Herdsman, 27.4.1964 and No. IV the same by S.A.M.I.P. Rome, in 1955. The Puglia bentonite's analysis was by P. Pastore (1941), (*Ind. meccan.*, *23*, 131), the Sardinian bentonite from the Sarcidano district by V. Cirilli and O. Amati (1942) and the bentonites of Melos VII and Kimolos VIII to X are given by way of comparison. The analysis of the Kimolian earth was given by Franzini *et al.* (1963) and by Robertson (1980) (X).

Fuller's earth from Pompeii, *Scottish Journal of Science*, (1967) *1* (1), 59–65. Analyst Dr. J. A. Clements, then of Robertson Research Co. Ltd.

(1942), *La Ricerca Scientifica*, p. 122

M. Franzini, R. Mazzuoli, M. Pratesi, L. Schiaffino, (1963), *Atti Soc. Toscana Sci. Natur., Pisa*, *70*, 49

R. H. S. Robertson, *Min. Polon.* (1980), *11* (1), 3–18

It is not possible from chemical analyses alone to tell where the Pompeian fuller's earth came from, though the low alumina content suggests Ponza. An x-ray spectrogram of II was interpreted as showing montmorillonite with 10 to 15% cristobalite; Ponza bentonite gave a

Kindly run by Ralph E. Grim

Table 1 Chemical analyses of some Mediterranean montmorillonites

	I	II	III	IV	V	VI	VII	VIII	IX	X
	Fuller's Earth		Bentonite			Bentonite		Kimolian	Earth	*Id*
	Pompeii		Ponza		Puglia	Sardinia	Melos	Nodules	Ground-mass	Purified
SiO_2	67·145	68·4	65·34	65·70	68·12	48·30	72·52	56·91	68·70	59·96
TiO_2	—	0·10	0·12	0·20	—	—	0·17	0·67	0·62	0·15
Al_2O_3	12·857	11·3	12·93	13·55	16·10	16.31	16·08	18·64	12·68	18·55
Fe_2O_3	2·107	0·87	1·96	1·13	1·65	2·6	0·93	1·86	1·39	1·56
FeO	—	0·15	0·75	0·31	—	—	—	—	—	—
MgO	1·822	1·52	3·49	0·89	2·23	5·29	0·82	2·49	2·44	4·51
CaO	6·412	1·00	2·08	1·67	3·68[a]	0·30	1·24	1·18	0·75	1·13
Na_2O	—	0·30	0·48	0·64	—	—	—	0·74	0·71	4·34
K_2O	2·253	0·50	0·20	0·84	—	1·20	1·70	0·96	0·30	0·18
CO_2	3·451	0·10	—	0·30	—	—	—	—	—	—
Cl	—	0·035	—	—	—	—	—	—	—	—
SO_3	—	0·12	0·15	traces	—	—	1·89	—	—	—
P_2O_5	—	—	—	—	—	—	—	—	—	0·86
H_2O +	3·953	5·38	7·24	9·35	7·93[b]	20·1[c]	4·57[b]	7·28	5·59	8·76
H_2O −	—	10·3	5·42	5·42	—	—	—	8·48	6·89	—
	100·000	100·08	100·00	100·00	99·71	94·10	99·92	99·21	100·07	99·90

[a] CaO in Puglia earth sometimes 6 to 7%
[b] Loss on ignition
[c] Loss at 120°C

similar pattern. Both were fairly well crystallised calcium-montmorillonites.

There was not enough difference in the ionic compositions of the montmorillonites of Ponza Pompeii and Kimolos (Robertson 1967, 1980) to prove the origin of the Pompeian earth, but thermal analysis showed great similarity between Pompeii and Ponza. Both of these were very different from Kimolian white and purplish, which I collected in 1961, thanks to Mr George Eliopoulos. There is therefore a high probability that some of the fuller's earth used in Pompeii came from Ponza. This conclusion will be discussed in the next chapter, on Pliny's description of fulling.

By Dr. Anna Langier-Kuźniarowa of the Geological Institute, Warsaw.

By courtesy of Silver & Baryte Ores Mining Co., Athens.

For curves of thermal analyses, see following page.

7ii

Fig. 18. (right). Differential thermograms and thermogravic curves of Pompeian fuller's earth compared with Ponza bentonite and two sorts of Cimolian earth, the white and the purple.

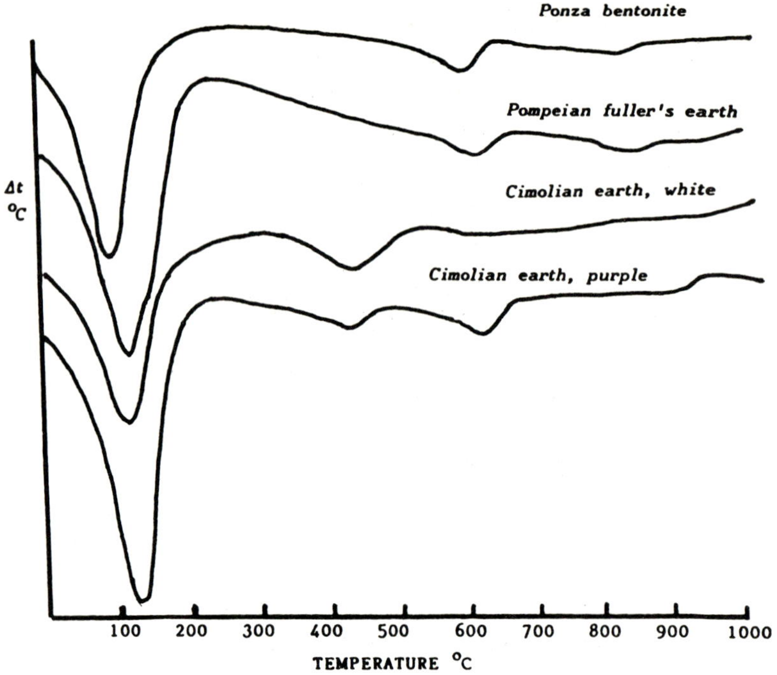

Fig. 18A. Differential thermal curves.

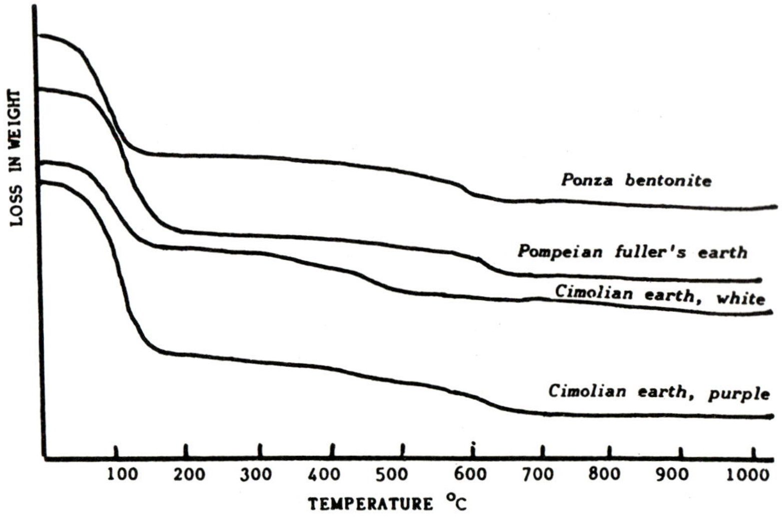

Fig. 18B. Thermogravic curves.

7iii Pliny, the Elder, and other authors.

> **Saxum**, *a great stone, a rock. Also a kind of earth that groweth by steping.*
>
> **ADAM LITTLETON, A Latine Dictionary, London 1678**

THE only detailed account of fulling in antiquity was given by Pliny the Elder in his monumental *Natural History*, which was begun about 77 AD. The passage which concerns us now is in Book 35.

196. Est et alius Cimoliae usus, in vestibus, nam Sarda, quae adfertur e Sardinia, candidis tantum adsumitur, inutilis versicoloribus, et est vilissima omnium Cimoliae generum; pretiosior Umbrica et quam vocant saxum.
197. Proprietas saxi quod crescit in macerando atque pondere emitur, illa mensura. Umbrica non nisi poliendis vestibus adsumitur. . . . . .
198. Ergo ordo hic est: primum abluitur vestis Sarda, dein sulphure suffitur, mox desquamatur Cimolia, quae est coloris veri; fucatus enim deprehenditur nigrescitque et funditur sulphure. Veros autem et pretiosus colores emollit Cimolia et quondam nitore exhilarat contristatos sulphure. Candidis vestibus saxum utilius a sulphure, inimicum coloribus. Graecia pro Cimolia Tymphaico utitur gypso.

This, in K. C. Bailey's translation, reads:

Cimolian creta *has another use, too – the whitening of clothes. The kind which comes from Sardinia, and is called* sarda, *being employed only for white garments and useless for coloured ones, is the least valuable of all kinds of Cimolia. The Umbrian sort and the one they call* saxum *are more highly prized.* Saxum *has the property of increasing in bulk when soaked in water and is purchased by weight, whereas Umbrian is sold by measure. The only use of Umbrian creta is for imparting lustre to garments.*

K. C. Bailey, *The Elder Pliny's Chapters on Chemical Subjects*, (1932), *2*.

This routine, therefore, must be observed: the garment is first washed with sarda, *them fumigated with sulphur, and finally scoured with Cimolian creta, provided that the dye is fast, for bad dye is detected by this procedure and blackens, and its colour is destroyed by sulphur. On the other hand, the valuable fast colours are softened by Cimolian creta, which freshens and brightens hues which have been dulled by sulphur. If the garments are white, it is better to follow the sulphur by* saxum, *but this cannot be used with colours. Instead of Cimolian earth, the Greeks use the gypsum of Tymphaea.*

Pliny, *Nat. Hist.*, 35, 196–198.

Matters are complicated a little here as regards *creta Cimolia* by uncertainty as to when Pliny is using the term generically and when specifically, but the nature of the specific earth is, as has been said, known. *Sarda* must have been of very similar composition, since it also is used for fulling and shares the Cimolian property of dye-adsorption. The significance of the term *gypsum* has been examined. Blackening would occur in sulphur vapour of any dyestuff containing an admixture of a lead pigment, such as red lead or ceruse. Umbrian earth we may judge from its use is a kaolin type of earth. (This practice, followed in Rome as in Greece, of whitening cloth by rubbing in a "pipe-clay" has had an amusing philological consequence. Such whitening was performed with particular ardour by candidates for offices – hence Persius in

Sardinian fuller's earth, a calcium montmorillonite, is again an article of commerce; see ITALY, Chapter 47.

Jus habet ille sui palpo quem ducit hiantem Cretata Ambitio refers to "chalked Ambition" – and from the whiteness of their garments came the description of them as *candidates*.)

Persius, *Satires*, 5, 177, translated rather obscurely by M. Madan, London, (1789).

candidati

Saxum has been variously identified. Hardouin and Beckmann both believed it to be common chalk, calcium carbonate. The properties of

Beckmann, *op cit.*, *2*, 105; followed by J. Bostock & H. T. Riley, London: H. G. Bohn. (1857), *6*, 300 note.

chalk, however, do not correspond at all closely with those given for *saxum* – it does not swell in water, nor is it injurious to dyes. K. C. Bailey held it to be rather quicklime, obtained by burning limestone – a much likelier identification, since lumps of quicklime are often hard and rock-like, it takes up water to form slaked lime which occupies a greater volume than the original quicklime, and it would destroy many dyestuffs. (Quicklime as Bailey points out, is an artificial product, not strictly to be classed as a variety of *creta*, but confusion of chalk and lime is not unusual even to-day.)

Robert H. S. Robertson, The fuller's earths of the Elder Pliny, *Classical Review*, 1949, *63*, 51–52.

A still more likely supposition, however, is that *saxum* was what we now know as bentonite, in lump form. This detergent clay has the property of swelling to many times its original volume if a dry lump is placed in water. The dried lumps of clay are quite hard and could reasonably be termed *saxum* in the trade, especially if other earths were more friable or powdered. It is also natural that such lump earth should be weighed out for sale, the softer Umbrian earth being sold by measure.

Even calcium montmorillonite swells to *some* extent in water, so the clay called *saxum* may have been calcium montmorillonite.

Bentonite, like English fuller's earth, as we have said, consists essentially of montmorillonite. Both, when fairly pure, have a high affinity for basic dyes.

The bentonite of Ponza was certainly known when the island's water galleries were cut. The meagre water supply of Ponza was collected in the porous rocks lying above the impervious bentonite, and the galleries actually penetrate the bentonite. It can be accepted that the Ponza fuller's earth was known and could have been shipped to Pompeii, before that city was ruined by earthquake in 63 AD and finally overwhelmed by the ashes of Vesuvius in 79 AD. Nero in 62 AD banished his wife Octavia to the neighbouring small island of Pandateria, now the penal colony Ventotene, and had her murdered there, in favour of the infamous Poppaea.

Since all the fuller's earths known to Pliny except *saxum* are named from their places of origin, we may suppose that there was some good commercial reason for withholding the place of origin of *saxum*. Perhaps the Emperor was not aware that quarrying was being conducted so near to the source of his palace's water supply.

Thanks to Warren E. Dawson, 1953.

Ovid, *Fasti*, iii, 821

Sir James Frazer, *Publii Ovidii Nasoni's Fastorum Libri Sex: The Fasti of Ovid*, 5 vols. London 1929.

This chapter may be completed by another reference to fullers in Rome: Ovid (43 BC–17 AD) wrote *Hanc cole, qui maculas laesis de vestibus aufers. Worship her [Minerva], thou who dost remove stains from damaged garments*. The commentary on this passage is as follows (vol. iii, p. 148)

Pliny, *N. H.*, 35, 143 "Officinam fullonis quinquatrus celebrantem." Novius, quoted by Nonius Marcellus, s.v. 'Sapivi", p. 817, ed Lindsay.

Ovid is here addressing the fullers. Pliny mentions a picture by an artist called Simus which represented a fuller's shop celebrating the Quinquatrus; and the comic playwright Novius alluded to a fuller at the Quinquatrus, a festival of Minerva of which two were celebrated at Rome: the greater on the 19th of March (the fifth day of the ides) and the lesser on the 13th of June. An inscription regarding a votive offering by fullers to Minerva has been found at Spoletium (Spoleto) in Umbria.

H. Dessau, *Inscriptiones Latinae Selectae*, No. 3127.

Soap

7iv

SOAP the Romans did not have – or at least did not recognise it for what it was until the latter days of the Empire. Pliny speaks of *sapo*, but by this he means a hair pomade, imported into Italy from Germany and Gaul. The tribes were said, probably inaccurately, to use it for colouring their hair. The *sapo* from Germany was made, he says, from tallow and ashes – the best from goat's tallow and ashes of the beech tree. Such a pomade, made from these raw materials would be what we now call soap and was in fact eventually manufactured in Rome. The first word of appreciation of its cleansing properties, however, comes only with Galen in the latter part of the 2nd century AD (he settled in Rome in 164), and then only in pharmacy. Sammonicus, who in the 3rd century AD recommends it for washing the body as part of the treatment of certain illnesses, does so in a manner implying that it was not anything like common in his day. During the Imperial period there lived a *negotiator artis saponariae* at Lyons, and at Grand (Vosges) a basrelief was found which seems to depict the workshop of a "soap-manufacturer" dedicated to Juno Saponaria – but there is no indication of course as to the use to which the product was primarily put.

Ger. *Seife*.

More likely they used it for washing their hair.

banker of the soapmakers' trade.

That quite effective detergents were available in Athens for cleansing of the person is clear from the quotation from Aristophanes. The Greeks and Romans of the earlier centuries showed considerable ingenuity also in mixing with oil for the purpose of cleaning the skin mechanical agents such as bran, sand, earths, ashes and pumice. But something of the necessities the Romans still laboured under in the baths is apparent in the message accompanying some strigils in Martial's *Epigrams: Pergamus has misit: curvo distringere ferro. Non tam saepe teret lintea fullo tibi* (*Pergamus sent these: scrape yourself with the curved iron, and the scourer will not so often have to cleanse your linen*).

Martial, *Epigrams*, xiv, 51

More on urine

7v

COLLECTION of urine was systematic. *Dolia curta* or *testae*, large earthenware vessels, were placed at street corners or in the narrow streets between the houses, and collected by the fullers when they had been filled by the passers-by. This had its disadvantages from the point of view of amenity, reflected in Martial's description of Thaïs, who *smells worse than an old jar of a covetous fuller broken in the middle of the street* (the rival subjects of comparison are rotten eggs, an amorous goat, a fetid jar of pickles and a dead dog floating in the Tiber.) After treatment with a depilatory and bean unguent, she smelt fine.

Lucretius, iv, 1026. (98–55 BC)

Martial, *Epigrams*, vi, 93. 1 London: Bohn's Ed., (1860), 302. As a 16th c. MS has it *Worse than a fuller's tub doth Thaïs stink, broke in the streets and leaking through each chink.* Thaïs was an Athenian courtesan who accompanied Alexander the Great to Asia.

Rome like the Eastern civilisations made shift however to accept such inconveniences, and so, though methods of collection varied, did Europe right up to the mid-19th century. In England stale urine, known as "old lant" was commonly mixed for fulling with pig's dung. So efficacious was this mixture held to be that when soap was first introduced for milling moleskins and corduroys the British workman became suspicious: he did not believe that cloths of this type could be durable unless the odour in preparation was overpowering. Urine was still in use for washing clothes at sea as late as 1875. F. T. Bullen writes in an account of a voyage in one of the New Bedford Whalers, which sometimes remained at sea for three or four years without putting into port:

Enzymes in dung gave limed leather before tanning a smooth silky grain. Synthetic "bates" are now used. Was there any such justification for bating woollen cloth?

F. T. Bullen, *The cruise of the Cachelot*, London, (1898), Ch. 4.

> *No water was allowed for washing except from the sea, and everybody knows, or should know, that neither flesh nor clothes can be cleansed with that. But a cask with a perforated top was lashed by the bowsprit and kept filled with urine, which I was solemnly assured by Goliath was the finest dirt extractor in the world for clothes.*

(The officers preferred apparently to use the lye of ashes from the galley fires.)

The fulling profession may have derived some benefit from the use of urine. The Roman members of it at any rate probably cheered themselves with the belief that this was so. *Male urine cures gout*, Pliny maintains stoutly – *witness the fullers, for instance, who for this reason, it is said, are never troubled with that disease.*

Pliny, *Nat. Hist.*, 28, 18.

The urine of the toper was no good to them, being too low in nitrogen content, and it was just the most inveterate drinkers who had most often occasion to use the urinals. There is a double social condemnation in Macrobius's scornful picture of the wine-bibbing dignitaries of his day:

Macrobius, III, 16, §15. Flourished ca 400 AD, just before the fall of Rome.

> *They sit at the gaming tables perfumed in a manner to which they have given much forethought, and surrounded by courtesans. When it is the tenth hour, they summon a slave to take them to the* comitium, *to find out what is happening in the forum: what the tribunes have decreed, what they have prohibited. Then they make for the* comitium, *for fear of involvement in personal responsibility for matters they have not thought about. On the way, there is not an alleyway at which they do not stop to add to the contents of the urine vessels: their bladders being ever full because of the quantity of wine they drink. They come with a bored air into the* comitium. *They order the trial to begin, the parties concerned state their case, the judge requires witnesses and goes out to urinate.*

This problem of quality was one Britain partly solved in the late 18th century when an instrument known as Twaddel's hydrometer came on to the market. It was invented by George Macintosh of Glasgow for the purpose of testing urine (in the first instance for use in manufacture of the dye cudbear.) Collectors of the liquid went out armed with pocket hydrometers, *to prevent being imposed upon by a spurious and inferior article,*

A. and N. L. Clow, Some chemical aspects of the industrial revolution in Scotland, *Aberdeen University Review*, (July 1940), 6 and make sure they got the "well-brewed", not the "toper's" variety.

7v

The Romans were well aware of the quality of urine of different animals. As Pliny noted, in Holland's incomparable translation, *as touching the camel, cammels pisse (they say) is passing good for Fullers to scour their wool withall* – and is good also for the cure of running sores. Barbarous nations were in the habit of keeping it till it was five years old.

Pliny, *Nat. Hist.*, 28, 26; Philemon Holland, London, (1601), vol. 2, p. 311, column C.

Whether the collected urine was of attested quality or no, the emperor Vespasian in the seventies of the first century put a tax on it. It is not clear how this was collected, whether it was paid by the fullers or was a kind of poll-tax, but, unpopular as it was, it was renewed by Vespasian's successors off and on for the next two hundred years. Suetonius records that Vespasian's son Titus at the time of the first imposition protested, declaring such a form of taxation a disgrace to a civilised community. Vespasian sent for one of the coins received in the first instalment of the revenue, applied it to Titus's nose, and enquired if the money stank, – an example of their sense of humour:

One of the sumptuary laws based on the Fannian law of 161 BC.

Suetonius *Vespasian*, 23
fl. 69–79 AD

> *Reprehendenti filio Tito, quod etiam urinae vectigal commentus effet, pecuniam ex prima pensione admovit ad nares, sciscitans num odore offenderetur: et illo negante, at qui inquit, e lotio est.*

According to Schaefer the Emperor Commodus again thought fit, in AD 190, to tax Roman urine collection because the *fullones* had become too prosperous. As far as is known fuller's earth escaped taxation.

G. Schaefer, some old-fashioned ways of cleaning textiles, *CIBA Review*, (April 1947) *56*, 2035.

A historical account of the technical uses of urine was given by Binz (1936).

A. Binz, Altes und Neues über die technische Verwendung des Harnes, *Angew. Chemie*, (1936), *49*, 355–60: see also *id.*, (1937), *50*, 120–1.

The ancient method of preparing woad was by fermentation in urine. Dyers did not always escape the odium experienced by fullers. In an Egyptian papyrus of ~ 2000 BC it was said of a dyer: *His hands stink, they have the odour of rotten fish, and he abhorreth the sight of all cloth.*

Earths in pharmacy.

7vi

> ***This seems a suitable place in which to remark that all earths should be washed freely with water, and set aside to settle, when they may be made into tablets. The vessels in which they are heated should be frequently shaken***
>
> **PLINY, Natural History, 35, 193.**

ALL medicinal earths were made into cakes or tablets, partly to remove grit, partly to justify a high price and partly to make it easy for the shop to store them.

7vi

If fulling had been practised for over 4000 years in ancient Greece there can be no doubt that fuller's earth and other clays had been used also for cosmetic and pharmaceutical purposes for a very long period before written records, during which time considerable expertise had evolved.

Pliny, *Nat Hist.*, 35, 194

Pliny gives a list of medicinal clays:

> . . . . *Chian earth, whose virtues are identical with those of Samian earth. Selusinian earth, which is as white as milk, and very readily forms a suspension in water, is similarly used. Pnigitis is very like Eretrian earth, but occurs in large lumps and is adhesive. Its medical properties resemble those of Cimolian earth, but it is less active.*

chapter 5 Theophrastus

or oven-dried?

There were two varieties of Samian earth, collyrium and aster, which we have met before. The former to be really good had to be fresh, remarkably smooth, and glutinous to the tongue; the second being of a more solid consistency, and white. Both were calcined and often washed before use; and both were prescribed for discharge of blood from the mouth and as an ingredient of desiccative plasters; also in the preparation of ophthalmic preparations, i.e., salves for the eyes.

Pliny, *N.H.*, 35, 56

John Beckmann, A history of inventions, discoveries, and origins, 4th ed., London: Henry G. Bohn, (1846), *2*, 224.

See chapter 19 v for scouring balls.

Chian earth was used for the same purposes as well as for use as a cosmetic for feminine skin. Beckmann, noting that Chian and similar earths were used in classical times instead of nitrum (sodium carbonate), said that this was still the case in the Levant in the 18th century when *a kind of reddish earth . . . that people of the first distinction never bathe without* was mixed with perfume and made into small balls that were allowed to disperse in the water.

Selusinian earth was also used as a face powder, as it has a milk-white colour and disperses very rapidly in water: suspended in milk, it is used for whitening plastered wall surfaces.

Pliny, *N.H.*, 35, 54 and 21.

or oven-dried?

Eretrian earth may be white or of an ashy colour, the latter being preferred in medicine, and the former presumably as a pigment. In medicine it is cooling and emollient. Calcined, it promotes the cicatrisation of wounds, is very useful as a desiccative, and especially good for pains in the head and for the detection of internal suppurations. If the earth when applied [to the chest] with water, does not dry quickly, the presence of pus is indicated. Eretrian earth should be of a soft consistency, and when rubbed upon copper it should leave a violet tint. Pnigitis is very similar to *eretria* but is found in larger masses and is of a glutinous consistency. The fact that these earths are rather like Cimolian earth but not so active suggests that they might be sepiolite, which does occur in Euboia, or palygorskite.

Both kinds are mentioned by Dioscorides 5, 172 Oribasius Bk 13, γῆ, Aetius 1, 2, 6. Marcellus Emp. 8, 5 Isidore 16, 1, 7, Galen XII, 178 ff. (ed. Kühn) Vitruvius 7, 7, 3.

Pliny still refers, as the Greek authors did, to the two kinds of Cimolian earth used in medicine, one white and the other of a purplish cast. Both kinds, moistened with vinegar, have the effect of dissipating swellings and checking discharges; of reducing tumours of the groin and parotid gland; and of a swollen spleen. It also removes pimples. He gives three recipes:

cimolian earth, 'foam of nitrum' cyprus oil vinegar	This mixture of synthetic bentonite, cyprus oil and vinegar was used for reducing swellings of the feet, care being taken to apply the lotion in the sun and at the end of 6 hours to wash it off with salt and water.
cimolian earth cyprus oil wax	This preparation was considered good for swellings of the testes.
cimolian earth,	especially the varieties from Thessaly and near Bubon in Lycia, when applied in the form of a liniment prevents excessive perspiration; taken with wine, in the bath, it removes pimples on the body.

Pliny, *N.H.*, ch 57, para 195.

A "moist plaster" is to contain:

(illegible)	6 obols	(67 grains)
Cimolian earth	2 drachms	(134 grains)
Pounded copper	2 drachms	(134 grains)
Iron Scale	2 drachms	(134 grains)
Wax	10 drachms	(670 grains)
Oil	1 sextarius	(1 pint)

Pliny, *N.H.*, 34, 155.

This is said to be effective for cleansing wounds and fistulas and for removing callosities. It would appear that the oil was used to soften the wax; the earth served to absorb poisons; and the metals were probably antiseptics.

"Samian stone", probably a compact form of kaolinite, is to be applied, as well as externally *in human milk to ophthalmic ulcers and for chronic watery discharge from the eyes* (where it would of course absorb the discharge), internally *for stomach disorders*; it also *relieves giddiness and restores mental balance* – the main and secondary effects of the excess acidity for which kaolin, as already mentioned, is prescribed to-day.

Pliny, *N.H.*, 36, 152.

The objective empirical nature of the Greco-Roman approach to these matters stands out in its true and really rather remarkable light when some of the vagaries are remembered of the mediaeval and indeed post-Renascence pharmacopoeias. As late as the end of the 18th century, the student of medicine was apt to find himself confronted with recipes made up of such ingredients as spider's web, pigeon's blood, juice of woodlice, human skull and concocted toad. The amazing thing is that the relatively sound understanding of the medicinal uses of clays survived till the first *Pharmacopoeia Londinensis* of 1618 which printed the Greco-Roman formulary for clays almost unchanged. The treatment of clays in future pharmacopoeias is discussed later.

According to a Syrian text, fuller's earth was prescribed for anal fistula. W. R. Dawson suspects that this was a Syrian translation of Galen's *De Locis Affectis*.

Sir E. A. Wallace Budge, *Syrian Anatomy, Pathology & Therapeutics*, vol. 2, pp 510, 515; Syrian text in vol. 1.
W. R. Dawson, personal communication.

7vii

Some fullers of antiquity and in hagiology

Plutarch, *Cicero.*

THEOPHRASTUS was not the only son of a fuller in the Greco-Roman world to rise above his original social station. Plutarch writes of Cicero – though without vouching for the accuracy of the report –

οἱ μὲν γὰρ ἐν κναφείῳ τινὶ καὶ γένεσθαι
καὶ τραφῆναι τὸν ἄνδρα λέγουσιν,

some would have him the son of a fuller, and educated in that trade,

Ammianus Marcellinus, *Roman History*, 14, 11, 44, translated by C. Yonge, London: Bohn's Edition, (1862).

and Marcellinus quotes in his *Roman History* as an illustration of the fickleness of fortune and the changes in life experienced by great persons the story of how Andriscus, fuller and son of a fuller of Adramyttium, was emboldened to declare himself a son of Perseus, last king of Macedonia, and, while the legitimate son was obliged to turn blacksmith to earn a living, in 149 BC he actually seized and held, if only for a year, the Macedonian throne.

Catholic Encyclopedia

A monophysite was a heretic who believes that there is only one nature in the person of Jesus Christ, a belief held in the Coptic, Armenian, Abyssinian and Jacobite (6th c.) churches.

A more remarkable story still is that of Peter Fullo, so-called (in Greek Γναφεύς) after the craft he originally practised in a Chalcedon monastery, who became intruding (because Monophysite) patriarch of Antioch – a rise it is true from the status of a fuller somewhat exceptionally placed, but as a saga of tenacity and mental and physical resilience hard to beat. Expelled first from his monastery for heretical doctrines and dissolute living, Petrus Fullensis, as he is also called, proceeded to Constantinople where by enlisting the good offices of the future Emperor Zeno he obtained an appointment as presbyter at the church of St. Bassa in Chalcedon. Driven thence by the populace, he accompanied Zeno to Antioch, incited the populace there against their patriarch, Martyrius, and on the latter's resignation in 470 usurped the see. Unqualified Monophysite acceptance was accorded him as introducer to the Trisagion of the words ὁ σταυρωθεὶς δι' ἡμᾶς ("who was crucified for us") in the monophysitic sense that the Father and the Holy Ghost were crucified with the Son. In 471 the Emperor deposed him, but he achieved a comeback usurpation in 476. Two years later he was deposed a second time, and banished. 485 saw him reinstated again, however, at the hands of the usurping Emperor Basilicus, and this time the see was his until his death (in 488).

Trisagion, an ancient hymn of the Oriental churches, beginning with a threefold invocation of God as holy.

The east window of Gloucester Cathedral shows St. James with a fuller's club.

Also known as St. James the Just. He was Jesus' brother.

AD 62.

The fuller's billet has also its place in history as the emblem of St. James the Less, traditionally a relative of Jesus (see Fig. 19). This beloved and pious first Bishop of Jerusalem, known as the "defence of the people" (whose knees tradition reports were "hard as a camel's, because he was always kneeling in the temple, asking forgiveness for the people") when the city was laid siege to by Vespasian, like Jeremiah in Nebuchadnezzar's time, opposed resistance, declaring that the Royal Door of Jesus was the open door of peace. A riot followed and he was thrown by the Jews from the pinnacle of the Temple and badly injured.

7vii

Fig. 19. (on left) St. James the Less carrying his Epistle and the Fuller's Beetle. Patron Saint of Dyers, Fullers and Druggists.

Burns, Oates and Washburn.

He was still praying for his murderers when he was finally killed by a fuller who dashed out his brains with his club. St. Blase, the Patron Saint of Woolcombers, was tortured during the persecutions of Diocletian with an iron comb.

Eusebius, *HE* II, 23.18.

The story of St. James the Less has been told many times. John de Trevisa's words may be quoted in their stark simplicity:

ðey smyte out his brayne wið a walkere his perch (Latin *pertica fullonis*)

St. Clement, more specifically, the patron saint of the fullers, before beginning a journey, put soft wool inside his shoes, and thus made a felted lining.

John de Trevisa, *Polychronion Ranulphi Higden*, tr. 1387, Rolls Series (1865–86).

K. G. Ponting, *The Woollen Industry of South-West England*, Bath: Adams & Dart, (1971), p. 10.

A note on beetles

7viii

"DEAF as a beetle" does not refer, as might be supposed, to the insect of that name, but is equivalent to "deaf as a post".

The beetle drawn in many representations of St James is long and heavy, like the Scotch beetle. More commonly beetles were not much

See ch. 30.

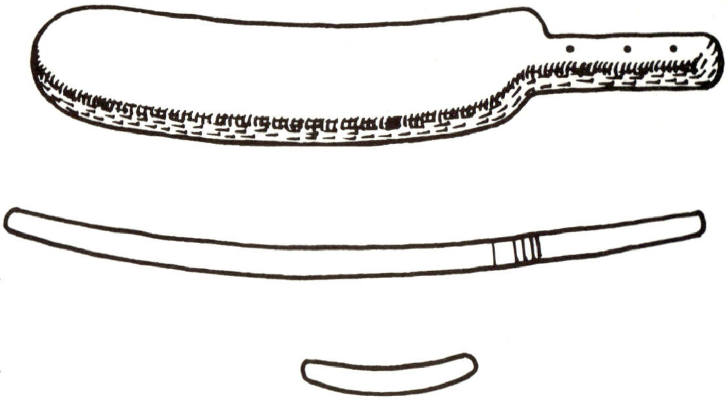

Fig. 20. A fulling bat formerly in Keighley Museum, adapted from a sketch by E. Kilburn Scott. Above, a perspective view; middle, a longitudinal cross-section; below, a transverse cross-section.

Fig. 21A. Fourteenth century Flemish fuller with two-handled beetle.

From the trade book of Ypres

Fig. 21B. Washerwoman at work with her beetle. Miniature in Queen Mary's psalter, 14th century, British Museum, London.

CIBA Review (1947), *5*, 2035

Fig. 22. Women washing at the river. In the 16th century the beetle was indispensible for washing. The ones shown here were used for washing shirts.

Woodcut by Hans Franck (d. 1526)
CIBA Review 5, 2037.

longer than a policeman's baton. In the Cliffe Castle art gallery and museum at Keighley there is a Fulling Bat or Battledore of black oak, about 440 mm long, 120 mm wide and only 10 mm thick: a thicker piece of wood must originally have been fastened to the underside of the implement.

For cleaning the insides of shirts a special short flat beetle was used, judging from a woodcut by Hans Franck (d. 1526). Fig. 22. An Irish beetle is shown in Fig. 69 and the long Scottish beetle in Fig. 66. As 'beetling machines' were used only for giving linen and some cotton fabrics a lustrous thready appearance, they are noted here only in passing.

A. J. Hall, *A Handbook of Textile Finishing*, London: The National Trade Press Ltd (1952), 131–5

By an extraordinary coincidence, *fullo* is also the name of a scarab-like beetle, covered with white spots and used as an amulet. Cotgrave (1611) in his French-English dictionary gives *foulon*, a fuller and second "the great, and stinking Droane, or dorre-Bee." Littleton (1678) tries "earwig or beetle" as the second meaning of *fullo*. Bostock and Riley suggest in a note that the name may possibly be derived from the Greek

Pliny, *Nat. Hist.*, 30, 11, 30/§100

Dictionary of the French & English Tongues (1611).

Latin Dictionary (1678) 1751 ed.

Pliny, *N. H.*, 5, p. 454 of their translation (1856).

7viii

φυλλὸν, a leaf. On the other hand Smith's Latin Dictionary (1912) recalls that Johann Christian Fabricius (1745-1808), the Danish entomologist, identified the beetle as *Melolontha fullo*.

7ix

Etymology of the fulling words

THE words used for naming the industry are so different in sound that it seems probable that they represent individual unit processes, the most characteristic of which would be the trampling of the cloth in a trough, or the beating of the cloth with a beetle, and the raising of the nap in the finishing of the cloth. The etymology of the Latin word *fullo*, *fullonis*, is generally regarded as uncertain, but it has a close similarity of sound with the Greek πλῦνός, a washing trough, as was first pointed out by Vossius (1662). The Greek word πλῦνειν properly meant to wash linen clothes rather than woollen ones, for laundering of which κναφεύειν was the more appropriate word. This word is known to have meant originally the carding or combing of the cloth with the teasel (κναφός) or comb. Similar words are the Welsh *cnaif*, fleece, the English *nap* of cloth, and the Lettish *knabt*, to pick or peck.

plunos plunein

G. J. Vossius, *Etymologicon*, Amsterdam 1662.

knapheuein (or gnapheuein).

The gestural theory of the origin of speech, propounded by Sir Richard Paget and developed by Alex. Jóhannesson, may be invoked here. In fullon-is and plun-os the action suggested by the mouth gesture is a pushing out (f or p), followed by a rounded motion (ul or lu) returning unabruptly to the top (n). This is a fairly accurate representation of the fulling action carried out whether with the hands or with the feet. The end of the cycle in either Greek or Latin is at the top of the movement.

R. Paget, This English London: Kegan Paul (1935).
A. Jóhannesson. *Nature* (1944), *153*, 171, & (1950), *166*, 60.

Batuo

IN Spanish the beetling action was chosen to describe the whole fulling process. In John Minshaeus' *The Guide into Tongues*, a fuller in Spanish is given as *Batanéro* and in Portuguese as *Batãaro*, and in his *Vocabularium Hispanicolatinum et Anglicum* he gives H. Batán de paños, L. mola fullonica, A. a mill to full cloath in. *Dicitur a verbo* batuo, *quod pannos ibi ut densior fiat, crebris ictibus* batuitur *i. conturnditur*, i.e. derived from batuo meaning to strike, beat or hit, to make cloth denser and thicker. Thus batañár paños is "to full or dresse cloath", and batañador is another word for a fuller. He also says in *The Guide* that *beetle* was derived from L. *batuere*. In fact it was derived from old English *bietel*, an ancient word in Germanic languages.

(Also known as Minsheu's Dictionary), John Browne, London, 1617.
H. Spanish
L. Latin
A. English

The word *beat* comes from OE béatan, whereas *bat*, thought by Littré to have come from the French *battre* to beat, may have had a much older, Celtic origin. *Batter* may be from the French, but all these words, whatever their etymology, are gesturally related.

Natta or nacca

THE Latin word *nacca*, used often derogatorily for a fuller, is something of a mystery. It may have meant simply a mean and *nasty* fellow and became associated by transference with one who toiled in a noisome industry, or it may be gesturally cognate with *nap*, or indeed with νάκη , a goat's skin or sheep's fleece. The alternative form *natta* services as a surname in Italy today.

Persius, 3, 31
Juvenal, *Satires*, viii, 95.
Horace, *Satires*, i, 6. 124.

Following the same principles, one can see that the other Greek word for a fuller στιβεύς bears a similar gestural relation to *stamp* or *step* and πιλόειν , to *felt or compress* [wool], is related both to *pile* and to *felt* (German *Filz*). ἡ σμηκτρίς, a late word for a kind of fuller's earth, is derived from σμῆμα, *that which is used for rubbing or cleansing*, and ultimately from σμάειν to *smear*, *anoint* or *cleanse* or its related form σμήχειν to wipe or wash away.

stibeus

piloein

smēktris

Nicochares (*Comicus*) *Herc.* 1; cf. Phrynics (*Comicus*), *Grammaticus*, ed. Lobeck, 253.

Thoughts on the meaning of *walking* and *tucking* are given in chapter 10 (placenames) and on how *smectite* came to be used as a clay mineral group name in chapter 70 iii.

8 ROMAN GAUL AND BRITAIN

8i Gaul

Fig. 23. THE GARB OF OLD GAUL
The Regimental Slow March of the Black Watch.

P. R. Drummond, *Perthshire in Bygone Days*, London: W. B. Whittingham & Co. (1879), 467–473.

In the garb of old Gaul, with the fire of old Rome,
From the heath-covered mountains of Scotia we come.

Translated from the Gaelic by officers of the 42nd Royal Highland (Black Watch) ca. 1763 and set to music by Baron John Reid, of Straloch.

CLOTH-MAKING in Western Europe of course long antedated the Roman Conquest. Wool was the primary material of clothing here as in Italy from earliest times, and quite complex weaves had been developed. Dyeing with woad and madder goes back to the neolithic period. Fulling techniques were probably primitive, however – Martial speaks of Gallic cloth in the first century AD as warm but "filthy", which meant probably that the grease had not been thoroughly scoured out. Retention of some of it in the wool may in fact have served a purpose in the cold and damp of the north-westerly climate, but the chances are that it was not a matter of choice. More advanced techniques must have filtered into the west of Europe to some extent through the activity of the Greeks. The Phocaeans of Ionia, more industrious and mercantile than the settlers of Sybaris, vied with the Etruscans and Phoenicians in trading with the far west of the Mediterranean. They regularly visited Tartessus in the south of Spain and traded with the Celts who lived about the Rhône, and when they founded the colony already referred to at Massalia they began making cloth there and the town became the centre of a number of smaller factories on the coasts of Catalonia and Provence.

Marseilles.

The Romans, as they set up army bases and installed administrative personnel, saw to it that cloth production in the provinces was stepped up technically to give them as nearly as possible woollens of the quality they obtained at home. And in order to cater for an expanding market at home, they introduced better quality breeding stock, and improved breeding methods, to obtain a finer and softer fleece, and, more immediately, set about teaching local labour their own fulling and finishing techniques, which of course included the use of high quality fuller's earth. The military at any rate required large amounts for laundering and set out systematically to discover suitable deposits. Nothing is known about these skilful prospectors whose discovery of fuller's earth proved to be of considerable economic benefit to the occupied countries.

The growth that now took place of a full-scale wool industry in the western provinces was more immediately of course a function of the *pax romana*, which, with the improved long-distance transport facilities, made possible the regional specialisation required for such development. Wool in any state was an excellent material for such traffic, being neither fragile nor perishable; and the needs of the military and other Romans on the spot, plus the fact that a cloth industry was one that could be developed from an already existing domestic industry with comparatively little expenditure of capital, meant that not more than a small proportion of it left the provinces in the raw state.

Cambridge Economic History of Europe, Cambridge University Press, (1952), *2*, 355 *et seq.*

Provisions relating to the cloth trade in Diocletian's Edict of 301 AD, in which he attempted to fix maximum prices and wages throughout the Roman dominions, show that by that date the wool products of the western provinces were of major importance in the economic life of the Empire as a whole, and the degree to which in Gaul, at least, cloth production had become the basis of the inhabitants' livelihood, is shown by the string of sepulchral monuments of the period to workers in the woollen industry. The manufacturing region stretched right across the country of the Belgae, from the Somme to the Moselle, through the modern Picardy, Artois, Western Belgium, Champagne and Lorraine, then occupied by the tribes of the Ambiani, Atrebates, Nervii, Remi, Treveri and Mediomatrici, with their

Edict. Imp. Diocl., 22, 1. **περὶ φουλλώνων** (note the Latin word used in Greek).

C.E.H., *2*, 361

Romanised cities of Ambiana (Amiens), Atrebates (Arras), Turnacum (Tournai), Remi (Rheims), Augusta Treverorum (Trèves), and Augustoduno (Metz).

Of the organisation of this industry we know very little. It is not even certain to what extent it was concentrated in the towns. Some of the clothing was probably exported in a semi-finished state, for there is no mention in Diocletian's edict of the price for fulling new Nervian cloaks (*birri*). But that some fulling and finishing was carried out in Gaul would seem to be indicated by the tomb at Sens (Fig. 25), which shows the fuller trampling cloth in a trough, precisely in the same attitude as the fuller shown in the painting at Pompeii, some washed cloth hanging on a bar behind him. And, above, steadied by his body, are depicted his three-foot-long shears (Fig. 24). Much dyeing was certainly done, though here the Gauls could not hope to compete with the Mediterranean lands since, much as they loved gaudy colours, they had neither the Tyrian purple nor the crimson *kermes*. Vegetable dyes, however, they cultivated extensively, and with these, relates Pliny, they imitated all the colours, even producing a substitute, if not a fast, Tyrian purple, *without running the risk of drowning by seeking the mollusc on the sea bottom: without exploiting the deep to multiply the attractions of courtesans*. The sepulchral monuments at Arlon in the heart of the cloth country, depicting men at work stirring a deep vat, have puzzled archaeologists, but there can be little doubt that in fact they represent dyers. From the monuments, indeed, it would seem as though fullers and dyers were in a position of special importance and were men of substance as compared with the other workers. This is to be expected in view of the capital equipment they required and the fact that they were concerned with the finishing processes of clothmaking. Even more important were the cloth merchants with their expert knowledge of the foreign markets on which the very life of the industry depended. Such were the Secundini, commemorated by the sumptuous monument at Igel near Trèves, whereon may be seen their bales of precious cloth in course of transport, first by barge along a river, then on mule-back over a lofty mountain range, doubtless up the Rhine, over the Alps, and into Italy itself.

C.E.H., *2*, 362

Charles Roach Smith, *Collecteana Antiqua*, London: T. Richards (1861) *5*, plate xx.

Kermes was prepared from the dried bodies of the females of a scale insect *Kermes ilices*, which live on certain oaks in the Mediterranean.

There were also the Romans' own imperial workshops or *gynaecia*, at Turnacum, Augusta Treverorum, Augustoduno and Remi – the latter to remain for centuries a centre of the industry. Cloth-making as a domestic or estate activity does not seem to have bulked very large in any of the provinces in early times but R.G. Collingwood and other economic historians have made it clear that the decline of the Empire, with diminished economic activity at the centre, did not immediately entail a corresponding decline on the periphery. Indeed, for a time trade was stimulated, as there was an increased demand to meet and provinces with a strong export trade found their markets abroad expanding as Italian production lagged. The general pattern outside the

There was also an Imperial gynaecium (factory employing women) at Winchester, see Chapter 8ii.

R. G. Collingwood, *Roman Britain* in T. Frank's *Economic survey of ancient Rome*, (1937), *3*, Baltimore.

Figs. 24 & 25. Cloth shears and tenting frame (on left); the Fuller of Sens (on right). These figures form part of a single bas-relief on a Gallo-Roman tomb at Sens.

C. Singer, A History of Technology, 2, 215

imperial cloth-factories, was one of committal to a public fullery such as probably existed at Vidi, a suburb of the Roman Lousonna (now Lausanne), and discussed under Switzerland.

Chapter 52.

Gallic cloaks and Gallic wool were finding a market in Rome in the first century of the Empire, but the wool was then rough and flocky and the fabrics woven from it coarse. Hence Gallic garments were at first the dress of the common folk, bearing the same relation, remarked Martial, to purple dyed garments made in Rome as did Arretine pottery to the finest crystal. Nevertheless they became increasingly popular, if only for the warmth and protection afforded by the stout cloaks which often had capacious hoods.

C.E.H., *2*, 361

Martial, *Epigrams*, vi, 11.

Charles Roach Smith, in his *Collectanea Antiqua*, gives an excellent

London: T. Richards, (1861), *5*, 168.

a) Trebellius Pollio, *Gallieni Duo*, 6
b) Vospiscus, *Carinus*, 20
c) Martial, *Epigrams*, xiv, 128
d) *id.*, iv, 19

M. & C. H. B. Quennell, *Everyday Life in Roman Britain*, London: B. T. Batsford, 1924, 41.

e) Juvenal, *Satires*, ix, v. 30

C.E.H., *2*, 361.

Paulus Orosius, fl. 415, a Christian priest.

J. Yates, *Textrinum Antiquorum*, London: Taylor & Watson, (1843), pt 1.

C.E.H. *2*, 361

A tenting frame, it is thought.

J. de Lapparent, *Comptes rendus*, (1932), *95*, 257.

description of the woollen manufacturers of Gaul. Unlike the *toga*, the Gaulish garments were tailored, and were more suitable for a colder climate. What might be termed the national garb was the *sagum*, a strong coarse cloak often portrayed in Romano-British monuments. But there were also local specialities. The *sagum* was particularly associated with the Atrebates[a] in the north, who were also noted for the *birrus*[b], a rather similar woollen garment. The Santones in the west were well-known makers of the *bardocucullus*[c], with a hood, and another warm garment called the *endromis* was made especially by the Sequani[d] (it is these which Martial describes as warm but filthy). (The hooded cloak was in use until the 14th century when it *was lengthened into the liripipe, hanging from the chaperon, and finished by snuggling down into a turban*. A hooded jacket was re-introduced in World War II and remains in use.) The *lacerna*, also coarse and warm, was more voluminous and could be worn over the *toga* as an overcoat; it was far from being fashionable wear[e], but was generally donned by all classes in cold and wet weather.

As the Roman sheep-farmers improved the quality of the fleece there was increasing export from Belgica to Italy of high-grade wool and also high-grade cloth, fetching as good a price as any. The quantity of wool produced was as notable as its quality, particularly in the region of the Atrebates. It is related by Orosius that one year real wool, mixed with rain, fell there from the clouds, and whatever the origin of his tale there can be no doubt that in the fourth century the wool of this district was both abundant and excellent. In the edict of Diocletian it was rated more highly than any other wool, even that of Tarentum. Garments made from it were more elaborate – Sidonius Apollinaris for instance mentions that attendants on Prince Sigismer at his marriage wore green *saga* with red borders – and were no longer scoffed at by fashionable Romans, but competed with the finest Italian clothing. Not only were they worn by Romans living in Gaul, but in Italy too they became the rage; an Atrebatic *sagum* was as indispensable in the fourth century for a well-to-do Roman as linen from Egypt. Indeed Nervian cloaks were so popular that they were being imitated in Asia Minor.

Looking again at the monuments preserved in the Mairie at Sens Fig. 24 and 25, one can see that the cloth being treated with the spring shears is hanging upon a frame, the horizontal bars of which are adjustable in height. This operation was not illustrated in the Pompeian murals, as stated above, though shears of the same type were found in the excavations. Absent from Pompeian representations with more reason was the trade of the *vestiarius* or tailor. A Gallo-Roman monument to a tailor, found near Bourges, shows a figure standing full-faced holding before him a garment with a fringe, similar to others appearing in some statues at Rouen.

What deposits of fuller's earth the Romans had found in Gaul is uncertain, but de Lapparent in an exhaustive survey in 1933 found only two areas in which montmorillonite earths occurred of commercially good quality – one on the geologically famous hill of St. Radegonde, near Apt, in Vaucluse (30 miles north, 60 east of Avignon) and the other at Montmorillon, Vienne (45 miles north, 30 west of Limoges), where the mineral montmorillonite was first discovered by Mauduyt. The historical picture is rounded off by the fact that all the well-known Cretaceous fuller's earths of England are of Aptian age – a time measurement taken from the deposit at Apt. Although some other earths have been worked in France for some centuries for use in fulling mills none of them was known to have a high montmorillonite content, although it is not impossible that some quite good fuller's earth discovered north of Paris in 1979 may have been used for fulling in ancient times. **8i**

Bull. Soc. ged. Fr. (1847), 2nd series, *4*, 168-170. See chapter 70ii.

See chapter 40 on France.

The earth at Ivry, near Paris, is halloysitic and the earth at Mormoiron near Carpentras in Vaucluse is similar to American fuller's earth in containing the clay mineral palygorskite; though it is underlain by calcium montmorillonite. The earth of Salinelles near Sommières, Gard, is sepiolite, but there is no evidence that this earth was worked for any purpose in antiquity. The wool industry which developed in Roman Gaul must have managed with inferior sources of fuller's earth – the quality of the wool probably being of greater importance.

Britain **8ii**

IN Britain as in Gaul the Romans found spinning and weaving already skilfully practised; some of the cloth produced had a check type of pattern, produced both with natural-coloured (black, brown and white; or grey made of mixed fibres spun together) and with dyed yarn.

Their own influence had to some extent preceded their main invasion. In the south of England there had been for a long time frequent commercial intercourse with Gaul, and there were tribes like the Atrebates and Belgae which had branches on both sides of the Channel – the closeness of contact is perhaps reflected in what seems to have been considerable similarity in the rather limited range of clothing styles in vogue in the two countries.

The wool industry in Britain, the last of the western provinces to be developed, seems to have grown more slowly than in Gaul. There is no record of export of cloth before the 4th century. By the time of Diocletian's Edict a Britannic *birrus* was on the market, fetching a good price, above that of the Canusian if below that of the Nervian *birrus*.

For a picture of a *birrus*, see mosaic of "Winter", The Roman Villa, Chedworth, Roger Goodburn, The National Trust, (1979), pl. 5.2.

For many miles around Winchester (Venta Belgarum) stretched rolling chalkland on which there were no great wolf-infested forests, but only occasional beech plantations consecrated by the Druids. Large flocks of sheep could be safely tended by few shepherds in this open country. Once the Belgae were sufficiently under control, the Romans were able to establish an imperial gynaecium at Winchester. There are traces also of what was probably a dyeworks at Silchester.

As noted by John Leland, King's Librarian in 1544.

Mineral prospecting went ahead to such purpose that in 49 AD, only six years after the conquest, pigs of British lead were being stamped with the name of the emperor Claudius. One can safely assume that several deposits of fuller's earth would have been known to the Romans not long after their arrival. Bath stone, the Upper Oolite, lying immediately above the fuller's earth, was used for a carved monument of 60 AD.

See B. Young, D. E. Highley & D. J. Morgan ...*calcium montmorillonite* ...*Tillington*, open-file report Inst. Geol. Sci. No. 1979/1.

Tillington is only 10 km from Stane Street, the road from Portsmouth to London.

The factory at Winchester possibly used at first some Oligocene or Eocene clays to be found between Winchester and *Portus Adurni* (Porchester) and *Portus Magnus* (Portsmouth). The Lower Greensand, which contains the now famous fuller's earth deposits of Nutfield, Maidstone and Woburn, also penetrated the land of the Belgae, and may have yielded fuller's earth at Tillington near Petworth some 32 miles away.

Somerset (Avon), Bath area

W. Philps, *The history and antiquities of Somersetshire*, London, (1836), 4 vols.

Tacitus, *Annals*, 12, 32.

See also Ivan D. Margery, *Roman roads in Britain*, I *South of the Foss Way – Bristol Channel*, London: Phoenix House Ltd., (1955).

CONSTRUCTION of roads and earthworks also led to the incidental discovery of fuller's earth deposits. When Ostorius Scapula having subdued the Belgae established in 47 AD his headquarters on a hill between Dunkerton and Wellow, six miles south of Bath, near the line of the Fosse Way, there can be little doubt that his men would have learned of or discovered the fuller's earth either near the encampment or at Odd Down just over three miles south of Bath, since the road and its two deep *fossae* cut right through the seam. At Wansdyke near Bath the Romano-British defence works terminated a few yards north of the fuller's earth outcrop.

Fuller's Earth and the Roman Occupation, *Laporte House Journal*, (1969).

Coffin 2nd c.

Villa 2nd or 3rd c.

There was an extensive settlement on Odd Down for nearly four centuries. Digging fuller's earth and farming were engaged in, but pottery, glass and iron seem also to have been made on an industrial scale, judging by the nature of the artefacts found. A Romano-British stone coffin was unearthed in 1969. It lay beneath the entrance to a Roman villa there.

The fuller's earth at Maidstone, Kent, almost certainly came to light in a road cutting, and about 1711 a Roman urn was dug up by workmen at the Grove, Maidstone, near the seam of the fuller's earth there; other urns were found in the 18th century, with other relics of antiquity and coins, both here and nearby at Vinters, most of the coins bearing the inscription of the Emperor Hadrian (117–138 AD), who came to Britain in the spring of 122. Doubtless the Kentish fuller's earth was taken at this early period to military laundries at Durobrivae (Rochester); and later possibly to some of the Roman villas which were built in the neighbourhood of Maidstone, though no traces of fuller's earth have been recorded in any of the excavations.

It is also worth recording that Magiovinium, a Roman encampment on Watling Street, near Bletchley, may have been put there because it is only two miles from the nearest fuller's earth at Woburn. The same deposit would have been accessible to the permanent settlement at Flitwick some eight and a half miles to the east.

Ostorius Scapula, once established so near to Bath, did not take long to be attracted to the hot springs, already used by the British. Within ten years of their arrival in Britain the Romans had begun to build the great series of baths in their New Town of Aquæ Sulis. The legions settled down to a comfortable existence in the area in amicable relationship with the civilised people around them and can have lost little time in seeing to it that the Cotswold wool of that time would be treated by the best Roman fulling processes using what is now called Bath fuller's earth.

Many pieces of Samian and British ware have been found in the Somerset parish of South Stoke within yards of the outcrop of fuller's earth, but evidence of a fullery has not yet come to light there either.

G. W. Hignett, Manor Farm, South Stoke, *private communication*.

There were many well-to-do landlords in the Cotswolds, living in comfortable villas. Many of them were Britons, who adopted Roman ways, spoke Latin and when they went to market at Corinium Dobunorum (Cirencester) enjoyed their entertainment at the amphitheatre and forum. (Such British often took to wearing the toga, which Tacitus found rather flattering. It is not clear whether the cloth they had it made from was produced in the Romans' workshops or their own.)

Agricola, c. 21 "Inde etiam habitus nostri honor et frequens toga."

One of the largest and most sumptuously appointed villas of the Roman period was built near Aquae Sulis at Wellow. No fewer than four fine tesselated pavements have been found here, as well as a sudatory. So far, though fuller's earth occurs nearby, there is no evidence of a fullery there either. Perhaps arrangements for fulling at Aquae Sulis had already been established by the Celts using the hot water at Bath. In any case the Wellow villas were too grand to be converted to commercial use.

J. Collinson, *History of Somersetshire*, Bath: R. Cruttwell, (1791), *3*, 325–6.

Gloucestershire (Chedworth)

SEVEN miles from Corinium, at Chedworth, is a Roman villa which was thought by Fox (1905) to have had one wing converted by one of its later owners into a fullery and dyehouse for the manufacture of white and coloured cloth – a view that was held until Sir Ian Richmond meticulously re-examined the site. He could find no trace of either fulling or dyeing; the rooms wrongly identified by Fox were all parts of bath-suites, which were redesigned on three occasions, on

G. E. Fox, Notes on some probable traces of Roman fulling in Britain. *Archaeologia*, (1905), *59*, 207–232.
The National Trust guide, 5th edition, (1953) follows Fox.
I. A. Richmond, Chedworth Roman Villa (1966), 15 pp,

8ii

following *Transactions* of the Bristol & Gloucestershire Archaeological Society (1959), *78*, 5–23 and 162–5 and The Roman Villa, Chedworth, The National Trust (1979) by Roger Goodburn, to whom I am deeply indebted for help in this chapter.

Mentioned in Chapter 7ii.

average about 70 years apart, to provide an undressing room, moist tepid room as well as hot steamy (Turkish) and dry hot (sauna) baths, followed by cold plunge baths – as in Finland. Fox had imagined that the steps shown in Fig. 26 led down to fulling troughs, but, as Richmond said, this could not be so since the muddy water would have spilled uselessly from the apsidal troughs into the rectangular space between them, since they had a common water level.

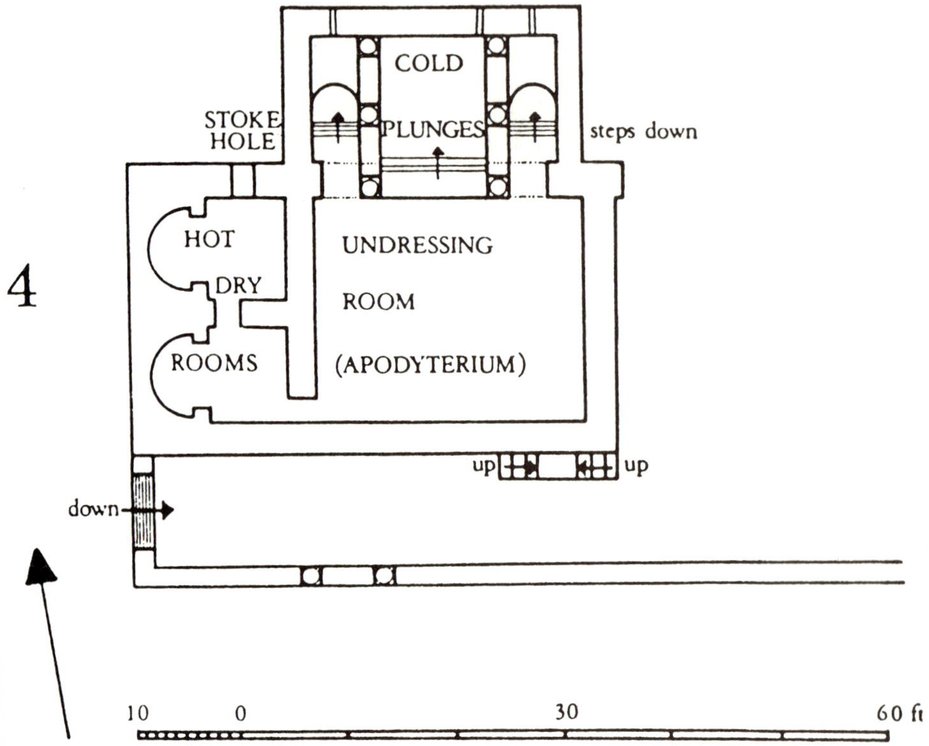

Fig. 26. The fourth stage of the baths in the northern range of buildings at Chedworth.

By courtesy of Roger Goodburn, 1979

Another reason why the round-ended troughs could not have been used for fulling is that the sides are 1.7 m apart. The unquestionable fulling troughs in the *fullonica* between Via Quinta and Via Sexta in Pompeii as measured by Fox in 1886, were 0.75 m long but only 0.6 m wide. No support can be given to the arms if the walls are much further apart than 0.6 m. (Fig. 27.)

Fox more correctly described this Pompeian fullery. *At the West end of the corridor on the North side is a raised platform (2). In this are constructed the treading places, six in number, two on one side, four on the other, of a narrow middle space. They consist of small rectangular compartments divided from each other by dwarf walls, the floor of each compartment being boldly dished, and in*

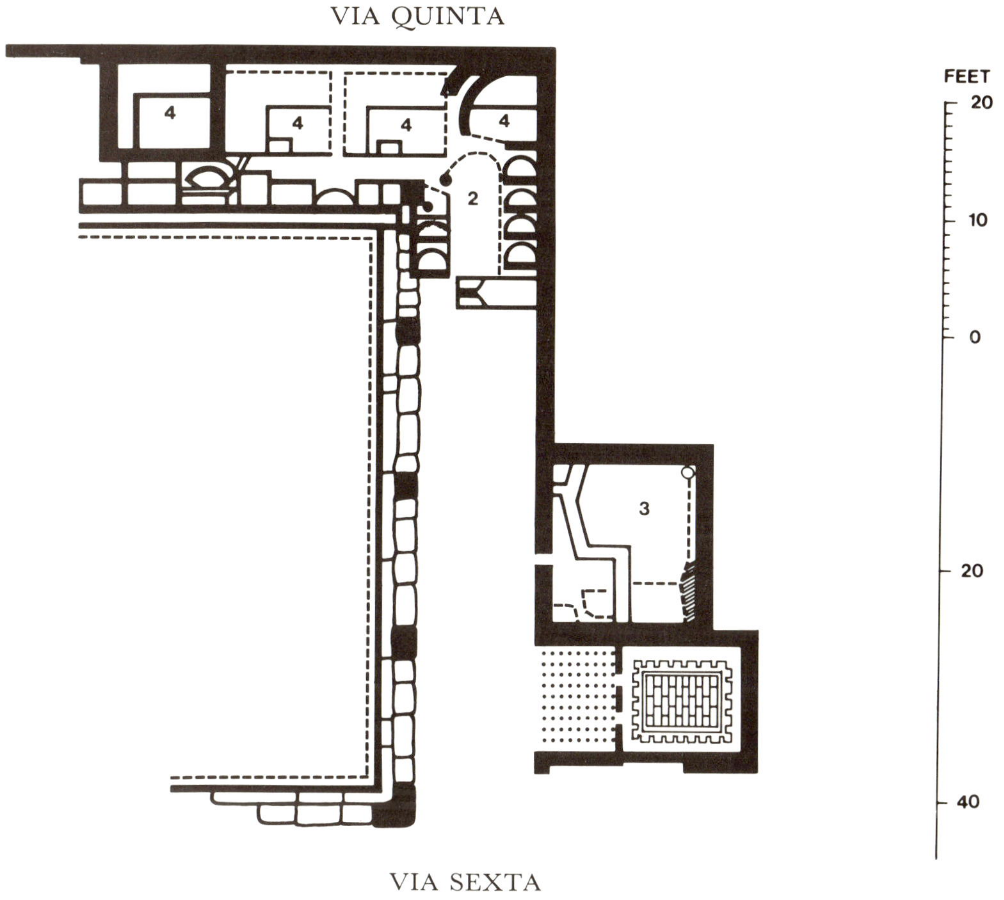

Fig. 27. Pompeian fulling troughs on each side of corridor 2 in house between Via Quinta and Via Sexta (Regio VI, Insula VIII, No. 20).

After G. E. Fox (1905), mens. et del. *1886*

8ii

Real Museo Borbonico, *4*, 13 et seq.

one corner, sunk in the masonry, is a large pot to contain the detergent in use. A supply of fuller's earth was found during excavation in room 3.

C. R. Smith, *Collecteana Antiqua*, (1861), *5*, 20, for a figure.

It is wrong to infer that scouring troughs were usually made of masonry. We have seen that the Cupids of the house of the Vettii are treading the cloth in a long wooden trough with high ends; and the fuller of Sens is at work in a square wooden receptacle with sides raised to a convenient height for resting the hands upon.

The reason why no traces of fulling apparatus have survived from Roman Britain must therefore lie in the use of wood for the troughs. This material was still used as long as fulling by foot was practised; a fuller in his tub is seen in Fig 40a (c. 1460 AD).

Surrey (Titsey)

G. E. Fox, *op. cit.*, 214.

D. C. Whimster, *The archaeology of Surrey*, London: Methuen, (1931). 151–2.

FOX was right to be on the lookout for "probable traces of Roman fulling in Britain", but his enthusiasm led him to two more false identifications, at Titsey and Darenth. The Roman villa at Titsey, excavated in 1864–5, lies close to the Harroway, parts of which were later called the Pilgrims' Way. The site is only eleven km distant from fuller's earth at Blechingley. Indeed it was this ancient trackway which gave access to the West of England after the Roman Conquest of 43 AD. After the

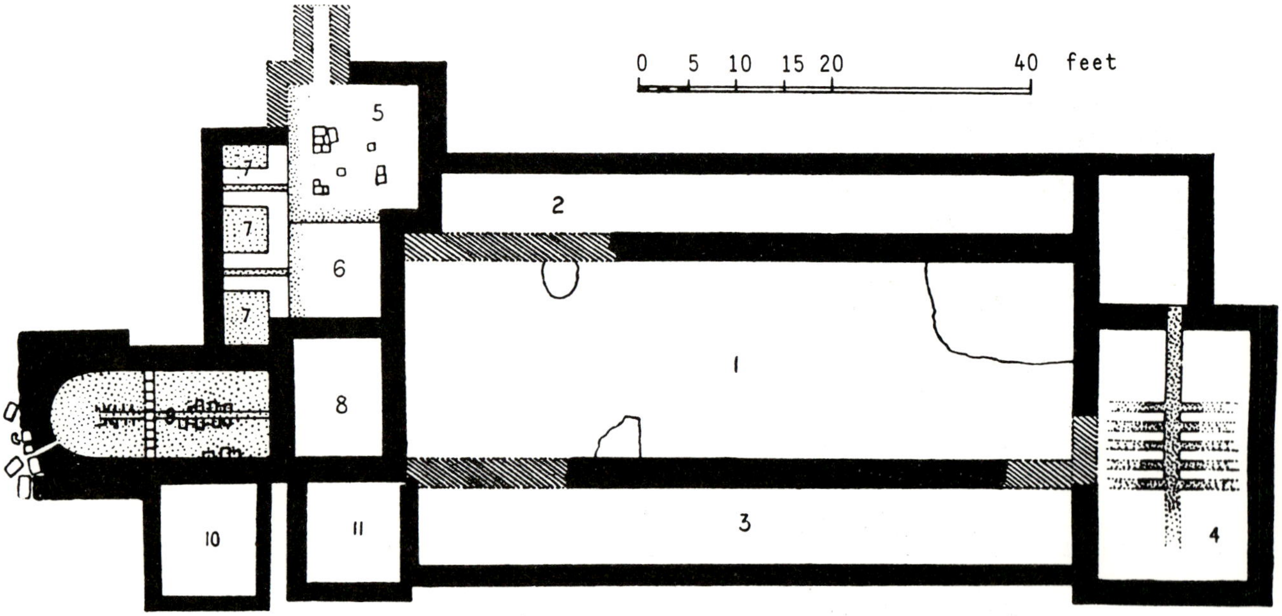

Fig. 28. Plan of a Roman building in Titsey Park, Surrey.

From G. E. Fox (1905)

new Roman road via London to Silchester was paved for wheeled traffic, the Harroway became a favoured site for Roman villas.

Fig. 28 shows the lower story of a "corridor house"; Fox supposed that the family lived in an upper wooden storey since the walls of the lower storey were 0.9 m thick and strong enough to support a second floor. The corridors 2 and 3 gave access to a suite of rooms in 1, and to the stone-built rooms at each end. The three small compartments 7, 7, 7, were all about 1.5 m long. The most northerly one has a width of 0.69 m, the next 1.42 m and the third 1.63 m. The depth of 0.95 m is a clue to Fox's error, for this is the right depth for baths. One might have been for a single person and the others for communal use.

The apsidal room 9 was thought by Fox to be a pool to allow fuller's earth to settle out from suspension, the dividing dwarf wall acting as a baffle, and the channel down the middle would facilitate cleaning out the sediment. He wrote: *these tanks have been called baths, but it would be difficult to find warm baths in any Roman villa in this country either so large or for cold baths so shallow* (0.3 m). Room 4 with its very unusual system of heating was thought to have been a drying room for cloth.

Note: Roman factories in Surrey included a tile yard at Reigate, a glass works at Chiddingfold, and a pottery at Farnham.

Surrey (Blechingley)

IN the parish of Blechingley itself a hypocaust was discovered in 1813 at Pendell Court (Fig. 29). The rest of this Roman villa has never been unearthed, but its owners must have known of the fuller's earth so close to them, and may have supplied the slave labour to work the mineral.

la Pende in 1259. (cf. pend, an archway).

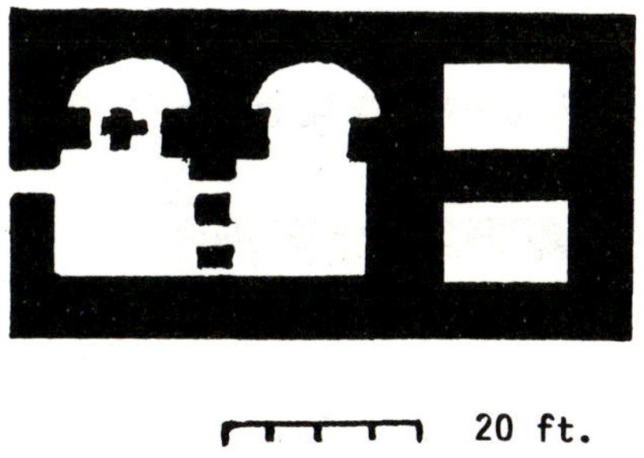

Fig. 29. Hypocaust discovered at Pendell, Blechingley.

From Uvedale Lambert (1921)

Kent (Darenth)

MORE extensive changes than at Titsey seem to have been made at a villa in what is now South Field on Court Lodge Farm, between Darenth and South Darenth in Kent. Here the whole place, in Fox's view, having become derelict, was at one time *converted entirely into a fairly elaborate fulling establishment. Later, a Romano-British family reconverted*

8ii

Now A229 and A2.

part of it into a dwelling house, complete with its suite of hot, tepid and cold baths, and kept one building only in operation as a fullery. The main detergent could have been fuller's earth transported by road to Rochester and then on by the Dartmouth road.

In Fox's opinion room 14 (Fig. 30) had been converted to a shallow trough serving as *the great treading place of the* fullonica, *in which the larger materials to be fulled could be trodden in the liquid detergent filling its lower portion and beaten and scrubbed when pulled up over the sloping sides.* This unusual structure was later converted to a hypocaust for heating a new suite of baths.

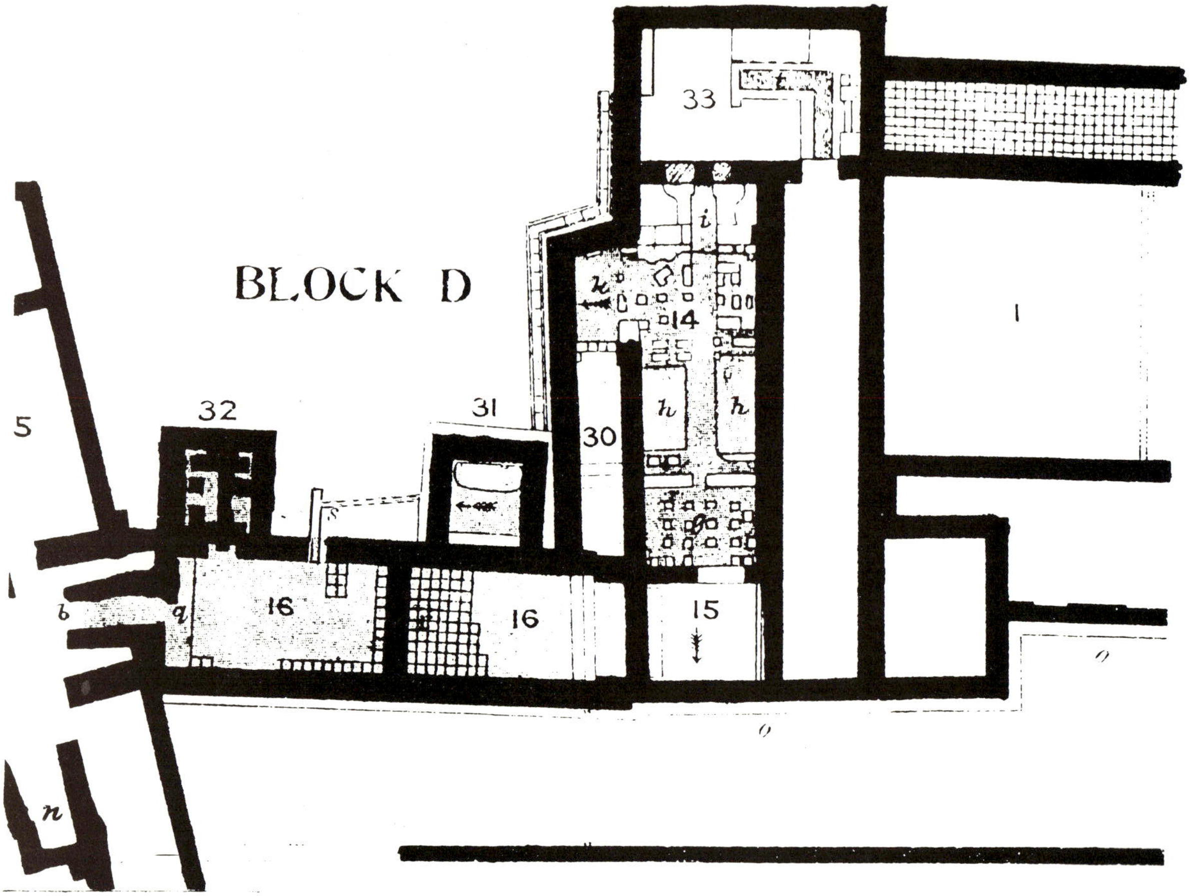

Fig. 30. The rooms in Darenth Roman villa known as Block D in which Fox wrongly thought room 14 had been a large shallow fulling place with sloping sides, h, h, coming out of the water.

From G. E. Fox (1905)

The site was re-excavated in recent times and the general points of Darenth were outlined by Alec Detsicas (1983). Again, no convincing evidence was found for stone fulling troughs in any of the buildings.

A. Detsicas, *The Cantiaci*, Alan Sutton (1983).

Sussex

FULLER'S earth was, however, discovered in 1949 in a Roman timber-lined pit at Chichester. The site excavated by Dr A. E. Wilson of Brighton Technical College was not intensively studied and no sample of the fuller's earth has survived.

The earth may have come from Tillington, 26 km away. See Chapter 28.

The excavation was at E. Palland House Garden, and was described in *Sussex Archaeological Collections*, (1951–2), *90*, 167–178.

THE DARK AGES

9

BLADUD, the King's son, was banished because he had a foul skin disease, and became a swine herd. The disease was transmitted to the swine, but, when they wallowed in the mud near the hot springs of Bath, they were cured. The prince did likewise and was cured.

Geoffrey of Monmouth ca 1139

Geoffrey of Monmouth, Historia Britonum, 1st ed. before 1139. First extant text 1147. The work is highly imaginative and not historical; but it did create the Arthurian legend.

THERE is no evidence for the legendary foundation of the city by Bladud in 863 BC, but the Roman name *Aquae Sulis* commemorates the native deity *Sul*, said to have been considered by the Romans as the counterpart of Minerva, who presided over the fullers. The name survives in the neighbouring Solsbury Hill.

The story of Bladud has no historical basis, but may have been put together from traditions, which Geoffrey richly embroidered. We are left though with some intriguing questions. Was the disease swine vesicular fever, a pustular skin disease? Can this be suffered by man? The waters of Bath *are very beneficial in cases of rheumatism, gout, neuralgia, sciatica, diseases of the liver, and cutaneous and scrofulous affections.* Fuller's earth is known to give relief in eczema and to reduce mustard gas blisters. Perhaps Bladud's affliction was eczematous. In any case wallowing in the mud was more likely to have therapeutical value than being touched by the monarch in cases of King's Evil.

Scrofula, a tuberculous condition.

I shall now return to more rigorously historical events.

Slavery wrote Engels *no longer paid; it was for that reason it died out. But in dying it left its poisoned sting – the stigma attached to the productive labour of freemen. This was the blind alley from which the Roman world had no way out . . . nothing could help here except a complete revolution.* That revolution, the work of the northern Barbarians, took place between AD 400 and 800, commented Benjamin Farrington. Engels continued *The social classes of the ninth century had been formed, not in the rottenness of a decaying civilization, but in the birthpangs of a new civilization.*

in *The Origin of the Family*

B. Farrington, *Greek Science*, 2, Pelican Books, A192, p. 167.

This new civilisation, arising out of the grave of slave society, soon flowered in a series of new inventions which transformed the economic basis of life. Although the distilling of fermented beverages to make spirits was practised in the East long before Christ, spirit of wine (brandy) was made by the Arabians, during the European Dark Age. Fermented grain was used for making spirits in Ireland and Scotland before the end of the first millenium. The modern harness of the saddle-horse dates from the 9th century; and the modern harness of the

It is naive to suppose that the science of the Renascence arose because Greek books from Constantinople arrived in western Europe.

9 Le Mercure de France, (May, 1932).

* probably earlier.

draft-animal in the 10th c. Des Noëttes attributes to the 12th c the watermill*, windmill, mechanical saw, forge with tilt-hammer, bellows with stiff boards and valve, window-glass, the domestic chimney, candle and taper, better paved roads and the wheelbarrow.

The watermill was known to antiquity but apparently little used. Its reintroduction for milling grain and then for fulling cloth took place gradually during the 10th to 12th centuries, a mechanisation as decisive in the Middle Ages as that of spinning and weaving in the 18th c.

Domesday Book 1086 AD.

Water-mills were widespread in England by the time of the Domesday Book, in which 5624 are recorded south of the Trent and Severn. Fulling-mills were known in Europe in the 11th century but the earliest reference to one in England is in 1175, when fulling and teazling mills were apparently worked at Berton in Gloucestershire. A survey made in 1185 of the Templars' English estates notes one at Newsham in Yorkshire, and another at Barton (presumably the Berton above) in the Cotswolds, near Temple Guiting between Cheltenham and Stow. The religious-military order of the Knights Templar was officially recognised at the Council of Troyes in January 1128; its Grand Master visited England later that year and within twelve months the Order held substantial estates in France, England, Scotland, Flanders, Spain and Portugal.

The Holy Blood and the Holy Grail, M. Baigent, R. Leigh & H. Lincoln, (1982), Book Club Associates and Jonathan Cape, pp. 38–41.

No doubt there was confusion when the Romans were recalled from Britain about 425 AD, but sheep flocks were still part of the scene as recorded by Bede in the 7th c. – it was while Cuthbert lay awake tending a flock that he had the vision that led him to enter a monastery.

St. Cuthbert, d. 687, bishop of Lindisfarne.

In the late 9th and 10th centuries, a clay rich in illite, less kaolinite and some interstratified illite-smectite, possibly Mercian Mudstone. may have been used by clothworkers for fulling and dyeing at the Viking site at Coppergate, York.

Allan Hall, Environmental Archaeology Unit, University of York, *correspondence*, (Jan. 1984) Mercia Mudstone formerly Keuper Marl.

The clay is mixed with dyeplants; and loomweights have been found nearby. The clay, occurring between two houses, was found within a pit used to dispose of a variety of débris.

Offa died in 796 AD.

Mercian cloth was famous, one of the few cloths of any importance at the time.

Certainly quite early in Anglo-Saxon times cloth was being exported. A famous letter from Charlemagne to King Offa complaining of a deterioration in the quality of English cloths bears witness to a well-established trade. By that time also woollen cloaks and tunics such as Anglo-Saxon clerics wore in England were being sent to English missionaries facing hard winters on the continent. Boniface, though he bought some clothing near the scene of his missionary labours in Frisia, wrote to the Abbot of Wearmouth: *If it be not too much trouble, pray send us a cloak, for this would be a great comfort on our journeys.* A Norwegian saga mentions English cloth among imports to Iceland at the end of the 10th century. Sweyn Olaf's son, cruising off Dublin in

1171, captured two ships out from England laden with English cloth, and returned in triumph to the Orkneys with his booty displayed sewn to his sails – his voyage ever after known as "the broadcloth cruise" (*skrno viking*).

There has been a suggestion that the so-called "Frisian cloaks" (*pallia fresonica*), which Charlemagne himself wore and which played an important part in the commerce of his day, finding their way to Norway and Southern Europe and with Charlemagne's embassies to Baghdad, may in fact have been made in England or Flanders and known as Frisian because Frisian traders marketed them.

How far the availability of good fuller's earth in England played a part in making English cloth so popular is unknown, though one must remember that the transport of fuller's earth for long distances, unless by sea, was difficult and costly.

Certainly the English cloth that was exported found favour abroad. There was also no lack of English merchants to sell it. Anglo-Saxons were to be found in the Carolingian period at the big European fairs – Charlemagne in his letter to Offa in fact promises them protection "in accordance with the ancient customs of trade" – and after a disturbed period in the 9th century references to them became increasingly frequent. In the Norwegian Sverrir's Saga King Sverrir commends in a speech the trade of the English who brought wheat and honey, fine flour and cloth.

or Sverre Saga c. 1186 AD. He died 1202 AD.

Cloth exports from Britain were, all the same, through the first part of the Middle Ages, small in comparison with those of Flanders, Brabant and later Italy. Even in the 13th century, after rapid expansion of the industry following the harnessing of water-power, production seems to have exceeded by relatively little the demands of the home market. The country remained a purveyor of wool rather than of cloth, and it was wool that the foreign merchants, who from early times frequented the islands, came through until the 14th c. to buy. Bede mentions Frisian merchants in York and London, and an enactment of King Aethelred about London tolls cites explicitly merchants from Flanders, Normandy, Francia and the German Empire. With Flanders in particular foundations were very early laid of the partnership, so characteristic of the European economy in the Middle Ages, whereby England supplied the wool that was the basis of the Flemish cloth industry.

Whether fuller's earth went out as well is not clear. No record of its export has come to light until the prohibition of it, late on in the Middle Ages. The very fact that after the prohibition foreign manufacturers made it worth while to smuggle the earth out to them, however, would seem to indicate a tried acquaintance.

See chapter 13 vi.

How cloth-making was organised in Britain in the early centuries is likewise problematic. Some was being done in the monasteries or convents – but the charters make it clear that most of the cloth for the

St. Gregory of Tours, (538–594), historian of the Franks.

pecten = reed or sley of a loom.

religious communities was ordinarily bought at markets, often at some distance. Some of the large estates probably supplied their own needs in this respect. It seems likely that the tradition of Roman times persisted of an industry of expert artisans, working, in some cases, doubtless for certain processes in their own homes, for a local if not an international market. Such at least was the case in France, where the independent artisan working on commission appears in the pages of Gregory of Tours – an *artifex lanariae* called in to manufacture the royal stock of wool – together with a manager of a corn mill, who also *pectinibus insedit lanasque composuit* (set up his looms and made cloth). The religious communities would have offered such workers the security of an admirably steady market.

It is such an independent *artisanat* that has, at all events by the time more information becomes available, in the early 12th century, either congregated or come into being in the towns.

The sources from which this information mainly derives, the first records of the guilds – the first of which were formed in England, significantly, by workers in the cloth trades – make it plain also that the places of their congregation, and of the growth of the newly-risen towns most concerned with cloth-manufacture, were in the first instance in some measure determined by location of the fuller's earth deposits.

In Roman Britain the *fullonicae* were presumably first sited near a deposit of fuller's earth and then rarely more than a day's waggon-journey away from one. And the considerations then operative were so still. Overnight stabling and foddering of a waggon-team, and its driver, was a costly proceeding. To relay with a second team involved maintenance expenses justified only by large-scale regular production. Water transport was hampered by the fact that sacks were rare and expensive, and the earth had to be moved either in wooden casks or loose in the waggon – transfer from the one vehicle to another being therefore very arduous or very time-consuming or both.

ANGLO-SAXON BRITAIN

10

"Hæthen cild bith ge-fullod"
A heathen child is christened
ALFRIC, Archbishop of Canterbury, *Paschal Homily*

Alfric died 1000 AD

THE only word resembling "full" that occurs in Anglo-Saxon is *fullian*, used only with the meaning of "to baptize". This sense continued into Middle English. *Fullian* is probably used figuratively meaning to whiten, cleanse or purify. We are reminded that in Egypt the word to purify also meant nitron. The word *fullo* and the late Latin *fullare* do however occur in monastic documents in religious establishments which operated fulleries, and *fullen* appears in Middle English. The early Latin documents often use the word *fullonia* for the place where cloth was fulled by foot but *molendinum fullericum* and variants when mechanical fulling mills had been installed. When water-mills are mentioned it was common to specify whether they are for grinding corn or for other purposes.

Century Dictionary, New York (1890).

Fullen in W. Langland, *Piers Plowman*, xv, 445 (Skeat)

Although the surname Fuller appears in the late 12th c., the root *full* does not appear in place-names of Anglo-Saxon origin, simply because the word rarely had this sense. Thus Fullerton, Fulstone, Fulbourne, Fulfen, Fullarton, Fulmer, and Fulshaw are thought to be derived from old forms like fuȝel meaning a fowl (bird). Some others mean foul or dirty, such as Fulford, Fulforth, Fulledge, Fulready, Fulwell and Fulwood. A few *ful*-names come from personal names, such as Fulham from Fulla's hamm, Fulking from Folka's people, Fulmodeston from Fulcmod's tūn, Fulscot, Fulstone and Fulstow from Fugol's cot, tān and stōw.

The Survey Gazetteer of the British Isles.

or fugel, cf G. Vogel

The Anglo-Saxon fulling words which have survived in place names are blǽcan, to bleach; túcian, to dress or finish cloth after weaving, especially to stretch on tenters; and wealcere, one who fulls cloth. More in explanation has to be given to these words.

Early forms of *bleach* are, besides blǽcan, *blechen* and *blech*. The word black has a difficult history. In Anglo-Saxon it was blæc or blac; as the word developed it was sometimes indistinguishable from blác meaning shining, pale. But in all later forms there is a *k*, never the ending *ch(e)* as in words derived from blǽc(e)an, to whiten linen or other cloth by washing and exposing to light.

Oxford English Dictionary

O.E.D. says "in its α- or older form".

Tuck in its original form had a long ōū; in later usage the ŏu sound

Fig. 31. Places of certain or possible fullonical significance in England

became short, and in its present southern usage it bears little resemblance to its older forms.

Walk in the sense "to full (cloth)" is not recorded in English before the 14th c, though prominent in other Teutonic languages, such as German *walken*, to full.

Reference in an Elizabethan statute to a "cloth-fuller, otherwise called Tucker or Walker" makes it clear that in some parts of the country "full" had by then taken preference.

5 Elizabeth c.4.23

The tuck-names

THERE is no doubt that all the tuck-names in the West of England, such as Tuck Hill, Tuckenhay, Tuckerton and Tuckington, are connected with the finishing of woollen cloth. There are dozens of such names.

Tukerton, Tokerton n.d. Buckland, Tokertone 1285 FA, TUN of the tuckers or fullers, E. Ekwall, see below.

The walk-names

BUT when it comes to the walk-names we find several which have undoubted connection with the walking of cloth, especially names like Walker Burn, which are modern. There is some doubt whether Walkerith could mean "the landing-place of the fullers" but Walkern, which was Walchra in the Domesday Book and the present form as early as 1222, could have been in Old English wealc-ærn or wealcera ærn "fulling-mill" or "fuller's house", though the expression is not found in the literature. As we shall see under Surrey, the place Walkhamstead was Wolnestede in the Domesday Book. This place is on the Lower Greensand and must surely be connected with fulling. (The past participle of *wealcan was wēolc*).

The Concise Oxford Dictionary of English Place-Names, Eilert Ekwall, 4th ed:, Oxford: Clarendon Press, (1960)

Now swallowed up in Godstone

The blech-names

FOR some unexplained reason all of the blech-names have been ascribed by most authorities to people called Blecca, Blæcci (a derivative of Blæcca) or Blecci, partly on the grounds that there was a man with a name like that in Lincoln. Uvedale Lambert was, it is believed, the first to have suggested an alternative origin in the Anglo-Saxon word blac or blæcan, pale or to bleach, an allusion to the earliest industrial occupation of places like Blechingley.

According to the *Place-Names of Surrey,* Cambridge: English Place Names Society (1934), *II*, the name Blæcca was borne by a 7th c. prefectus at Lincoln.

Uvedale Lambert, Blechingley, London: Mitchell Hughes & Clarke (1921).

The earliest forms of many blech-names support Lambert's idea, as may be seen below:

Bletchingdon, Ox. Blecesdone DB, Blechesdune c 1130 Oxf., Blechesden 1197 FF, Blechingdon 1279 Abbr.

Blaisdon, Gl. Blechedon 1186 P, Blechesdon 1200, 1202 Cur, 1220 Fees.

DB = Domesday Book.

All from Eilert Ekwall, *op. cit.*, who gives references.

Blechingley, Sr. Blachingelei DB, Blecchingeleg 1195 P (In 1680 gazetted as Bleachley)

Bletchley, BK. Blechelai 1106-9 Fr, c 1155 Oxf, Blechelee 1212 FF

Bletsoe, Bd. Blechesho, Blacheshou, DB, Blechesho 1199 FF, Blechenesho 1247, Ass.

Bletchley Sa. Blecheslee 1222, Bleccheleg 1254 Eyton, Bleccheleg 1255. RH

Now all of these are closer philologically to Anglo-Saxon blæcan, which, as we know, became blechen or blech in the course of time. Three of these places are within five miles of fuller's earth deposits and the others are within practical distance except Bletchley in Shropshire (Sa). One can only surmise that once the tribe of Blæcca had been proposed, all subsequent authorities gratefully accepted the easy invention. One can add still more early blech-names such as Bletchingley, Sx (Blecchingelegh) and Blachford; Dev (Blacheorda), I do also agree with Lambert that the late, inserted t is "odious and otiose".

To be more precise, Blechingley, Sr, straddles the fuller's earth outcrop; Bletsoe, Bd, only five miles from the Wavendon deposits, is on the Cornbrash; and so is Bletchingdon, Ox, where there is some calcitic fuller's earth in the Wychwood Beds. According to the Domesday surveyors there were at Bletsoe in 1086 *2 bordars and 2 serfs and a moiety of a mill* [*din' molin*] [*worth*] *10 shillings.* This is only 20 miles from Wavendon, but there may have been nearer deposits of calcitic earth.

The question arises whether a similar etymological significance attaches to the name of Witney, which like Winchester remained a cloth-making town right through from Roman times and from the 17th c has been famous for its blankets. There were two fulling mills there by 1223 AD. And Witney, Surrey, is on the Lower Greensand.

10
Adam, *Index Villaris*, (1680)

For Blackheath, Sy formerly Blackedfeld, see Chapter 21.

Ox = Oxfordshire
Gl = Gloucestershire
Sr = Surrey
Bk = Buckinghamshire
Bd = Bedfordshire
Sa = Salop
Sx = Sussex

Geol. of the Country around Witney, Geol. Surv.Explan. Sheet 236, (1946), 67–69

Note: the felt-names deserve also to be investigated, e.g.,
Felting
Feltsted
Feltham
Feltwell

THE EARLY MIDDLE AGES 1000–1300 AD

11

"Must a name mean something?" Alice asked, doubtfully. "Of course it must", Humpty-Dumpty said with a short laugh. "My name means the shape I am, and a good handsome shape it is, too. With a name like yours you might be any shape, almost".

'LEWIS CARROLL', Through the Looking Glass, (1871)

Quoted from *This English*, by Sir Richard Paget, Bart., (author of Human Speech), London: Kegan Paul... (1935)

Surnames

11i

WHEN surnames were becoming fixed, about the 12th–13th centuries, they were often taken from a man's trade. Where the occupation was carried out by women in every home, like spinning, the surname Spinner was rarely adopted, and Spinster possibly never. Weaving was a more specialized trade for men, so Webbers and Weavers are more common surnames, like Dyer. I do not know why in Scotland particularly the feminine form Webster should have become so common, but it can be said that the trade which treated the work of many weavers was far more distinctive; and so it comes about that for every Weaver in London 1956–1965 there were 1.9 Tuckers, 3.6 Fullers and 8.5 Walkers. Tuckers are common in the West of England, Fullers in the South East and Walkers in the North and in Scotland.

London Telephone Directories

If we apply the gestural theory of the origin of words we can see that these names describe different stages in the preparation of cloth.

Sir Richard Paget, *op. cit.*

Weaver. (OE wefer); to weave is to push(w) a thin (thread) through the warp (ve). The ee sound also indicates the thin, flat shape of cloth on the loom.

cf sieve, which also has the 'woven' character.

Fuller; the gestural meaning has been given in chapter 7ix, but *Walker* is closely similar, a pushing away (w) in a rounded or cyclic manner (l) and ending in the case of walk with an abrupt stop (k); the k here may represent a return to the top of the cycle.

Tucker is not the tenting operation as has been suggested because tent belongs to the series tit, tot, tight and tart, all of which are something between two sharply defined limits (t-t). The n in tent is the insubstantial cloth lying between the solid tenting frame. (A tent is also cloth stretched over a withy frame). The original pronunciation tōukere shows the relationship with German Tuch, cloth. Gesturally the tucker

11i

is rolling a long web of cloth into a cylindrical (ou) roll. The movement is not unlike *take*, the hand reaches out to its limit (t) and brings in an object towards the body (k).

Draper. The draper showed his cloth by putting a length over one shoulder. The word means the falling down of a flat material to the ground (p). A very small thing falling to the ground is a drip. The French *drap* has the same meaning.

11ii

Encyclopædia Britannica, 14th ed., WOLF and EXTINCT BRITISH ANIMALS

Wolves

THE wolf was the greatest devastator of sheep flocks. Wolves catch their prey by running it down in open chase, which their speed and endurance enable them to do.

Wolf-hunting was a favourite pursuit of the ancient Britons as well as the Anglo-Saxons, and it was not until the reign of Henry VII (1485–1509) that wolves seem to have become extinct in England. One of the reasons for the rise of the English wool industry in the 15th century was the extermination of the wolf; another was freedom from wars which plagued the Continent, the English Channel providing protection from both.

In fact the wolf was exterminated in the *South* of England soon after the Norman Conquest and large flocks of sheep could thrive in safety.

~1275–~1355.

David Robertson, *A Brief Account of the Clan Donnachaidh*, Glasgow (1894), p. 16, quoting Nisbet's Heraldry (1816), i, p. 324.

In Scotland the wolf maintained its hold much longer. Donnachadh Reamhair, a huge man, who had led the clan in support of Robert the Bruce at Bannockburn (1314), was chiefly instrumental in clearing the Atholl Highlands of wolves; and 3 wolves' heads are on his armorial bearings. Sir Ewen Cameron of Lochiel is said to have killed the last wolf in 1680, but wolves probably survived in Sutherland for a few decades longer. In Ireland there is a tradition that the wolf became extinct as late as 1766.

11iii

The fulling mill

"His luddokys (loins) thai lowke like walk-mylne clogges".
TOWNLEY MYSTERIES, ca. 1460, xxx, 313.

FULLING was first mechanised in France early in the 12th century, or possibly earlier, as an adaptation, it is thought, of a corn mill. Instead of the rotary movement of the mill-stone, a stamping or pressing action had to be simulated. In its simplest form two stamps or stocks were set in a strong frame; each was provided with a tappet for lifting it. The axle, powered by a waterwheel, bore a number of cams acting on tappets on the stocks alternately. The web of cloth would have been slowly drawn through the mill if the stocks were vertical and therefore not able to slide the cloth forward.

Fig. 32. The vertical stamping mill for ores.
A. Mortar
B. Upright posts
C. Cross-beams
D. Stamps
E. Their heads
F. Axle (cam-shaft)
G. Tooth of the stamp (tappet)
H. Teeth of axle (cams)

Georgius Agricola, *De Re Metallica*, (1556), from the Hoover edition, (1950), 284

Fig. 33. Millwrights making the various parts of a stamping mill.
A. Stamp
B. Stem cut out in lower part
C. Shoe
D. The other shoe, barbed and grooved
E. Quadrangular iron band
F. Wedge
G. Tappit
H. Angular cam-shaft
I. Cams
K. Pair of compasses

Id., 285

11iii

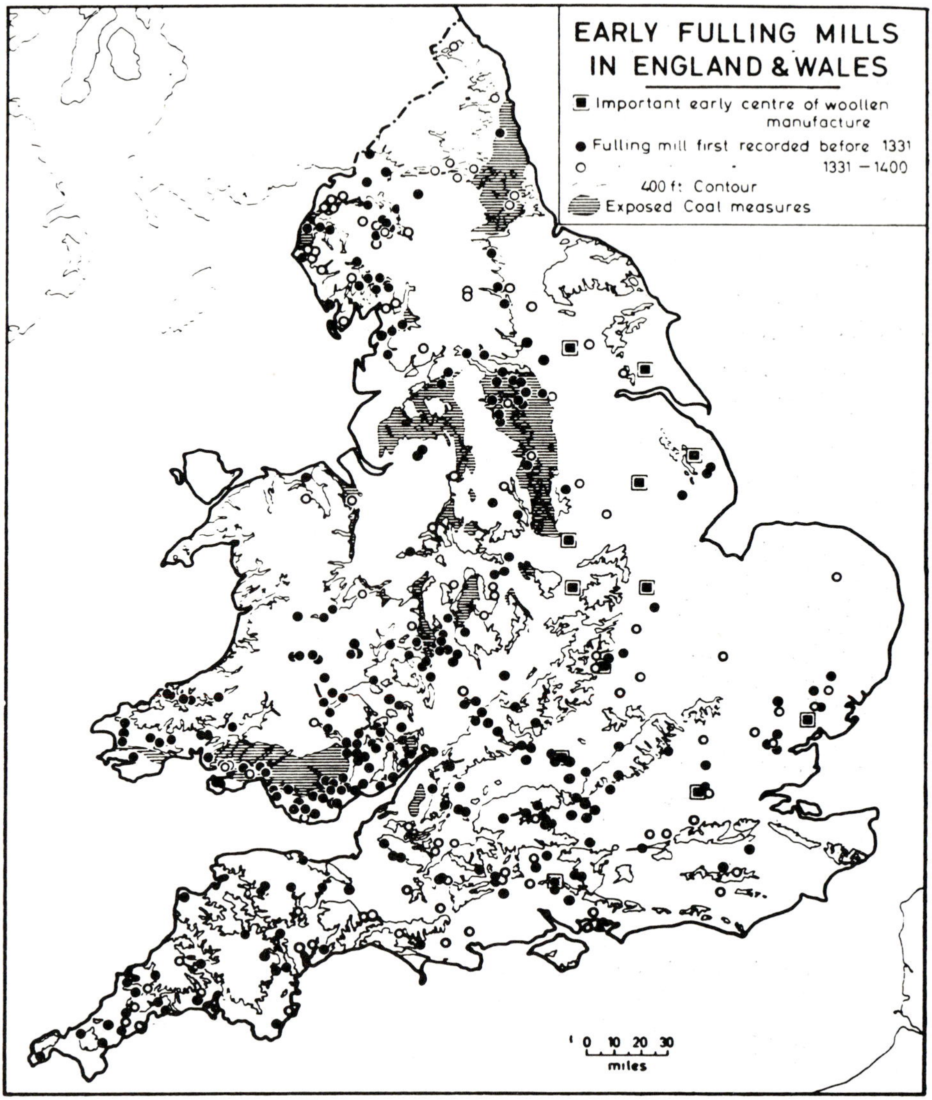

Fig. 34. This map, which is compiled by R. A. Pelham from a wide variety of manuscript and printed sources, indicates roughly the extent to which fulling had already become mechanised before the Flemish immigration of 1331, and also the degree of early mills in the West Riding, where the industry is centred today; another is the general absence of symbols in the worsted region of Norfolk. The important early centres of woollen manufacture shown are: York, Beverley, Lincoln, Louth, Nottingham, Leicester, Huntingdon, Northampton, Colchester, Oxford, Winchester, and London.

There are no early pictures of this simple, though at the time revolutionary device. Some idea of it can be got from Agricola's pictures of ore stamping mills (1556). A fulling mill would be similar but the heads of the stocks would be rounded at the corners. Another picture shows all the pieces which went to make a mill and millwrights at work, Figs. 32 and 33.

Georgius Agricola, *De Re Metallica,* translated from the first Latin edition of 1556 by Herbert C. Hoover and Lou H. Hoover, Dover Publications: New York 1950, pages 284 & 285.

In Paris and elsewhere in France in the early 12th century it was the fullers who caused cloth to be put through its final processes, fulling and shearing; they could therefore sell it to the general public to the annoyance of the weavers who had hitherto supplied the market. The term "drapers" in the 13th century became a synonym for fullers, who had by this time got the sale of cloth into their own hands.

The cloth industry in England developed rapidly after the introduction of the fulling mill. Besides the early fulling mills of Barton and Newsham, mentioned above, one was built by Winchcomb Abbey at Cleveley near Eustone in North Oxfordshire. This was working by 1185 to 1195. Other fulling mills were built at Bourton-on-the-Water (1206) and Brightwell (1208); and there were two at Witney in 1223. They were all built by ecclesiastical landlords. The Church was an important initiator of industrial progress in the medieval period.

Newsham in Yorkshire

The Brightwell mill cost more than £9 to make and erect.

> *The invention of the fulling mill . . . and its spread in the 13th century throughout the country districts, led to a shift of the cloth industry from the old-established urban centres to the rural areas, a movement marked enough to be called an early industrial revolution*, see Fig. 34.

As a result there are innumerable places called walkmill through the country. One could not log them all. In Shropshire alone, by way of example, there were:

At least, as the *Gentleman's Magazine* (1742) puts it, they were to be found "from Devon to Lancs."

Thanks to Sir Hugh Rankin, Bart., for posing a query to the *Shrewsbury Chronicle* in 1940.

Names connected with fulling in Shropshire

Walkmill, 3 miles W. N. W. of Bettws-y-cryn	
Walkmill, E. S. E. of Chirbury	This would have supplied Chirbury Priory in its export of cloth or "retainers of Montgomery Castle."
Walkmill, Clun . . .	This would have supplied Clun Castle, i.e. Barony of Clun "retainers".
Walkmill, Hockleton, Chirbury (Distinct from above) . . .	This is very close to the "china clay" pit on Corndon. It would have supplied Chirbury Priory or "retainers of Montgomery Castle."
Walkmill, ½ mile W. of Wentnor . . . Walkmill, 1 mile N. W. of Leebotwood . . .	Wentnor and Leebotwood were no doubt associated with Cardingmill Valley of Church Stretton and Cardington village. These would have dealt in good medium wool, i.e. in those days Longmynd and Morfe Forest breeds, the progenitors of the Salop breed.
Walkmill Bridge (River Tern), Market Drayton.	

11iv

Guilds

THERE were no cloth-makers' guilds before 1093, but by 1130–31, the date of the earliest extant record of the Royal Exchequer, weavers' guilds had already been established in London (three), Lincoln (two), Oxford (four), Huntingdon (five), Nottingham (one), Winchester (six), which later also had a fullers' guild. At this time (1130) the weavers' guild in Oxford paid £6 to the Crown, but in the reign of Edward I they petitioned the Crown for a reduction of their licence money to £2.2s on the ground that the number of weavers had dropped from 60 to 15. A little later there were only 7 weavers left, all poor, and the fee was reduced to 6s 8d. In 1328, even they were all dead and had no successors; the cloth industry had ceased to exist in Oxford, and gone away to flourish elsewhere.

Edward I (1272–1307)

The guilds were a monopoly it is true but they controlled the proper apprenticeship of those that entered the mystery or craft; and they maintained the quality of the cloth, appointing 'alnagers' to check it.

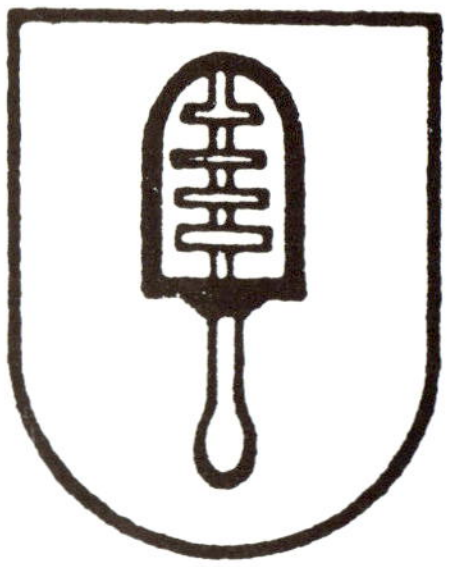

Fig. 35A. Arms of the Fullers of St-Trond, c 1350.

Fig. 35C. *Labor Labori Laborem Addit.* Arms of Tuckers' Hall, belonging to Guild of Weavers, Fullers and Shearmen, Exeter. Incorporated in 1490.

Fig. 35B. Tuckers' Hall, Exeter. From a photograph by R.H.S.R.

Import and export control

11v

AS a protective measure, import of cloth was prohibited in 1271, but this was repealed in 1274, possibly for fear of reprisals. In 1273 an embargo was placed on wool exports – a measure taken in the first instance as a result of military hostilities with Flanders, but the possibilities of using export control as a commercial weapon were seen, as is shown by the issue of export licences. The French could play the same game. Philip III, the "Bold", King of France (1270–1285), in 1277 forbade all export of wool to protect the French trades. In later centuries further measures were taken to promote the woollen industry not only for the sake of employment but more as a source of wealth for the King.

Ashley gives 1275 for the Statute of Marlborough imposing customs dues on wool.

See his statue on the tomb in the abbey of St. Denis. *Ciba Review*, 6, p. 2375

The use of Fuller's earth in 12th–13th century London

11vi

FULLER'S earth is first mentioned in a document dated 1295. Details are given in the section on Woburn, Bedfordshire.

Geoffrey Egan of the Museum of London excavated a series of hearths beside which fuller's earth was found in a remarkably pure condition – much cleaner than the Pompeian samples. I infer from this that the fuller's earth was part of a stock held ready for use after being dried by the fire; exactly as was found in Cyprus (5000 BC). A very good analysis of the 12th to 13th century material by Dr. Paul Henderson shows that the fuller's earth almost certainly came from Surrey. Because of the great interest of this find the chemical analyses are given below (Table 2).

of the British Museum (National History)

Although fulling machines had been invented by that time, it is reasonable to suppose that treading or beetling were still the normal procedure for use in small workshops, to clean and thicken fabrics, to launder excessively dirty clothes and for removing excess dye in dyehouses. The main stock of fuller's earth may have been warehoused in Tooley St., Southwark, as it was centuries later, for this was a convenient entrepot whether the earth came from Surrey or from Kent.

R.H.S. Robertson, letter to Mr. G. Egan 3.2.1982

Table 2 Chemical analyses of 12–13th century fuller's earths

		Beechfield Pit	Museum of London	Patteson Court Pit
	TiO_2 %	0·62	0·575	0·59
	Al_2O_3	13·5	14·8	13·62
total	Fe_2O_3	6·29	5·33	{4·66 1·20 FeO}
	MnO	0·0036	0·0454	0·01
	MgO	3·22	3·05	3·17
	CaO	2·20	2·08	2·65
	Na_2O	0·25	0·14	0·23
	K_2O	0·418	0·477	0·54
	p.p.m.			
	Be	2·3	1·5	
	Cr	<10·8	<10·8	
	Li	42	37	46
	Ni	<10·5	25	<10·5
	Cu	10	14	
	Zn	110	105	125
	V	30	<10	20
	Yb	5·7	2·8	
	Sr	50	40	85
	Ba	120	<20	455
	Rb	10	10	10
	P_2O_5			0·11
	CO_2			0·37

There may also have been smaller supplies held in drychandlers' shops throughout the City of London for sale as a household cleaner and possibly for use in medicines.

The Gentlemen's Magazine, (1742)

In 1208 the Guild of Fullers and Dyers in London complained *that persons sent cloth out of city to be fulled in mills instead of being treated in customary manner by the feet.*

Fig. 36. Flemish miniature of dyers at work, from a French version of *De Propietatibus Rerum* by Bartholomaeus Anglicus. Bruges, 1482. Dyehouses changed little until the 19th century. When modern technology took over, the problems of continuous textile processing were solved within a short space of time – with revolutionary technical and economic results. Miniature by courtesy of the British Museum, London.

CIBA Review, 1959, 11, *No. 131, 4.*

12i

THE LATER MIDDLE AGES

Cloth that cometh fro the weuyng · is nouȝt comly to were,
Tyl it is fulled under fote · or in fullyng-stokkes,
Wasshen wel with water · and with taseles cracched,
Ytouked, and ytented · and vnder tailloures hande.
And so it fareth by a barne · that borne is of wombe,
Til it be crystened in Crystes name · and confermed of the bisshop,
It is hethene as to heueneward · and helpelees to the soule.

W. LANGLAND, Piers the Plowman, ca. 1377–1379, xv, lines 444–450.

The Vision of William concerning Piers the Plowman, by William Langland, Text B, edited by Rev. Walter W. Skeat, Oxford: Clarenden Press, (1886), Vol. 1, p. 466.

IN this well-known passage, Langland is no longer able to make use of the word fulled to mean christened. A modern translation runs: *Before newly-woven cloth is fit to wear, it has to be cleaned in fulling-mills or trodden underfoot in water; and it does not come to the tailor until it is combed with teazles, fluffed out, and thoroughly stretched. So also a new-born child, until it is christened in the name of Christ and confirmed by the Bishop, is still a heathen.*

Langland, Piers the Ploughman, a new translation by J F Goodridge, The Penguin Classics, L87 (1959).

Langland often draws upon the cloth industry for his allegories. Of *Avarice* he says *Then I went to school with the drapers, and was shown how to stretch the selvedge and make the cloth look longer. My chief lesson was with the best, striped stuff – how to pierce it with pack-needles and join the strips together and lay them in a press, till ten or eleven yards were stretched into thirteen.*

id., The Confessions of the Seven Deadly Sins and the Search for Truth, v, p 99.

Eileen Power reminded us how, as early as the 14th century *honest burgesses climbed upon wool into the ranks of the nobility, only outstripped in their progress there by the dishonest ones.*

Spectator, (29.1.1960), 142.

Other references of Langland to the wool industry include:

> *Wisdom and Intelligence are not worth a straw nowadays, unless they are carded over with covetousness, as clothiers comb their wool. A man has only to plot and scheme . . . and his advice is sought by great men everywhere. Then he . . . twists the truth as he pleases.*

Piers Plowman x, Goodridge, p. 151.

Conscience tells Haukyn how to clean his suit of Baptism with Contrition . . . (it was to be) *scrubbed with the soap of sickness which penetrates very deeply.*

id., xiv, p. 204

Do-better will boil and beat it till it is as bright as scarlet, and dye it with a pinch of goodwill.

id., xiv, p. 205

Elsewhere he speaks of a laundry as a symbol of self-examination and penance; and of the rough wool chafing his skin.

Langland was born about 1332, probably at Cleobury Mortimer in Shropshire; his father was Stacy (Eustace) de Rokayle, who held land under the Despensers at Shipton-under-Wychwood in Oxfordshire not far from Witney and the calcareous fuller's earth of there and Bletchingdon.

id., p. 9

For an account of the remarkable rise of the English wool industry in this period one should turn to Eleonora Carus-Wilson's chapter in *The Cambridge Economic History of Europe, II* "Trade and Industry in the Middle Ages."

C.E.H. II, Cambridge University Press, (1952), 355–428; bibliography, 560–561

The fourteenth and fifteenth centuries saw a complete redistribution of a cloth industry which now depended on water-power. Surrey, for instance, an area possessing fuller's earth in plenty, was almost deserted by cloth-makers. The centres were now the West Riding of Yorkshire, the Southern Marches of Wales, the Cotswolds, the Mendips, the valleys of the Berkshire and Wiltshire downs, the Lake District, Devon and Cornwall.

In the inventory of Peter Baker, fuller, made in 1433 in Winchester his fuller's earth was entered as "item fullers clei pretium xvi d." *Survey of Medieval Winchester*, by Derek Keene, Clarendon Press: Oxford (1985)

In all these areas by the 14th century water-power was being actively exploited, if not to its fullest possible extent.

What abundance of water-power meant to the British cloth industry

Fig. 37. Cloth finishing in a monastery; (right) removing imperfections from the fabric; (left) stretching cloth on a tenterframe. From an Italian manuscript, 1421.

Fig. 38. Fuller trampling cloth in the vat. From the painted window of the Clothiers' Guild, Semur-en-Auxois Cathedral, Côte d'Or, c 1460.

Richard Hackluyt, *Voyages* . . ., London: (1598), *2*, 163.

For continental cloth-making operations, see Figs. 37–40

cf. *Fullers earth is best in Great Britain and abundancy,* Edward Chamberlayne, "Anglia notitia; or the present state of Great Britain, 1667". (1684), *1*, 7.

was not exaggerated by Hackluyt:

. . . *And where the Clothiers in Flanders by the flatness of their rivers cannot make Walkmilles for their clothes, but are forced to thicken and dresse all their clothes by the foot and by the labour of men, whereby their clothes are raised to an higher price, we in England have in all shires scores of milles upon falling rivers. And these rivers being in temperate zones are not dried up in summer with drought and heat as the rivers be in Spaine and in hotter regions of the world; so as our milles may go and worke at all times, and dresse clothes cheaply.*

Nevertheless, although the importance of fuller's earth to the cloth industry has been exaggerated, its ready accessibility did have some effect on the type of cloth made in different parts of England. In the West of England where fuller's earth was most easily obtained the most heavily fulled cloths were made, beginning with broadcloth, often black, an important men's suiting for centuries.

In the same area were made well-fulled cloth dyed with "Bristol Red", no doubt madder or kermes imported at that port. Such cloth was later used by the army, whose famous "Red Coats" made conspicuous targets until the Crimean war (1853–56). Another cloth of the same sort was baize, though not so heavily fulled. It could be of any colour, but was familiar in Victorian times for lining doors, especially in the servants' quarters, or the pews of churches. In billiard table cloth the original woven threads cannot be seen at all, so heavily fulled it has been.

In other parts of England where fuller's earth was expensive, owing to the difficulties of road transport, the wool cloth industry developed in a different way. In East Anglia wool was usually combed, not carded, and the fine yarn so produced made a denser, more compact fabric, which required only a light fulling, such as wolseys (or serges).

Flemish weavers had settled in Kendal in 1331.

In the West Riding of Yorkshire the woollen industry was founded on mountain breeds of sheep grazing the Pennines. In Cumbria wool from another breed of sheep gave "Kendal Green" its main characteristics. Kerseys, a kind of narrow cloth, woven from long wool and usually ribbed, were made, along with other cheap and light fabrics in Devon, Dorset and Cornwall, as well as in the West Riding, Cumbria and the Welsh marches.

E. Carus-Wilson, *C.E.H.* (1952), 412

Even a journey, to fetch fuller's earth, of 25 miles each way required at least two nights away from home. In the manor of Minchinhampton where there was a concentration of industrial workers, seven tenants, one of whom rented the manorial fulling mill, paid for leave to dig fuller's earth, presumably from Bath, or even locally.

Nottingham, Rec. II (1403).

Far from the deposits of southern England fuller's earth was occasionally sought. At Nottingham there is a record of 'Unum stryk de walkerherth' (1403). There are more records of other branches of the fuller's art, such as *No walker off the Cite of Conventre shall Rakke no Clothe on the Tey(n)tur that shall be solde for wette-clothe.*

Coventry Leet Book, 172.

Fig. 39. Raising the cloth before cropping. From the Clothiers' Window at Semur-en-Auxois, c 1460.

Fig. 40. Shearing (or cropping) cloth. From the Clothiers' Window at Semur-en-Auxois, c 1460.

As Carus-Wilson remarks, it is not surprising that Chaucer, who had lived through so notable a development of England's economy and had himself been a collector of customs, included among his Canterbury pilgrims a vigorous west country clothier, the much travelled goodwife living near Bath, and wrote of her:

Of clooth-making she hadde swiche an haunt
She passed hem of Ypres and of Gaunt.

The author's, or in this case perhaps the excise officer's, eye for detail comes out in the description of the goodwife's Sunday *cover-chiefs*, that *ful fine were of ground* (i.e. full thick were of texture); and of which he *durste swere they weyeden ten pound* – typifying the heavily fulled, close-textured cloth already characteristic of the west country.

At first much wool and little cloth was exported; but in this period, after the cloth makers had left the towns and cities – and the restrictive practices of the guilds – and set up their mills where overshot water wheels could be installed, the cloth industry was greatly encouraged by sundry fiscal measures and by the banning of the export of wool, except under licence.

C.E.H., 2, 416

Geoffrey Chaucer, *Prologue,* c. 1386: lines 447–8.

haunt = skill.

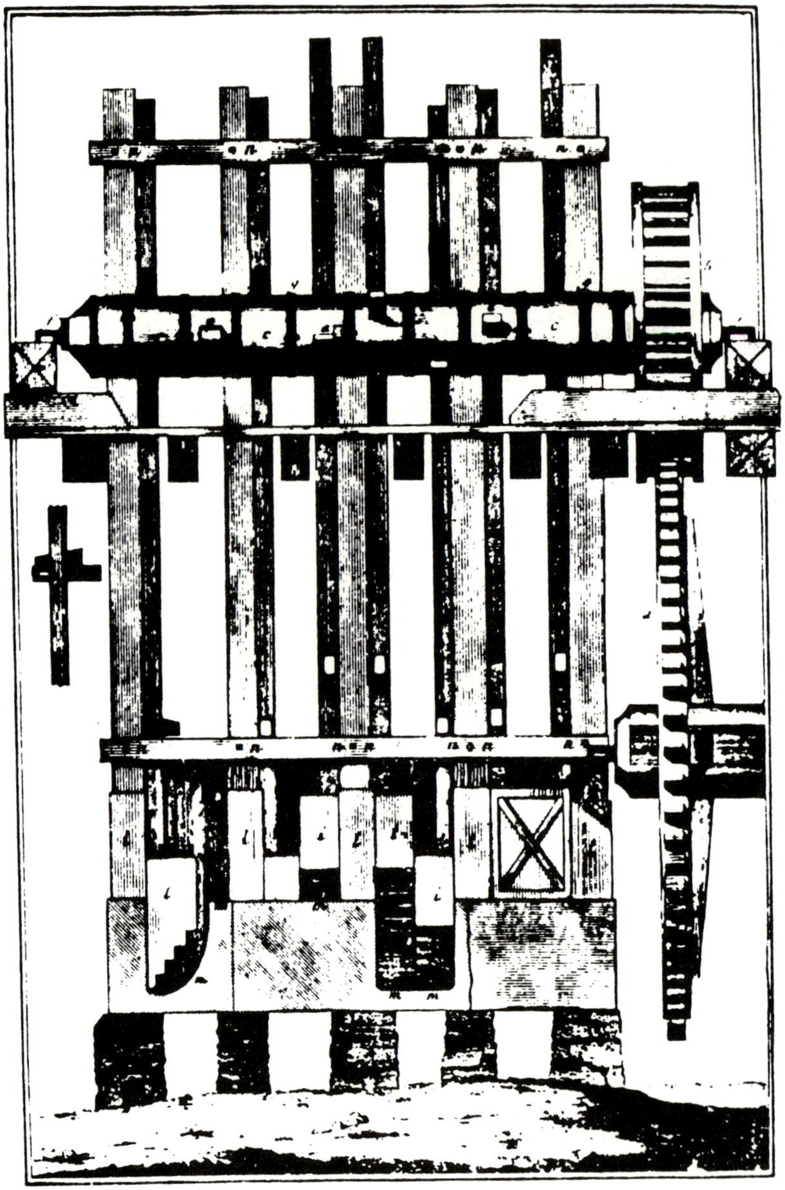

Fig. 41A. Diagram of 'falling' stocks from Diderot: *Encyclopédie* (1773).

Fig. 41B. Diagram of 'driving' stocks from Diderot: *Encyclopédie* (1773).

The fulling mill, again.

See Figs. 41 A & B.

THE fulling mill may have been improved before 1500 AD and assumed its modern form, at least in principle. Instead of dropping vertically the stocks were hinged at the top and were raised through an arc by means of a tappit wheel – instead of the tappits being fixed into the axle-tree. As Kenneth G. Ponting puts it *there were in fact two rather different kinds of stocks in use – driving stocks and fulling stocks. The heads of the driving stocks came down in a slanting direction. Instead of pounding the cloth, they threw it up and swirled it so that it fell back again as they rose. The heads of the fulling stocks came down almost perpendicularly and pounded the cloth. The driving stocks really performed a scouring or cleaning process which removed the oil added to assist in spinning.*

The Woollen Industry of South-West England, Bath: Adams & Dart, (1971), p. 38.

Fig. 42A A diagrammatic representation of a fuller's equipment, V. H. von Hove (1628–1698)

In the more complicated, and therefore later, driving stocks the stocks were always in pairs. A primitive form of fulling stocks is in an engraving by Frederick Hendrick von Hove (1628–1698). The mechanism cannot be made out; both stocks are in phase, whereas one should be up and the other down. Fig. 42A.

CIBA Review, (78), 2859

The earliest known picture of the driving stock is by V. Zonka (1607). Part of the water-wheel is shown and two swinging stocks. The tappit-wheel is so badly drawn one cannot make out the mechanism.

Novo teatro di macchine et edificii, Padua: Bertelli, (1607), 69–75. Fig. 42B.

Let us return to the 15th century. It was when Leonardo da Vinci's father took him to the workshop of Andrea Verrocchio that his early genius was confirmed. He amazed Verrocchio with his drawings, paintings and sculpture. He was the first to propose a canalised River Arno linking Pisa to Florence, and made designs for mills, fulling machines, and other engines powered by water (~1468).

Vasari's *Life of the Painters,* (1568), in Ludwig Goldscheider's *Leonardo da Vinci,* Phaidon Press, 3rd ed., (1948), p. 6.

Fulling mills did not escape "Luddite" attention. In 1485 one was attacked and looted at Bisley, Gloucestershire. Works already quoted should be consulted for the stirring social history of fullers and other clothworkers.

Fulling was of course not the only textile process to be mechanised in this period. The spinning wheel was invented about 1300 AD and

Fig. 42B (on left) Water-driven mill as illustrated by V. Zonca (1607)

British Library

V. Zonca, *Nuovo tcatro di macchine* ed *edificii*. Padua: Pietro Benteli, 1607, 69–76.

teaseling had been mechanised ('gigmills') before 1435 AD. Hand-shearing, however, remained a craft of great skill.

Fulling mills were probably not very large installations even in the 17th century as one may judge by the juxtaposition of "fulling mylnes, sythe mills, cutlers mylnes" (Fitzherbert, 1523). Apart from the various detergents, little has been written about the nature of the other materials used in this craft. A misprint in "The Golden Fleece" (1656) conceals the meaning of *clothing* [*i.e. cloth*] *cannot be scoured from the Seame and Oyls, where with they must necessarily be wrought.* What is meant may be "Sesame (and other) Oyls."

John Fitzherbert, *A newe tracte or treatyse moost profytable for all husbande men*, (1523), 9 b.

The Golden Fleece, by W. S. (1656).

12ii

Nicolaus Pevsner, *The Buildings of England; S. and W. Somerset*. Penguin Books (1958, reprinted 1976), 294.

A fuller and his tools are shown in a wood-carving at the end of a church bench at Spaxton, five miles from Bridgwater, West Somerset. Fig. 43.

Fig. 43. Renaissance church furnishing, Spaxton: bench end with a fuller and his tools.

After Nicolaus Pevsner

THE PERIOD 1500–1700

13

Household cleaning

13i

A profitable boke declaring dyuers approuved remedies to take out spottes and staines . . . with diuers colours how to die velvets and silkes

translated out of Dutch by L. Mascall, 1583.

1583 (1588 ed.) A 11 b.

IN this early book on household cleaning, Mascall recommends one to *take a quantity of walkars claye, called Fullars earth*. This theme will be continued in chapter 19.

Pharmacy: an unpleasant antidotal experiment

13ii

IN 1581 a prisoner condemned to death was given the option of ingesting a fatal dose of the highly poisonous mercuric chloride followed by a suspension of Terra Sigillata. He accepted this offer and lived. The story was considered by Thompson (1912) and Sir Hans Sloane before him as an authentic account. As we are unlikely to test again the therapeutic value of clays in so direct a manner, the account is given in full.

$HgCl_2$

C.J.S. Thompson *Int. Congr. Med.* (1912), Sect. 23, 433.

We Wolfgangus Earle of Hohenloe, Lorde of Langenburg, &c. Do openlie make knowen unto all men by these my Letters Testimoniall, that there came lately before me at Langenburg, my wellbeloved friende, Andreas Bertholdus of Oschatz, and declared unto mee that he had a most excellent kinde of Terra Sigillata, which was not alonely of great force against sundrie diseases: but also a most undoubted remedy against all manner of venemous poisons, as had beene prooved by sundrie witnesses upon a greate number of dogges, which made me also desirous to see the triall of it. It happened at the same time, that one called Wendel Thumblardt was by our Lieutenant of Langenburg for certaine fellonies imprisoned, who being examined by our Justices, confessed himself guilty of a great number of robberies: And therefore brought to the barre was condemned to be hanged. Being yet deteined in prison, and coming to his ears that there was such a medicine, so soveraigne against sundrie sicknesses, and the most deadly poisons, [he] has made humble request as well by his parents, as by other his friends, of

A. Bertholdus, *The wonderful and strange effect and vertues of a new Terra Sigillata lately found in Germany . . .* London, 1587.

The original account was *Terræ Sigillatæ, nuper in Germania repertae*, [Frankfurt], 1583.

which there were present no small number, desiring for the mercie of God, and respect of his poor life, that being thus condemned, hee might have given unto him the most deadly poison that might be devised, whereby a perfit triall might bee had of the worthines of medicinable earth. And in this respect, not onely for this pittiful request of his: but also for the commoditie and benefit of all Christendome (if so be the medicine proove answerable to the report), pardoning the offender, we graunted his life upon that condition. Therefore the day of the date of these presents, in the presence of our selfe, and our wellbeloved Cosin the Countie George Friderick of Hohenloe, and Lord in Langenburg, and in the presence of all our Nobilities and Commons, the said patient received a dram and a half of Mercurie Sublimate, mingled with Conserve of Roses, and immediately after it he drank a dram of the Terra Sigillata in olde wine. And albeit the poison did in the judgement of our learned Phisition George Pistor, Doctor in Phisicke, and John Lutzen our Apothecarie, who were both by him all the while, extremely torment and vexe him: yet in the end the medicine prevailing overcame it, whereby the poore wretch was delivered, and being restored to his health was committed to his parents. Whereas therefore the foresaid Andrew Berthold, hath humbly required to have our Letters Testimoniall for his farther credite, wee have thought good for the furtherance and advancement of the truth, to graunt him these our Letters, signed with our seale Manuell. Given at Langenburg the 25. of Januarie, in the yeare of our Lord. 1581.

D. A. K. Black, a revaluation of terra sigillata, *Lancet*, (27.10.1956), 883–884.

T. Sollmann, *Manual of Pharmacology,* 7th ed., Philadelphia, (1948).

D. A. K. Black (1956) commented on this strange event that *if we dismiss the Conserve of Roses and the olde wine as active agents . . . we are left with the possibility that* terra sigillata *did in fact protect against a dose of 6 grains (sic) of mercuric chloride, a substance which has generally caused death when given in amounts of more than two grains (Sollmann, 1948), and has often been fatal in much smaller amounts.* Actually the dose was 1.5 drams or 90 grains. If the clay were illite and had a cation exchange capacity of 0.3 milliequivalents per gram, the dose should have been not less than 65.24 grams; but if the terra sigillata were calcium montmorillonite with a CEC of 0.7 meq/g, the dose would have been 28.76 g – not far from an ounce, not a dram. I suspect that the wretch was given an ounce of the clay and that the clay newly found (nuper) in Germany was a calcium montmorillonite or Walkererde – see Fig. 45A.

Jessie M. Sweet. *op cit.*

Now in Poland.

During the 16th century medicinal clays had been discovered in Prussian Silesia, at Kupferberg, near Hischberg; and also the districts round Striegau and Liegnitz. Andreas Bertholdus considered the Kupferberg clay to be more effective than the Lemnian, and better I suspect than the clays of Striegau and Liegnitz, which were kaolinitic weathering products of basic igneous rocks. They may have been halloysitic like the occurrence at Dunino. It is known that smectites are also formed in this area, when the weathering conditions were different.

G. Brown, Report No. 39, (8.12.1960), acknowledgement to Rothamsted Experimental Station, Harpenden, Herts.

G. Brown examined powder from seven troches in the Sloane Collection and the British Museum in London. Two were of identical composition and presumably came from the same batch. One sample had illite, two had disordered kaolinite, one had well crystallised kaolinite, another an unspecified type of kaolinite and a single example had montmorillonite as the dominant clay mineral.

Sample Sloane 148/11.

Hematite was always present in the kaolinitic clays, but not in the illite or montmorillonite. It is safe to assume that the genuine Lemnian earth was essentially disordered kaolinite, hematite, and quartz with a small proportion of illite and goethite. However, a sample chemically analysed by Thompson (1912) seems to have contained, roughly, calcite 40, quartz 24, illite 20, kaolinite 13 and moisture 3 per cent.

C. J. S. Thompson, *Int. Congr. Medicine*, Sect. 23, (1912), p. 433; my interpretation.

The seal of Bertholdus of whom we have read was a *scutchin wherein is portrayed the Sunne, the halfe Moone, and five Starres. In place of the helmet (are) these Letters, A.B. and on the other side V.O.* A similar seal is figured in Museum Wormianum and reproduced by Sweet, (Fig. 44). *Others showed three towers and crossed keys with the words* Terra: Sig. Strigonien *(Striegau), (Fig. 44B), and also a two-headed eagle with the words* Terra Sigillata *surrounding it (Fig. 44C). Another is marked* Terra Sigillata 16*N*32/*S. Pauli Terra Adam. Axungia Soles (Solis) Terra (Fig. 44D).*

Ole Worm, Museum Wormianum ... Lugduni Batavorum, 1655.

There is also a specimen of terra sigillata silesiaca in the Woodwardian Collection (collected ca. 1695–1727) at Cambridge University, but in the Collections of Materia Medica made by J. Poynter (in ca 1700 and presented to Oxford in 1720) and by Dr. James Oglander (presented 1729), both now in the Museum of the History of Science, Old Ashmolean Building, Oxford, there are some pale buff troches marked *Terra Sigilada*/Montaȝui with crossed keys above the legend.

Woodward was sent by Scheuchzer a specimen of T. sigill. Laubacensis. It is brown and shiny and has white specks – an altered igneous rock, Laubach is in Hessen, NE of Frankfurt/Main.

Fig. 44. Seventeenth century sealed earths. (a) Probably Terra Sigillata Silesiaca from Prussian Silesia between 1587 and 1655 (Reproduced from *Museum Wormianum*, 1655). (b) Terra Sigillata Silesiaca from Striegau, Prussian Silesia before 1655 (Reproduced from *Museum Wormianum*, 1655). (c) Terra Sigillata Silesiaca from Germany before 1655 (Reproduced from *Museum Wormianum* 1655. (d) St Paul's Earth or Axungia Solis from Prussica Silesia, 1632 (Reproduced from M. B. Valentin, *Museum Museorum*, 1704).

13ii

Poynter thought this earth came from Constantinople but the spelling and possibly the name suggest a Spanish origin, the Silesian logo having been inserted to confuse the public.

An unsealed troche is labelled *Ter. Cimol. & pulv./Belgic* in the Oglander Collection, and similarly prepared earths at that time are known to have come from Austria, Italy and elsewhere. Another German sort was *Terra miraculosa Saxoniæ* or Sächsiche Wundererde (C. Richter, 1732). Jung and Neuhaus found that a grey-violet tablet of this therapeutical clay was two thirds kaolinite, with quartz, mica and limonite.

H. Jung & L. Neuhaus, *Chem. Erde,* (1945), *15,* 389–394. *Min. Abstr. 11,* 174

13iii

Social History

Forsooth

"For soth his name is Povertie, for he and his cosyn Necessitie hath brought us to this day," said the spokesman of the Suffolk rebels in 1525, when asked who was the captain of the four thousand men who had risen in arms, "The clothmakers", he continued, "have put all these people and a farre greater nomber from worke".

Cambridge Economic History, 2, 427.

THESE words describe the sad effect of war taxation which threatened at that time to cripple the industry. The plaint is echoed in Shakespeare's Hentry VIII:

My emphasis. R.H.S.R.

Norfolk: *Not* **almost** *appears;*
It **doth** *appear: for upon these taxations,*
The clothiers all, not able to maintain
The many to them 'longing, have put off
The spinsters, carders, fullers, weavers; who,
Unfit for other life, compelled by hunger
And lack of other means, in desperate manner
Daring the event to the teeth, are all in uproar,
And danger serves among them.

W. Shakespeare, Henry VIII, 1.2 ca. 1520–1523.

The fullers had cause to be in uproar on many occasions during the long crescendo of the English woollen industry. Discontent had arisen when in times of war or trade recession, unemployment rose sharply, especially in places where labour was organised in what almost amounted to factories – a trend which led to the construction of some huge mills centuries later. The woollen and worsted trades were however usually highly profitable as one may judge from the hundreds of beautiful Perpendicular churches in rural England. Henry VIII raised money by the illegal devices of "benevolences" and forced loans when Parliament was not functioning, and England was becoming poorer and more discontented.

Dissention was even more frequent and violent on the Continent. In 1302 the armed craftsmen of Bruges and Ypres won a decisive victory over the army of French nobles near Courtrai. As a result their municipal privileges were acknowledged. In Ghent the armed force of the Guilds was organised according to town quarters, 23 of which were in the hands of the weavers and 18 of the fullers.

CIBA Review, (77), 2822.

In 1378 there was a mass rising in Florence and in the following year a worse one in Ypres, Ghent and Bruges. Dissention was a cause of the decline of the industry in Flanders and Tuscany. England profited by their distress – and so did Germany and Holland. The cloth industry moved out into small towns in Italy and Flanders, and Spanish wool replaced English in the 15th century. Nevertheless when the export of wool was prohibited and Flemish cloth workers had often been encouraged to settle, woollen cloth became England's most important export. Fully finished cloth of the highest quality was hard to sell abroad; what fetched a ready sale for centuries was white broadcloth, well fulled. This was dyed and finished on the Continent where a variety of fabrics was made. The West of England was the main beneficiary of this excellent compromise and the fuller's earth in Somerset must have been in demand throughout the Middle Ages.

E Carus-Wilson, The Woollen Industry, The Cambridge Economic History, ii, (1952).

For the distribution of cloth *types* in Britain see Thomas Fuller, "Church History of Britain", (1655): *CIBA Review 2*, 498.

How much earth was used to make a length of broadcloth is not known. We know the numbers exported, and if we assume 15 lb of fuller's earth per cloth we arrive at a very crude estimate of the annual requirement for the export trade. If the home demand were four times the exports in 1392–95 and three times the exports in 1550, a table can be constructed, and a graph of production (Table 3; Fig. 45).

In France in 1751, 10 lb of soap was the recommended weight for fulling a piece of white cloth, 45 ells long, and 15 lb for coloured cloth.

As white broadcloths were made mostly in the West of England we may assume that Bath fuller's earth was mainly used there. The output had fallen to 2000 tons of Bath earth in 1744; there may have been a maximum about 1600.

Table 3 Commercial data on fuller's earth 1347–1550

Date	Number of cloths exported	Fuller's earth used for exported cloth Tons	*Total* export ratio	Fuller's earth used for all white broadcloths Tons
1347–48	4,422	30	×5·0	150
1392–95	43,000	300	×5·0	1500
~1450	54,000	360	×4·7	1700
~1483	63,000	420	×4·5	1900
~1520	84,000	570	×4·2	2400
~1550	122,000	810	×4·0	3250

Based on figures given by E. Carus-Wilson, *op cit.*

13iii

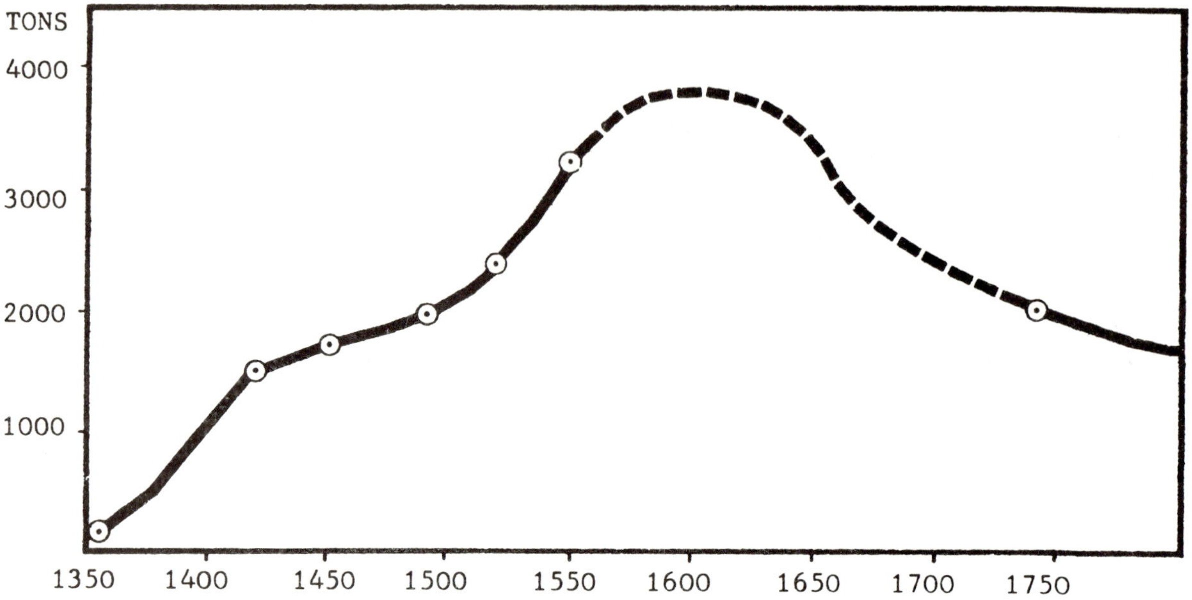

Fig. 45. Indication of output of fuller's earth for cleaning white broadcloth, 1350–1750.

13iv

The cult of nastiness

B. Gott MS in the Brotherton Library, Leeds.

THE trampling of pig's dung into cloth was recorded as early as 1613. B. Gott's MS "Prices and pieces" (1800–25) gives some particularly revolting recipes. Sometimes fuller's earth was mixed with the dung. In France apparently sheep's dung was preferred. The horrid truth is that man in various periods has found virtue in loathsome mixtures – perhaps in some half-understood way he believed in a law of opposites: a filthily defiled thing would become exceedingly clean. There was magic in disgusting recipes. Paracelsus lambasted them in medicine. Cullen two hundred years afterwards was still cleaning up the pharmacopoeia. Who cleaned up fulling we do not know, but the malodorous processes gradually died out – not always with the workman's immediate approval for he had often come to regard a nasty process as an effective one. Similarly hygienic education had made the use of dung in foundries rather rare by 1900. One suspects that some masters made but little effort to eliminate degrading processes in the hey-day of industrial exploitation.

According to Caroline Davidson (1982), the poor used dung or urine for washing their clothes; and when they eventually took to using soap some objected to its smell!

13iv

A Woman's Work is Never Done, Caroline Davidson (1982), Chatto & Windus, p. 152.

Import and export control

13v

. . . it is possible and probable that other parts of the world may produce Fullers' Earth, but neither in such fineness nor abundance as in England

"The Golden Fleece" W. S. Gentleman 1656

W. S. stands (oddly) for John Sergeant.

WHEN an export duty was imposed on cloth in 1347 it was less than 2 per cent, as against one of thirty-three per cent on wool. The 14th century saw also the forthright measure of prohibition in 1376 – a reflection as much of the Crown's acumen as of the growing power of the clothiers – of all export of unfulled cloth. And legislation right through to the reign of Elizabeth was "tightening loopholes" in the wording of the acts; for instance in 1487 Henry VII forbade the export of any stuff which was not fulled, teazled or shorn. Only certain cheap-quality cloths were to be shipped abroad. The Merchant Adventurers in 1593 found these restrictions irksome, but the decree regarding exports remained in force until 1707, after which year half-finished goods were charged with export dues, while finished stuffs could be sent abroad duty-free.

H. Wechsler, Textile Trades under Queen Elizabeth, *CIBA Review*, (Feb. 1950), (78), 2859–60.

The cloth industry was controlled in other ways. A Restoration Parliament went so far as to ordain that everyone who died must be buried in English cloth. Alexander Pope (1732) voices the unpopularity of this Act:

(1660)

Odious! in woollen! 'twould a saint provoke,
Were the last words that poor Narcissa spoke.

A. Pope, *Moral Essays*, (1732), i.

Owling

The owling trade is in a manner supprest by the diligence of the officers appointed for that purpose.
NARCISSUS LUTTRELL, A brief historical relation of state affairs, 1678–1714 (1857).

A NATURAL consequence of the prohibition of the export of wool, wool-fells or unwashed fleeces, and even live sheep, and fuller's earth, was a well-organised and profitable smuggling industry, known as "owling" either from its nocturnality or as a corruption of a dialect form of "wool".

George P. R. James, *The smuggler, a tale* (1845), p. 8.

The risks of this illicit traffic were heavy: *The Walker and Fuller shall truely walke fulle thikke and werke every webbe of wollen yerne"* required for exportation, ran the decree, and the penalty for non-compliance was forfeit of the value of the unwrought goods he tried to export. And *we do . . . streightly charge . . . that no manner of Fulling clay be . . . exported.* These embargoes were repeated in 1693–4, and again in 1720 when the *London Gazette* refers to the forbidden export of "any Fuller's Earth, or Fulling Clay."

Act 1688, c.12, of last year of James II. Repeated in 12 Car. II, c. 32 (1672).

London Gazette, (1688), No. 2338/1.

In 4 and 5 W. and M., c. 24, sect. 10, 1693–4.

London Gazette, (1720), No. 5853, 1.

In 1739 the forfeit was three shillings per pound weight of the goods. Despite petitioning by the Merchant Adventurers of the Privy Council, however, and legislation for prevention of *exportation or carrying out by sea or land (out of any of His Majesty's aforesaid dominions) of wool, woolfelts, woollen yarn, and fuller's earth, and the confiscation of the vessel that shall carry out any of the same in it*, and increasing inducements offered to informers – in 1739 these were to receive three-fourths of the penalties – owling continued. Even customs officials apparently took a hand in it, for the 1739 Act stipulated a penalty of £300 for bribing them. And when not amenable they were liable to be given a rough time of it. The government was driven to include in the same Act provision that

Act 12 George II, c 12, sect. 14, (1739).

> *if any officers of the Customs, Excise or Salt, or any other Persons that shall act in aid or assistance of any above officers, in putting this act in execution, shall be hindered, opposed, obstructed, molested, wounded, or beaten, in seizing any Wool . . . or any other species of goods before enumerated* [*these included fuller's earth*] *by any Person or Persons whatever, either in the Day or Night, by Land or Water, the Persons that shall so hinder, obstruct, molest or wound such officers; or any person using offensive weapons, or wearing any mask or disguise, who shall attempt to rescue any Wool or the goods aforesaid, such Person or offenders that shall be convicted of any of the said offenses, shall be convicted to be transported to some of His Majesties Colonies and Plantations in America, for such terms or such Coutt shall think fit, not exceeding seven years, according to an Act of George I . . . and*

[Probably 'coust' = cost]

if any such offender shall return to Great Britain and Ireland before the Expiration of his time, he or they shall suffer as Felons, and have execution awarded against them, as persons attainted of Felony, without benefit of Clergy.

How little the measures took effect is clear from the reports of travellers.

Mr. Mynn states (1621), that he saw lately the manufacturies of the Netherlands, and their materials; saw there English wool and Fuller's earth, brought out of England, plentiful as here, and that *they can make no cloth without it* – a slight exaggeration.

Several prosecutions, some resulting in very heavy penalties and humiliation, were made in the 1620s. They are described under *Kent*, (Chapter 20). Apart from secret shipment there were less extreme ways of circumventing the regulations. Licences could be obtained to ship fuller's earth to other parts of England, as from Kent to Lincoln or Yorkshire, and opportunity could then be taken to divert it. Nevertheless there was distress in 1622, particularly in Suffolk, because licences were not given when urgently required. Overland transport was then costing £6 a ton, as against 2s a ton by sea.

Of course the State had to protect its greatest source of income from overseas, for the products of the cloth industry amounted to two-thirds of all exports. Protecting the staple industry was vital for maintaining the country's position in the power structure of Europe and the social structure within. Enterprising people often took a narrower view and used illegal means to enrich themselves. One way was to pass fuller's earth off as potter's clay, but in the case of Edmund Warner the officials made a big mistake.

13vi

Victoria County History, *Kent*.

A. Toynbee, *Industrial Revolution*, quoting Charles Davenant on Gain in Trade (1699).

Trial of Edmund Warner

13vii

A BRIEF Account of the Evidence given on behalf of Edmund Warner, *at a* Tryal *had at the Bar the 24th of* November *1693, before the Four Barons of the* Exchequer, *relating to a Parcel of* Potters Clay, *seized by the* Custom-House *Officers, under the pretence of its being* Fuller's Earth runs as follows:

William Riddal *swore that he was Stewart and Tenant to the said* Warner *for above 20 years, in all which time the said* Warner *constantly sold to the Potters in* London *considerable quantities of the same sort of clay now in dispute, for the making of White and Painted Earthenware; That he never knew or heard that the least Parcel of it was ever sold as Fullers-Earth, or put to any other use than making Earthenware, excepting some small quantities yearly in the Neighbourhood for daubing of Houses, which may be alledged as a good Argument that the said Clay is not of the nature of Fuller's Earth; for, whereas this makes the best Clay Wall in* England, *if it were of the nature of Fuller's Earth, upon the first wet weather it would all fall to the Ground; He likewise farther deposed, That the first time the*

Reprinted in the *Catalogue of British Pottery and Porcelain*, London: Museum of Practical Geology, 2nd ed., **1871**, and 3rd ed., **1876**, Appendix B.

An early example of an expert witness. He makes a clear distinction between kaolin and montmorillonite.

said Warner *shipt the said Clay for* Holland, *it was seized by the* Custom-House *Officers as Fullers Earth; but upon Tryal of it, they were convinced of their Error, and cleared the same.*

William Knight, Thomas Harper, Henry de Wilde, John Robins, Moses Johnson, *are all Potters in* London, *and swore that they had all seen the said* Warner*'s Clay, which was seized, that it was really Potter's Clay, and of the same sort which they had constantly bought of him some for above 25 years, others ever since they were Traders, that it is of a quite different nature from Fullers Earth; for the Said* Warner*'s Clay tho' never so often dissolved in Water, may be brought into a Body again, and will work, like Wax, into any shape, whereas Fullers Earth being once dissolved is never to be got into any Body, but when it is dry crumbles like Sand, and all the art of Man can never make a Pot of it.*

Hen. de Wilde, *one of the said Potters farther deposed, That about five Years since he shipt some of the said* Warner*'s Clay for* Pensilvania, *where his Son had set up a Pot-House, that the Customs-House Officers did then likewise seize it as Fullers Earth, but upon Trial of it found they were mistaken, and cleared the same.*

Benjamin Furly, *Merchant in* Rotterdam, *to whom the said* Warner *consigned his Clay,* John Sonman, *Servant to the said* Furly, *were, by virtue of a Commission from the Court of* Exchequer, *examined upon Oath, (by Commissioners appointed for that purpose) to several Interrogatories, and their Depositions being read in Court, did plainly prove that all the said* Warner*'s Clay which was sent for* Holland *was really sold to Potters for making of Earthen-ware. That they never heard, nor do believe that any part of it was ever used for the Fulling Trade, or is any way proper for it; that they sold the said* Warner*'s Clay from 25 to 30 Guilders a Last, when they could buy Fuller's-Earth at 7 Guilders a Last.*

Note the much higher price of potter's clay.

Adrian Van Arde, *sworn Measurer of all the Earth and Clay imported to* Rotterdam, *was examined by vertue of the said Commission, and Deposed, That for many Years which he had been in that Office, he never knew or heard of any* English *Fullers-Earth imported into* Holland, *that they were so plentifully supplied with that sort of Earth from* Flanders; *that it is commonly bought and sold at Seven Guilders a Last, and had an Hundred Last by him to sell at the same Rate.*

Ten or Twelve Potters in Holland were likewise examined by virtue of the said Commission, and deposed severally, that they had bought considerable quantities of the said Warner's clay of Mr. Benj. Furly, and paid for the same from 25 to 30 Guilders a Last; that they used the same in making Earthen-Ware, but do not believe it any way useful for the Fulling Trade, nor never heard that any of it was ever sold for that use.

Colonel Holt, *a Member of Parliament, swore that he (being very lately in* Holland*), made the strictest inquiries into the Truth of this Business, that he could, First amongst the Potters, who all declared to the same Effect, as in their Depositions before mentioned, that he likewise took some of the* *said* Warner*'s Clay, and got a Fuller to try it, whether it were any way useful for their Trade, who answered it was not? nor could he use it if he might have it for Nothing; after which the Colonel asked him, From whence they were supplied with Fuller's-Earth, and at what Rates? The Fullers answered, generally from* Flanders, *at 6½, and 7 Guilders* per *Last; and sometimes from* Rosendale, *in the* States *Dominions, at 10 Guilders* per *Last; that being much dryer, and wasted less by lying; at which the Colonel (being unwilling to believe upon the Fuller's bare word) caused to be drawn up in the form of an Affidavit, and the Fuller before a Magistrate, swore to the truth of it, which Affidavit the Colonel hath to produce.*

Mr. Edward Paget, *a Divine, swore, that he being in* Holland *at the same time, made the like Enquiry amongst the Fullers at* Layden, *the chief cloathing Town in Holland, who all gave him the same Information, that the Fuller at* Rotterdam *had done the Colonel above; and for his better satisfaction, Three of the Chiefest went with him before a Magistrate, and swore to the Truth of what they Affirmed; after which he went to* Delfe *amongst the Potters, who all affirmed the same thing, that the Ten or Twelve had sworn in their aforementioned Depositions, the Truth of which they attested by a Certificate, under the Hands of above Twenty, which Affidavits, and Certificate, the said* Paget *hath to produce.*

Rich. Cutler, Matthew Hanson, *Ship-brokers, swore, They had hired many Ships for the said* Warner, *to load Clay for* Rotterdam*; and the general Price of the Fraight, and Primage paid for the same, was 15 Guilders, 8 Stivers* per *Last; and sometimes 17 Guilders, 12 Stivers* per *Last.*

Will. Read, Tho. Wood, John Saliby, *Mariners, swore, They had the same Prices above mentioned, for several Fraights which they carried, and saw delivered to the Potters in* Holland.

Hump. Bellomy, Richard Hopkins, Timothy Fowler, . . . Shepherd, *Worcester, are all Buyers, and Sellers of Fuller's-Earth, and swore they could well judge of the same, having dealt therein from Ten to above Twenty Five years, that they all took Samples of the said* Warner*'s Clay, out of the King's Celler, and tryed all the Experiments to judge of it, that they were capable of; and possitively affirmed that it was no Fullers-Earth; neither could they find anything of the Nature of Fullers-Earth in it; that for their Use they would not give a Shilling for 100 Last, for they should never be able to sell it.*

Alderman Bearcraft, Tho. Cooksey, John Wynn, Rowl. Hancock, Gloucestershire; Rich. Osborne, Tho. Woorrel, John Chilton, Sam. Beadle, Essex; Simon Meazy, John Lawson, John Backer, John Peartree, Suffolk; John Clarke, Senior, John Clarke, Junior, Will. Baines, Tho. Baines, Dan. Baines, John Smith, John Carter, Dan. Wenden, Sam. Wrinch, London; Dan. Barrs, John St. Berry, *are all Clothiers, Fullers, and Cloth-Workers; being the most eminent in their several Places; as well as for Honesty, as great Traders, and good estates; they all swore they had made*

13vii

As A. L. Gutmann puts it in *CIBA Review*, (Oct 1938), (14), 484, [in Flanders] "The clayey soil also provided excellent fulling material." ("Serviceable" would have been a better word than "excellent").

Rosendaal in Brabant between Bergen op Zoom and Breda.

Perpetuana, a durable fabric of wool manufactured in England since the 16th c. "The sober Perpetuana-suited Puritan", Thomas Dekker, *The Seven Deadly Sinnes of London* (1606), 27

the most impartial Experiments they could of the said Warner*'s Clay, now in dispute, being the same taken out of the King's Celler, that they tried it upon several sorts of Goods, as several sorts of* Cloth, Bays, Cloth-Serges, *and* Perpetuanaes, *and every one possitively affirmed that it was no Fuller's-Earth, but on the contrary was an absolute Enemy to the Woollen Manufacture; for instead of scouring, it fix'd the grease in the Grownd of all the Goods, that were done with it, and instead of makeing them White, it absolutely stained them Yellow, all which was apparent to the most common Eye, in all the above said several sorts of Goods, produced in Court; They likewise declared upon Oath, that they were all Strangers to the said* Warner*; and that it was as prejudicial to their interest, to have Fuller's-Earth transported, as any men's, and therefore could have no Inducement to favour the said* Warner, *beyond the Merits of his cause, to the Violation of their Consciences.*

This is so just a Recital of the Evidence on behalf of the said Warner, that he challanges the greatest of his Enemies to detect of him of the least Falsehood therein; and if his Adversaries had been as fair in representing theirs, there would have been no need of this: for whereas in their printed Papers they make their Witnesses to affirm the said Warner's clay was tried in all Experiments, without any mixture, there was not above One or Two, but upon Cross Examination at the Tryal, owned they used either Segg and Hogs dung, or Soap and Gauls with it, which would have scowered any Cloath better without the said Clay than with it.

A segg was a bull castrated when adult.

It may not be amiss likewise to observe that for a whole Year in which the Custom-House Officers have been so very diligent in spreading the Fame of the said Warner's Clay to be the best Fuller's-Earth in England, he hath not been able to sell the least quantity of it as such, tho' he has proffered it to all Men for a very little more than one Third of what Fuller's-Earth is generally sold for.

Needless to say Warner won his case.

There is some doubt about the date of repeal of the Acts prohibiting export of wool and fuller's earth. Adam Smith's 'Wealth of Nations' refers to the withdrawal of the prohibition in 1721; but Chesterfield refers in 1738 to *Fuller's-Earth, the export of which is strictly prohibited*. The embargo was still in force in 1739 as we have seen, but was not finally repealed until 1823.

3 Geo. IV, c. 107.

More on owling will be found in Chapter 20, Kent.

Newton's rash experiment

To ☿ I added by degrees fullers earth grinding them together till y^e earth drank up all the ☿, wch was almost done wth twice as much earth, & very well wth thrice as much, that is wth 60 gr/g.

☿ = mercury

IN THIS experiment carried out on 29 February 1683/4, he might have thought of using fuller's earth because of the publicity attending Warner's trial only three months before.

The unusually high concentration in Newton's hair of several heavy metals known to produce nervous symptoms is strong circumstantial evidence in favour of the hypothesis that Newton's so-called "nervous breakdown" of 1692/93 was due principally to poisoning by the metals which he used so frequently and with such cavalier disregard for his own safety in his chemical experiments.

P. E. Spargo and C. A. Pounds, "Newton's Derangement of the Intellect: New Light on an Old Problem", *Notes & Records Roy. Soc. London,* (1979), *34* (1), 11–32.

Not only did Newton make a measurement of the absorptive capacity of fuller's earth but he carried out an early thermogravic experiment:

The whole weighed 7.5 gr, there being lost 5 gr in ye grinding. Perhaps some moisture might exhale from ye earth. To this I poured spirit of ♁ [antimony] *19 gr. In a gentle heat of digestion for a day or two some moisture came over. Then distilling in naked fire there came over first much moisture then some ☿ running. No sp^t of ♁ arose. At length almost in a red heat arose a white salt. And increasing ye fire to a red heat and continuing it 3 or 4 hours there arose fumes continually but at last more slowly than at first by much. This salt (some of it) stood melted in ye neck for some time & I am apt to think it dissolved some of the ☿ there. In a while it coagulated & some of it next ye fire sublimed before it melted again. Perhaps it was incrassated by ye dissolution of some ☿. The matter in ye bottom looked redder than fullers earth & weighed 43 gr & on a red hot iron did not smoake. The sublimed salt & ☿ together weighed 26 gr beside a grain or two left in ye retort neck ffullers earth 60 gr after being well dryed in ye fire in a fire shovel not red hot weighed 43½ gr. The salt was very ponderous. Its tast strong sourish ungrateful & tasting something like sublimate.*

From these figures one can show that the fuller's earth must have contained about 18% moisture, which is possible if it had been stored in an unheated shed.

For the sake of interest a few more chemical symbols are given in Fig. 47. It will be seen that the symbol for clay, argilla, is made by adding ears to the symbol for earth, terra. For an absorbent earth, a dot is added.

13viii

Fig. 46. Chemical symbols used in the 17th and 18th centuries.
The last symbol is a French symbol for absorbent earth of 1718 (Ch. 40). The previous symbol is a Mediaeval one for Armenian earth (Ch. 17i). The previous six symbols (from "talck" to "allum") are "Characters Chymical" from *Lexicon Technicum*, John Harris, 1704.

THE PERIOD 1700–1800

14

Money, wife, is the fuller's earth for reputations; there is not a spot or stain but what it can take out.
JOHN GAY, The Beggar's Opera, 1727 (1728), i, 1.

Peacham's words were omitted in the 1940 revival at the Royal Theatre, Haymarket, London.

FULLER'S earth must have been a very well known article for it to be used in metaphor; and have remained in the public mind for Horatio Smith to use the simile

. . . *like fuller's earth, defiling for the moment but purifying in the end.*

An early example by Eachard, (1670), is given in Ch. 4.

Horatio Smith, *The tin trumpet, or heads and tails,* 1836 (1876).

John Dyer published one of the most successful epic poems – "The Fleece" – in 1757. Over 6000 lines long it deals with every aspect of the treatment of wool to make cloth. He mentions the exotic origin of some of the materials used in dyeing in some lines which head the section on Woburn fuller's earth. Elsewhere he writes

Chapter 23

Next from the slackened beam the woof unrolled
Near some clear gliding river, Aire or Stroud,
Is by the noisy fulling mill received,
Where tumbling waters turn enormous wheels,
Where hammers, rising and descending,
Learn to imitate the industry of man.

John Dyer, *The Fleece,* (1757), III, 570

Dr. Johnson had spoken slightingly of Dyer's *Fleece*, and said the subject could not be made poetical. *How could one write poetically of serges and druggets? And yet*, he said, *you would hear many people talk of that excellent poem*, "The Fleece" *One might write* The Parsley bed, a Poem, *or* The Cabbage Garden, a Poem, *as well as* The Fleece.

Boswell: *The Ominous Years 1774–1776,* edited by C. Ryskamp and F. A. Pottle, William Heinemann: London (1963), pp. 284–285.

. . . *Thy oil-imbibing earth,*
The Fullers skill assisting, safe defies
All foreign rivals in the clothier's art.

Robert Dodsley, *Agriculture*, ch. 3, (after 1764), the first four lines he had lifted from *The Fleece*.

Literary references to fuller's earth are sporadic compared with the frequent use, even today, of phrases which come down to us from the great days of wool. G.M. Trevelyan lists "thread of discourse", "spin a yarn", "unravel a mystery", "web of life", "fine-drawn", "homespun", "tease" and the classifying of all unmarried women as "spinsters". Others are "on tenterhooks", "dyed in the wool", "tissue of lies".

George Macaulay Trevelyan, *English Social History,* (1944).

The fulling mill was often alluded to because one couldn't be unaware of its characteristic noise.

(1) Tobias Smollett, *Roderick Random*, xvi, 97 (1748).

(2) Bishop Thomas Percy, *Reliques of ancient English poetry*, 'Thomas Cromwell', (1765).

Quarter Session Order Books U, p. 141, Pontefract, (1 May 1739).

Voltaire, *Lettres Philosophiques,* Dixième lettre: sur le commerce. Basil Blackwell: Oxford (1976), 28–29.

S. Hales, "Vegetable Staticks", (1729).

R. C. Mackenzie, "De Calore : Prelude to Thermal Analysis", *Thermochimica Acta,* (1984), *73*, 251–306 (pp. 278–280).

Published in London. Guide Book, Dept. of Archæology, University of Durham (1980). Guide Book, Museum of the Lancashire Textile Industry, Higher Mill, Helmshore, BB4 4NP.

1 *My heart went knock, knock . . . like a fulling-mill.*

2 *Thou myghtest have learned thy cloth to flocke upon thy greasy fullers stocke.*

During the second quarter of the 18th century the output of cloth was increasing, and the work of the fuller grew heavier. So they worked on Sundays. Hence *milling from 12 Saturday night to 12 Sunday night* was prohibited under penalty in certain areas.

The importance of wool and fuller's earth to the British economy must not be exaggerated. Voltaire was surprised to find that Britain's international influence was based on such meagre natural resources. As he put it (1726) *La postérité apprendra peut-être avec surprise qu'une petite île, qui n'a de soi-même qu'un peu de plomb, de l'étain, de la terre à foulon et de la laine grossière, est devenue par son commerce assez puissante pour envoyer . . . trois flottes à la fois en trois extrémités du monde. (Posterity will perhaps be surpised to learn that a small island, which possesses only a little lead, tin, fuller's earth and coarse wool, has become by its commerce powerful enough to dispatch at one time three separate fleets to the extremities of the world).*

In the world of science, the Rev. Stephen Hales (1729) followed Newton's lead in examining the volatiles evolved on heating numerous – this time non-toxic – materials and weighing the residues. Among these were *Horn,* calculus humanus, *Oystershell, Oak . . . Honey . . . Sugar, Amber, Coal, Earth, Walton mineral, sea Salt, Salt-petre . . . Minium.* Of these Mackenzie reckons that Walton was probably Walkthon, German for fuller's earth.

Towards the end of the 18th century advances in mechanisation vitally affected the cloth industry and accelerated the industrial revolution which had begun with the introduction of the fulling mill – the Spinning Jenny patented by James Hargreaves in 1770, the Yarn Spinning or Water Frame (because powered by a water-wheel) by Richard Arkwright in 1771 and the Power Loom in 1784 (preceded by the luckless John Kay's Flying Shuttle of 1733). These inventions, applied to both wool and cotton, especially in the north of England, were the foundation of British industrial supremacy.

During the whole of this century roads were almost indescribably bad. As mechanisation proceeded the demand for fuller's earth increased, and interest was shown in the deposits at Woburn as we shall see later.

There are many books about the growth of the wool industry in Yorkshire. One may refer to Ralph Thoresby's *Ducatus Leodinensis, or the Topography . . . of Leeds,* (1725), and The Heaton Collection of Account Books, (1796), held in the Cartwright Museum, Bradford. Heaton had his own fulling mill and used soap for the scouring. In 1792 a 600 year old meal mill in Durham City was leased for the carding and cleaning of cloth; this Old Fulling Mill is now a Museum of Archæology. A more complete fulling mill may be seen at Helmshore in Lancashire, between Bury and Accrington. This was built in 1789

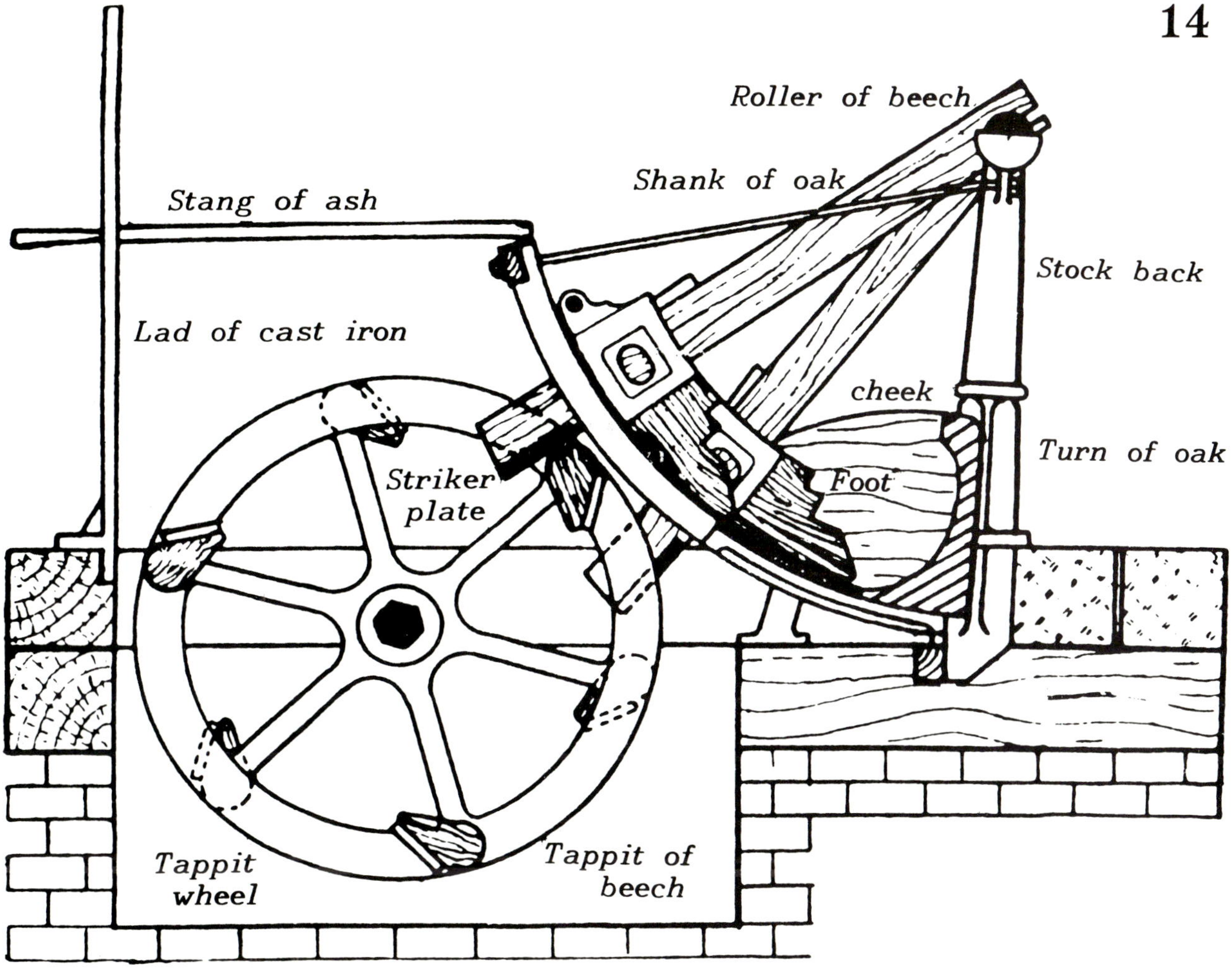

Fig. 47. Fulling stock, showing names of parts and materials of which it was made. NOTE – the shaft runs at about 13½ rpm, so that each foot strikes 40 blows per minute. Weight with shank and striker plate = about 2½ cwt, each foot.

From E. Kilburn Scott

and rebuilt in 1860 after a fire. The Lancashire County Museum Service took over Higher Mill, abandoned in 1967, and put most of the plant into working order.

15 THE NINETEENTH CENTURY

15i Competition from soap

Some old-fashioned ways of cleaning textiles, by G. Schaefer, *CIBA Review*, (April 1947), (56), 2036.

AS early as the 16th century soap was used for the first time in the cloth finishing trade. In the 18th century it was used all over Europe for the scouring of wool, by hosiers, fullers, blanket-makers and wool-workers, at least to some extent.

James Malcolm, *A compendium of modern husbandry,* London: C. & R. Baldwin (1805), *1*, 42–46.

At the beginning of the 19th century soap was becoming a serious rival to fuller's earth. Indeed, in 1805 Malcolm foretold that the use of fuller's earth on any scale in the cloth industry would be soon a thing of the past.

Archibald Clow & Nan L. Clow, *The Chemical Revolution,* London: The Batchworth Press, (1952), 72 . . .

id., p. 83.

id., p. 107.

That soap had not come hitherto into more extensive use was due, as already stated, to its price. The alkali required for its manufacture had to be obtained either from barilla, which in addition to shipment costs carried a high import duty, or kelp – ashes of seaweed – likewise not a cheap item. The price of kelp rose to a peak in 1798, fell and rose to a very high point in 1808. Barilla showed much the same pattern; and both fell to uneconomic levels by 1826. In a series of Acts from 1805 to 1810 the British government rescinded the barilla import tax. Destitution came to 80,000 seaweed burners in the Western Isles and seaboard of Scotland. But with duty-free barilla and kelp prices down to a quarter soap became for industrial processes an economic proposition. Even more helpful to the soap industry was the removal of the tax on salt and sulphur, which allowed synthetic soda to be made at a fairly low cost. Nicolas Leblanc patented his method of making soda on 25th September 1791; and it was at one time thought that the founding of James Muspratt's works near Liverpool in 1823 was the pioneer in Britain. The Clows trace the earlier development of this industry and show how synthetic soda brought the price down long before Muspratt's enterprise.

In 1806 there was a Parliamentary Committee on wool trade statutes, and in the following year there was

47 Georgii III Regis, ch. xlix, (25 April 1807).

GEORGII III. REGIS.

CAP. XLIX.

An Act for permitting the Exportation of Fullers Earth, Fulling Clay, and Tobacco Pipe Clay, to any Place in Possession of His Majesty.

[25th *April* 1807.]

Export to foreign countries was however still prohibited. In 1814 came the repeal of the Wool Acts, but the yield from soap duties reached the huge figure of £1,200,000 in 1827. In 1822 the value of the commercial soap output in England was £89,000,000, whereas ten years later it had risen to £119,000,000; the most important centres being in London and Liverpool. Much smaller quantities were made in Newcastle, Bristol, Brentford, Frodsham and Glasgow.

15i

See ch. 13 vii.

CIBA Review, p. 2047.

Effects on demand for fuller's earth in the cloth industry were gradual. There were some who retained a preference for it in any case:

> *Some manufacturers in Gloucestershire,* writes Malcolm, *as well as chemists with whom I have covered the subject, affect to say, that soap will answer the purpose equally well, but I have no hesitation in saying that the French are of a very different way of thinking; for as they manufacture soap of the finest and best quality so much cheaper than we do, they would certainly be at no loss in fulling their cloths so as to discharge all those offensive substances equally well as we do with fuller's earth. But I have seen a beautiful cloth exhibited by one of the principal manufacturers at the Louvre, on the first complimentary day at Paris, who assured me, after a great deal of conference upon the subject, that it was the fuller's earth only which they wanted, and they could then equal, if not surpass, our finest and best cloths.*

J. Malcolm, *op. cit.*, *1*, 43.

That was still in 1805, before soap prices had begun to drop anywhere in Europe.

The Surrey Iron Railway

15ii

ANOTHER remarkable development, also noted by Malcolm (1805), was the building of the Surrey Iron Railway, which provided the Surrey deposits with direct transport to Wharves at Wandsworth in London from which it was shipped to Yorkshire and elsewhere.

J. Malcolm, *op. cit.*, *1*, 29.

Plans for the first part of this, "the first public railway in the world" – proposed as an alternative to an original scheme for a canal, rejected as causing too much inconvenience to many mills sited on the river Wandle, were laid before Parliament in 1801, and in 1803 a line was opened for traffic linking Wandsworth and Croydon. Interest had been so much roused that in that same year a new company was formed to construct a continuation from Croydon to Godstone via Merstham. In fact this never got beyond Merstham, but for the purposes of fuller's earth transport this was a good enough terminus, and sidings were specially built there for its loading.

Surrey Iron Railway: An Act for making and maintaining a railway from the town of Wandsworth to the town of Croydon. 41 Geo. III, c. 33, (1801).

This was called the Croydon, Merstham & Godstone Railway.

F. G. Bing, in his history of the railway, comments on the curiosity of the fact that *those acquainted with the achievements of this railway had no fear of comparison with the method of steam traction which was being experimentally tried by Stephenson at about the same time. One champion of the horse traction in use on the Surrey Iron Railway exclaimed: Does anybody in possession of his*

The Grand Surrey Iron Railway, Croydon: Roffey & Clarke Ltd. for Croydon Public Libraries Committee, (1931), 7.

Fig.48. The Surrey Iron Railway 1803–48

Croydon Public Library

senses expect that this snorting, spluttering, hideous iron machine, belching forth smoke and steam, can ever accomplish such a draught as is easily undertaken by the horse on the Croydon iron road?

James Malcolm, op.cit., (1805), *3*, 446.

One horse drew 37 tons of stone for six miles along the line in two hours without evincing any symptoms of distress or fatigue – *to prove which, for the last ½ mile of his journey, 17 additional tons were given him, making the whole 54 tons, with which he proceeded without difficulty.*

Such feats notwithstanding, it is doubtful if the fuller's earth producers reaped much economic benefit. The tolls they paid on the public line are laid down in the railway company's regulations.

O. Manning and W. Bray, *History and Antiquities of Surrey,* (1804–1814), *2*, 266 [1809].

These charges do not look particularly heavy, but according to Manning and Bray, in 1809 the cost of transport by common carrier to London, where it was sold at the wharves in Tooley Street for from 24/- to 26/- a ton, was 18/- to 20/- a ton, and the considerable quantity then being carried by the Iron Railway was sold at little if anything above that carriage cost. The latter had after all to be taken first by waggon for loading at Merstham. (Mr. Lewis F. Grece, whose family had been long in the fuller's earth business, and who found in a search

SURREY
Iron Railway.

The COMMITTEE of the SURREY IRON RAILWAY COMPANY,

HEREBY, GIVE NOTICE, That the BASON at *Wandsworth*, and the Railway therefrom up to *Croydon* and *Carſhalton*, is now open for the Uſe of the Public, on Payment of the following Tolls, *viz.*

For all Coals entering into or going out of their Bason at Wandsworth,	*per Chaldron,*	3d.
For all other Goods entering into or going out of their Bason at Wandsworth - -	*per Ton,*	3d.

For all GOODS carried on the said RAILWAY, as follows, viz.

For Dung, - -	*per Ton, per Mile,*	1d.
For Lime, and all Manures, (except Dung,) Lime-ſtone, Chalk, Clay, Breeze, Aſhes, Sand, Bricks, Stone, Flints, and Fuller's Earth,	*per Ton, per Mile,*	2d.
For Coals, - -	*per Chald. per Mile,*	3d.
And, For all other Goods, -	*per Ton, per Mile,*	3d.

By ORDER of the COMMITTEE,

W. B. LUTTLY,
Clerk of the Company.

Wandsworth, June 1, 1804.

BROOKE, PRINTER, No. 35, PATERNOSTER-ROW, LONDON.

Fig. 48B. *From a poster of the period*

The terminus was at a small dock at Frying Pan Creek on the Thames at Wandsworth.

A chaldron was about 25½ cwt.

With acknowledgment to the Science Museum and the Railway Museum at York.

through papers a number of receipted accounts of the business, *including some of the old iron railway from the Jolliffe Arms at Merstham to Croydon* (where earth was sometimes unloaded on to barges on the Croydon Canal), writes: *The bags of fuller's earth used to be taken by six draught horses to the Jolliffe Arms. I myself remember old Tom Packham, my Grandfather's chief Stableman, and his talking of his splendid teams of grey draught horses.* There was then the transfer again to boats.

All told, there cannot have been benefits to the users. And the railway itself did not last long in spite of conversion to a pneumatic system latterly. *An Act authorising the dissolution of the earlier company and the sale of its properties received the Royal assent on August 3rd, 1846. Traffic on the Wandsworth to Croydon section ceased on August 31st, 1846, and after*

15ii

For further information see Charles E. Lee, papers presented to the Newcomen Society 11 Dec: 1940 and 11 Jan: 1950, the latter with C. E. C. Townsend. Lee also wrote a brochure The World's First Public Railway in 1931.

a career lasting forty-three years the railway was finally dismantled in 1848. It would appear that the line was never adequately protected or fenced along the greater part of its route, and that the track suffered much from being torn up in certain places by competing conveyances. As the Croydon to Merstham section had never been a financial success it probably passed away with few regrets, and before many years became the butt of humorists, who sarcastically compared its achievements with those of the steam railway. One needs scarcely to be reminded that the "Stockton and Darlington" (1821) was the first public railway in the world to be worked by locomotives. One pessimist in 1805 thought that *it is not probable that railways will ever come into general use.*

15iii

Improvements in transport

= lumps

UNTIL about 1840 fuller's earth clods were usually sent by road either loose or in casks, linen sacks being far too expensive to be used; but as the Dundee jute industry got under way, cheap sacks became available. The saving of labour costs in loading and unloading the earth must have been considerable.

Jeremy Greenwood's Fullers Earth in Surrey 1500–1900, Reigate (1983) traces the history of transport of fuller's earth.

The fuller's earth industry had already benefited by the repeal of the Act prohibiting its export in 1825, and an American market was opened up. Road transport was feasible since the turnpike road was built ca 1770 and remained the preferred mode of carriage even after Redhill was reached by the steam railway in 1841. Nevertheless within a few years it was possible to send the earth by rail almost all over the country. Certainly there was a definite rise in fuller's earth sales towards 1850. The American trade ceased during the Civil War, but was not insignificant thereafter.

15iv

Nineteenth century fulling: processes and materials

A LONG-ESTABLISHED empirical tradition discriminated between different types of fuller's earth for different fulling purposes.

James Malcolm, *op. cit.*, (1805), *1*, 44.

> *The sorts are distinguished by the names of the blue earth and the yellow earth, and have each their respective appropriations; the blue earth is required by the manufacturers of Leeds etc. where our second cloths are made; the yellow is preferred by the clothiers in Halifax, Huddersfield, and that neighbourhood for the manufacture of kerseymeres. That this distinction is invariably observed I am assured on the authority of Mr. Silva, a gentleman of veracity, and of consequence as a merchant; who is perhaps the largest shipper of earth in London; and to whom I am indebted for much of the information respecting this earth. The broadcloths are therefore fulled with the blue, and the kerseymeres and superfines with the yellow.*

So much for quality preferences – but we have failed to unearth figures relating to the weight of fuller's earth required to treat a given length of a given cloth. Colin Mackenzie (1825) gave the following instructions "To full cloths, woollens, etc."

Colin Mackenzie, *Five Thousand Receipts*, 5th ed., (1825), London: G.B. Whittaker, p. 125.

> *The method of fulling with soap is this; a coloured cloth, of about 45 ells, is to be laid in the usual manner in the trough of a fulling mill, without first soaking it in water, as is commonly practised in many places. To full this trough of cloth, 15lb of soap are required, half of which is to be melted in two pails of river, or spring water, made as hot as the hand can well bear it. This solution is to be poured by little and little upon the cloth, in proportion as it is laid in the trough; after which it is to be taken out and stretched. This done, the cloth is immediately returned into the same trough, without any new soap, and there fulled for two hours more. Then taking out, it is wrung well, to express all the grease and filth. After the second fulling, the remainder of the soap is dissolved in as in the former, and cast four different times on the cloth, remembering to take out the cloth every two hours to stretch it, and undo the plaits and wrinkles it has acquired in the trough. When sufficiently fulled, and brought to the quality and thickness required, scour it in hot water, keeping it in the trough till it is quite clean. As to white cloths, as these full more easily and in less time than coloured ones, a third part of the soap may be spared.*

56.25 yards = 49.2 metres.

15 lb = 6.8 kg

Possibly
4 gallons = 15.14 litres

Unfortunately he failed to tell us the width of the cloth in his remarkably ungrammatical account. If white cloth one ell wide required 15 lb of soap to full a 45 ell length, some 21 to 27¾ lb would have been needed for a broadcloth. If soap acted more quickly in the milling machine, fuller's earth would have been at a disadvantage. On the other hand in the slow process of the 16th century and earlier it might not have been so very different in its effectiveness; but the wool then was less greasy than later on (see chapter 16 ii).

originally 6 feet to
4 feet 6 inches

See Chapter 13iii

Nor does Mackenzie mention the use of fuller's earth under the heading "To bleach wool". *Scouring, to free it from grease, by immersion in an ammoniacal ley (5 parts of river water and one of stale urine); 20 minutes in this mixture at 56°; drained; rinsed in running water; this operation is repeated twice or even thrice . . . In some places, scouring is performed with water slightly impregnated with soap; and, indeed, for valuable articles, this process is preferable, but it is too expensive for articles of less value.* Sulphur dioxide in water may be used for bleaching wool and silk, as in Roman times.

op. cit., p. 174

ley (sic); but read lye.
56°F = 13.3° C

In 1834, the rotary mill, Fig. 49, in which pressure was exerted between rollers, was invented by John Dyer and improved in 1841. Mr. Chevalier of Leeds announced in the *Leeds Mercury* of 21 October 1843 that he had acquired the rights from Mr. Dyer who no longer had any. E. Kilburn Scott, a descendant of the Kilburn millwright family, gives a picture of a typical mill of this type, which was still in use in the 1930s, albeit with improvements. (Fig. 50).

Patent (25 March 1834),
J. Dyer of Trowbridge

From E. Kilburn Scott, Early Cloth Fulling and its Machinery, *Trans. Newcomen Society*, (1931–2), *12*, 31–52. (p. 45).

By 1850 handloom weaving had virtually died, but animal and vegetable dyes were still in use. The sale of fuller's earth to Yorkshire

Trowbridge, October 17, 1843.

JOHN DYER begs to call the Attention of Cloth Manufacturers to an Advertisement of Mr. CHEVALIER'S, in the Leeds Newspapers of last Week, in which the Public is informed that J. DYER had SOLD his PATENT RIGHT in the FULLING MACHINES, in March, 1834: but Mr. Chevalier does not state the fact, of which he must be acquainted, that the Party who Purchased such Right, subsequently granted to J. DYER, on certain Conditions, the SOLE RIGHT of MAKING and SELLING these MACHINES during the Terms of the Patent Right, and which exclusive privilege will be maintained by every lawful means. Further information may be had on application at Mr. J. Jones's Counting House, 24, Park Cross Street, Leeds.

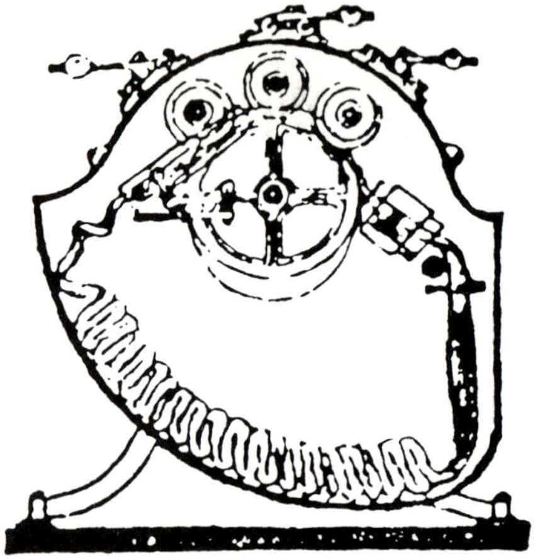

NEW PATENT FULLING MACHINE.

MR. CHEVALIER begs to inform the Cloth Manufacturers and others, that Mr. John Dyer, of Trowbridge, Original Patentee, has SOLD HIS PATENT RIGHT the 25th March, 1834, and consequently *cannot enter into any Arrangements* with any one; and for the Information of Manufacturers and others, *his Sealed Patent is in the Hands of Mr. Chevalier*, so that he is PROPRIETOR OF JOHN DYER'S PATENT, and also of the New Patent taken out in 1841.

CAUTION.—Any Person infringing on the above Patent will be prosecuted.

Machines can be tried on the Premises, and every satisfaction given to the Purchaser, by applying at the Manufactory, Dewsbury Road, Leeds.

Fig. 49. Dyer's and Chevalier's advertisement, 1843, of the New Patent Fulling Machine, which replaced fulling stocks. The poor quality reproduction is due to its being a photograph from a regional newspaper, not now available.

For dyestuffs see *CIBA Review*, volume *1*, 1–173, 206–240, 278–302, 309–312, 406–410.

was much reduced before the new type of mill was widely adopted, but the industry was revolutionised after 1856 when William Perkin synthesized the first synthetic dyestuff, mauvine, and went into production in 1857. The decline of cochineal began with the introduction of aniline red by W. von Hofmann in 1858. Azo dyestuffs from coal tar naphthalene came on to the market in 1880 and sealed the fate of natural animal dyes.

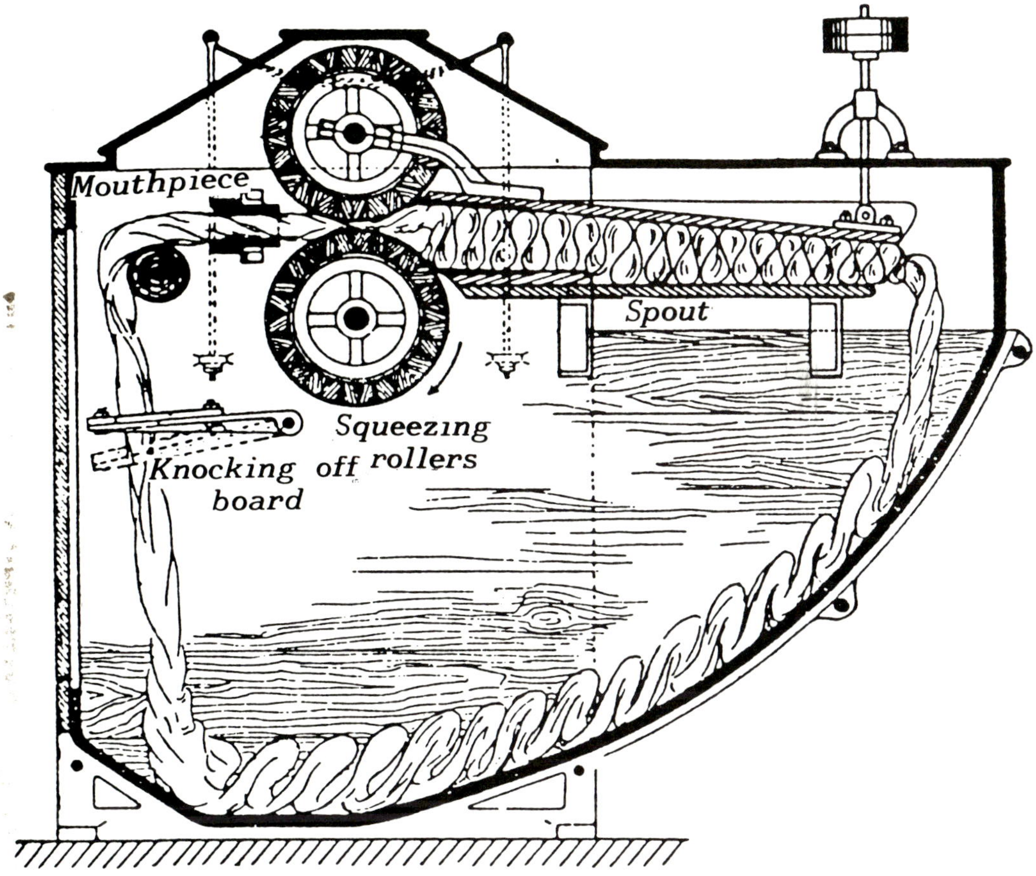

Fig. 50. Rotary Milling Machine and names of parts. The spout here is shown longer than usual.

E. Kilburn Scott (1931–32)

In the 19th century the two main uses for fuller's earth were for earthing after dyeing and in the rinse after milling, not after raising as is commonly stated. *When natural dyes were the only ones available, all dyed material had, like the indigo pieces of today (1950), to be given an earthing if fastness to rubbing was to be obtained.* Fuller's earth was also useful for treating brightly coloured blankets, for example, since the synthetic dyes were at first not resistant to soap or alkali.

C. O. Clark, Ancient and modern in milling and scouring, *J. Soc. Dyers & Colourists.* (1950), *66*, 187–190.

By 1870 the British production of soda exceeded that of all other countries. Hard, industrial soap reached a maximum production in 1895. And the nineteenth century saw the final triumph of the factory system.

15iv

Note on industrial archaeology

(1) 'Armley Mills', The Leeds Industrial Museum booklet, after 1981, p. 3.
(2) *J. Ind. Arch.* (1964), *1(1)*.

AT Armley Mill, Canal Road, Leeds, may be seen (1) two pairs of stocks made by J. Pickles of Bramley ca 1837 and (2) two pairs of stocks (not erected) made by William Kilburn of Leeds sometime between 1810 and 1850.

The Museum of the Lancashire Textile Industry, Higher Mill, Helmshore and the Museum in the Old Fulling Mill at Durham also exhibit fulling machines (see chapter 14).

FULLING IN THE TWENTIETH CENTURY

16i

ONE might have expected fuller's earth to have a new lease of life when it was discovered by Herbert à Brassard (1949) that it improved the milling and scouring of wool pieces with secondary alkyl sulphates. Clark refers to this as another *obvious method of use being ignored in the history of fuller's earth. In the literature there is just one isolated reference to the fact that fuller's earth is an admirable scouring agent to use in an acetic acid bath.*

H. à Brassard, talk to Huddersfield section of the Society of Dyers & Colourists, (March, 1949), in C. O. Clark, *op. cit.*, p. 187 and 190.

The use of fuller's earth was not doomed wholly because of technical faults in its performance, but, because in 1912 a Batley Carr firm ceased to make brightly coloured woollen rugs which they had made since about 1852, they could no longer find any cheap land on which to dump the sludge. One ton of dry fuller's earth gives 10 tons of sludge deposited in the settling tanks; and recovery of indigo from it proved to be unfeasible. Roberts Beaumont also speaks of the use of fuller's earth after dyeing with logwood black.

Roberts Beaumont, *The Finishing of Textile Fabrics*, London: Scott, Greenwood, 2nd ed., (1926) [1st 1909] p. 67.

Even in 1911, Murphy comments that for scouring worsteds [at least those few types which respond to the treatment] a very small quantity of white soap, aided by a little pearl ash, and ammonia, will suffice. Fuller's earth is also a very good detergent for worsteds and for clearing out soaps and dyeing substances generally.

Wm. S. Murphy, *The Textile Industries*, London: Gresham, (1911), *8*, 103 et seq.

Modern woad- or indigo-dyed material in its lighter shades requires no earthing, but in the dark blues familiar to us as the serges for school uniforms, fuller's earth is used to remove the surplus dye in what was still an essential finishing operation at least until World War II. In the two wars fuller's earth was used, to save soap, in the treatment of coarse blankets and in the earthing of dark blue naval and other uniforms. It was Hargreaves (1932) who showed that the addition of acetic acid in the treatment of indigo-dyed goods with fuller's earth gave a good effect. The last days of this application are described in Chapter 38.

P. W. Hargreaves, *Dyers Calico Printer,* (1932), *67*, 5313.

Changes in wool fineness

16ii

RYDER (1960) examined follicle remains in British parchments from the 12th to the 19th century. He concludes that *hairy sheep [having both coarse and fine fibres] were apparently present in all periods, whereas most of the fine-wool samples came from the medieval period, and most of the medium-wool*

Follicle remains in some British parchments, M. L. Ryder, *Nature,* (9 July 1960), *187*, 130–132;

16ii

but see also "Sheepbreeds in history", in Year Book of the National Sheep Breeders' Association (1960).

P. J. Bowden, *Econ. History Review*, (1956), ii, *9*, 44.

samples came from the sixteenth century onward, those of the eighteenth century being the coarsest. The findings therefore support historical records which mention the extreme fineness of medieval wool, and also work which has shown evidence of an increased supply of long wool during the 16th and 17th centuries. There was evidence, however, of the existence of medium to fine-to-medium wool sheep in the Middle Ages that could have been the ancestors of the long-wool sheep, the origin of which has recently been much discussed. It is interesting that Ryder concluded that Soay sheep were used in Scotland, because it had been thought that the Soay type was widespread there until the 18th century. The tartan cloth used for Highland dress was much softer than the long-wool cloth used today.

The earlier types of sheep were quite small and yielded only two pounds of fine wool which contained far less wool grease than now. By breeding, the long hairs were eliminated and the wool became coarser and longer, so that a sheep bore over ten pounds of wool containing a far higher proportion of wool grease. Over the centuries the need for scouring and fulling agents such as fuller's earth gradually increased.

16iii

Laundering

Kingzett's *Chemical Encyclopaedia* (1923). Worked first by C. M. Powder Co. Ltd., latterly by Treamble Mines Ltd.

Report by British Laundry Research Association, ~ April 1951.

SINCE 1923 at least, "fuller's earth" has been claimed to exist at Treamble, Perranporth, Cornwall. The clay was formed by the alteration of slate containing quartz, mica clay and kaolinite, and the industrial product IGOR was sold to some laundries for removing oil and grease from heavily soiled dungarees and the like. It was not a good detergent for normal loads, so the company, who had sold only 20 tons in 1950, went into voluntary liquidation the following year. This clay should not be called a "fuller's earth", which is more effective for scouring and fulling and is far less gritty. Indeed fuller's earth is better than soap or solvents for cleaning heavily oil-soiled textiles, such as dirty mats from motor cars. It is now known that it helps to prevent redeposition of soil when present in a complex household laundry detergent.

PHARMACY

17

The best cure for aeroplane sickness is equal parts of Fuller's Earth and Petroleum.
THE WEEK-END BOOK, 1931, as an example for the Parlour Game SALTED ALMONDS.

London: The Nonesuch Press, 3rd ed., (1931), 341.

Bole Armeny

17i

THE use of fuller's earth in babycare has made it a figure of fun, but for centuries clays have been used in pharmacy and cosmetics in a variety of useful ways.

Means had existed since the Middle Ages for the importation and distribution of pharmaceuticals, including clays. In and around London supply was in the hands of the members of the Grocers' Company, founded in 1373 and deriving from the even older Guild of Pepperers. (The name comes from *en gros*, meaning wholesale or in bulk).

J. C. Hanbury, Early developments in the fine chemical industry, *Chem. & Ind.*, (1952), 300–308.

One of the most important medicinal earths was bole armeniac, often corrupted to bole ammoniak and the like. As Sir John Pettus (1683) put it *I find these words of kin ... viz. Amoniack, Armenick, and Armoniack.* It was a pale red-coloured earth, which originally came from Armenia, but by 1776 from Bohemia, Silesia, Saxony, Hungary, Transylvania, Norway, and elsewhere. The French employed chiefly the bole of Blois. *Bole Armenic is smooth and slippery to the touch, somewhat glossy, pure from sand, or any perceptible gritty matter. It readily crumbles betwixt the fingers, and adheres to the tongue. Softened with water, it forms a smooth paste; diluted with a larger quantity of water, it remains for a considerable time suspended.*

Ἀρμενία

Fleta minor: the laws of art and nature in assaying metals, 1683 (1686).

The Rev. Samuel Ward, Vicar of Cotterstock cum Glapthorne, *The Natural History of Waters, Earths, Fossils, Minerals and Vegetables*, (1776), **12**, 33–36 London: F. Newbery.

Ward noted that the iron in it could be reduced to magnetite and that it contains a sulphate. The earth was used medicinally as a styptic, both internally and externally; *a virtue which seems to proceed from its ferrugineous impregnation: to the antipestilential and alexipharmic qualities, for which it has long been celebrated, it has no pretence; nor have the absorbent ones, which some ascribe to it, any just foundation.* The medicinal grade was made by washing, settling, drying and forming into rolls or cakes.

Alexipharmic: preserving from the effects of poison or counteracting them, from ἀλέξειν to ward off and φάρμακον poison.

In Chaucer's *Canon Yeoman's Prologue and Tale* (c. 1386), the Alchemist's laboratory contained "bole armoniac, verdigrees, boras". The main uses for the clay at that time were as an antidote to poison – including food poisoning and for the staunching of blood. Some of the other medicaments of the time were listed in Chaucer's description of the unpleasantly pustulous Somnour:

line 238

Prologue to the *Canterbury Tales*, (1386), lines 629–633.

Ther nas quik-silver, litarge, ne brimstoon,
Boras, ceruce, ne oille of tartre noon,
Ne oynement that wolde clense and byte,
That hym mighte helpen of his whelkes white.

whelkes = pustules

♀ *or* AB

The mediaeval symbol for bolus was a sphere capping a cross, and early imports included both *terra sigillata* and *Bolus armenus*, also known as *Bolus orientalis*. Not even the great physician Galen (~130–200 AD) seems to have been sure of the latter's provenance. Indeed during a "great and cruel plague" Galen was brought a pale-coloured earth from another district in Armenia, bordering on the old province of Cappadocia. This seemed to him different from the ordinary Bolus armenus and more like Samian Aster, a kaolin, but it was found most beneficial in its effects.

Other symbols have been given in Ch. 13 viii.

Pills of *Bolus Alba* had been prescribed in England from a very early date as a remedy for gastric disorders. From the early 17th century we find references to *Bolus Armenus Alba*, which might have been the clay shown to Galen. But the name *armenus* was no proof of origin. In 1776 white bole was imported *ready washed, and formed into large rolls or cylindric globes, not of a chalky whiteness, but rather of an ash-colour. Formerly it came only from Tuscany, or the island Elva; at present it is supplied chiefly from Norway, the island Bornholm, or places nearer home. It differs from the two preceding in containing no manifest irony matter, and consequently in wanting their astringency.*

Samuel Ward, *op cit.*

Gr. Αἰθαλία ; Lat. Ilva; later Elva; now Elba: The "smoky island" named from its iron smelting furnaces.

Andrew Boorde (1547) gives an early prescription *Take of Terre sigillata, of boole Armoniake, of eche an unce.* William Ward or Warde recommended Bolearmenicke in "The secretes of Alexis of Piemont containing remedies against diseases", (1558); so did Humphrey Lloyd or Llwyd in "The treasury of health conteyning many profitable medycines gathered out of Hypocrates, Galen and Avycen", by one Petrus Hispanus (1550): *take of . . . Mastycke, Dragon's bloud, bole Amonike new, of eche like quantity.* (Dragon's blood was the resinous exudate of a palm tree).

The breviary of healthe, 1547, (1552).

From Ἀλέξησις warding off. Tr. 1558 (the first part) p. 40 b.

Tr. 1550, *Treasury of Health*, O iii.

Bernard Palissy (1580) tried to explain the pharmaceutical action of Armenian earth and terra lemnia, *sealed earth has no power against poison except because of the action of congelative water**. He describes these clays, as well as *sanguine (or bloodstone), orcane (red ochre), and black stone, as being no more than petrified desiccative and astringent rocks.* He noted that boliarmeny stuck to the tongue.

B. Palissy (1510–1589), *A lively dialogue between theory and practice* (1580), pp. 176, 224–7, 246–7.

* water of crystallisation as opposed to "exhalative water" = moisture.

Thomas Cogan (1584) prescribed *coriander, ginger, bole Armoniacke, of each a drum.* Edward Topsell (1607) was perhaps the first to note that *plaister is made thereof with bole-Armorick.*

The haven of health, (1584) [1636]. Possibly a dram.

The history of foure-footed beasts (1607).

The egregious Gervase Markham (1610) also described a veterinary application in *take of bolearmony a quarterne* – say 114 grams – and in the *English Housewife* (1625) II; [1668] 41 he writes of a *defensitive plaister.*

Markham's maisterpiece or what cloth a horse-man lacke, containing all possible
(contd.)

Far more fanciful and uncritical was the doctor turned author

William Salmon (1710), whose Houshold Companion should be seen to be believed.

> 94. Bole Armoniack: To prepare this, you need no more than moisten it with May-dew, or any other Dew not too grofs, and dry it in the fhade.
>
> *Rain Water will do as well, and then you may give it from a fcruple to half a dram, againft Heart-burnings, and vehement Pains of the Stomach.*

John Quincy includes (1718) Bole Armonick, a natural earth, in his *Pharmacopoeia officinalis et extemporanea* or *a compleat English dispensatory*.

This clay continued in use throughout the 18th century. In 1758 and 1771 a "defensitive" composed of Bole Armenia was mentioned, and George Smith in the *The laboratory or school of arts* 1799 gives a recipe, presumably for a plaster, of *bole armenic, parched barley, etc.*

The poor hygienic conditions in the preparation of food led inevitably to frequent bouts of illness. Armenian bole is ... prescribed internally against diarrhoeas (1727–51). There were those who became so frightened of food poisoning that they kept a whole armoury of medicines, including clays, to try to ward off these unpleasant complaints.

I shall not deal at length with the other uses of Armenian earth. The redder sorts, containing more hematite, were used for marking sheep and were called *redding or ruddle*, and, I suspect, as an ingredient of face powder or rouge. More certain is that *Margaritone was ... the first that devised laying* Gold or Gilding *upon Bole Armoniacke to be burnished*, according to Peacham (1634).

The history of this class of clays ends on a superlative note for it is asserted in *Fraser's Magazine* for 1832 that *the best toothpowder in the world is Armenian bole.*

17i

(contd.) *knowledge whatsoever which doth belong to any smith, farrier or horse-leech* (1610).

William Salmon, M.D., *The Family Dictionary: or Houfhold Companion*, London, printed for H. Rhodes, 1710.

J. S. *LeDran's Observ. Surg.* 1758 (1771) 94 It should be *defensative.*

Chamber's *Cyclopaedia*, (1727–1751).

Henry Peacham the younger, *The compleat gentleman*, *13* (1634), 139 [other editions 1622 and 1661].

(1832) *6*, 714.

The growth of the pharmacopoeia

17ii

IT IS not known at what period English fuller's earth became so common a feature of the druggist's stock as to merit its own drawer labelled TERR. FULLON.

Something has already been said of the vagaries of the medieval pharmacopoeia. Although the London College of Physicians was established in 1511, it was not until 1618 that there appeared the *Pharmacopoeia Londinensis*, first of the regional compilations that served until the British Pharmacopoeia was brought out in 1864. In 1617 it was decided by James I that grocers, having no training in the properties

J. C. Hanbury *op cit.*

Pharmacopoeia Londinensis of 1618, George Urdang. New York: Schumann's (1944).

of drugs, were not fit and proper persons to have charge of selling them, and the privilege passed to the apothecaries. Effects of this measure however were not altogether what had been intended. Through the 17th century a bitter warfare was conducted between the physicians and the apothecaries, whom the former accused – with some exaggeration, possibly – of having the best of both worlds in a manner bound to excite antagonism; but also of excessive profiteering. Apart from this, they showed in their use of the so-called remedial drugs in vogue at the time a penchant for the more particularly revolting, employing every sort of human and animal organ, not to mention their excreta. Such choice of commodities as moss grown on the skull of a hanged felon were sometimes prescribed for a wide variety of ills.

From *alchemists*. At first, *chimists* or chymists, retaining the Gr. root as in χυμεία infusion. The modern spelling from ~1700 is based on Gr. χημία or χημεία, meaning chemistry. The word is gesturally connected with χέω, I pour.

About the year 1697 the physicians finally kicked over the traces and began to employ their own dispensers, who became known as chemists and druggists. Not only were these more reasonable in their charges, but a trend set in of steadily increasing discrimination in the use of drugs, which were eventually confined largely to vegetable extracts, more or less wholesome, and to relatively pure chemical substances, which began to figure in their formulations.

As far as clays are concerned the Greco-Roman formulary remained almost unaltered, the result of uncritical copying through the centuries. A typical passage reads:

ὀξύκρατον a mixture of vinegar and water.

Cimolia, Terra Saponaria, Creta Cimolia, Smectis, Fullers Earth. It is cleansing and discutient, refrigerates, digests and works according to the nature of the things with which it is mixed. Applyed with Oxycrate, it is very good against burns and scalds. Dioscorides saith, that both the White and Purplish helps swellings behind the ears, and other knots, tumors in the Testicles, and other swellings; as also St. Anthony's fire. The best is the cold and fat, which is good against the Gout and falling of the hair.

Pharmacopoeia Londinensis, London: Thomas Newcomb, (1677).

At least, fuller's earth is recognised as being in the same category as Creta Cimolia. In the 1677 edition of the *Pharmacopoeia Londinensis Terra Cimolia* (candida and purpurascens), *Terra Lemnia* and *Terra Samia* still stand side by side with *Terra Silesiaca* and *Bolus Armenia* and *Bolus Bohemica.* The clays were in the class known as simples and had, as we now know, some virtues, but the craze for elaborate formulations of umpteen mainly useless ingredients was accelerating towards a peak of stupidity. Nicolas Culpeper 1679 tried his best to comment helpfully on the "Colledge"-approved recipes by following each one of them by his often astringent and certainly refreshing commentary.

Pharmacopoeia Londinensis; The *London Dispensatory* with notes by Nicolas Culpeper, (1679).

Trochisci Terra Lemnia, 137, Or, Troches of earth of Lemnos.

Colledge.

Take of Earth of Lemnos, Bole-armenick, Acacia, Hypocistis, Gum Arabick tosted, Dragons blood, white Starch, red Roses, Rose-seeds, Lap Hematitis, red Coral, Amber, Balaustines, Spodium, Purslain-seeds, a little tosted, Olibanum, Harts-horn burnt, Cypress Nuts, Saffron, of each two drams; black Poppy-seeds, Tragacanth, Pearls, of each one dram and a half; Opium prepared one dram: with Juyce of Plantane make it into Troches.

The inspissated juice from the fruit of *Cytinus hypocistis* (Rafflesiaceae).
Dragon's blood a bright red gum or resin.
Balaustines, dried pomegranate flowers.
Spodium, bone-black.
Olibanum, *Boswellia* resin, also used in incense.

Culpeper.

Indeed in external applications, if an Inflammation, or Fever be, I think it better with Opium, than without, else better without than with it. It was invented to stop blood in any part of the body, and for it 'tis excellent. Well then, for the bloody flux, take half a dram of them inwardly (being beaten into powder) in red wine every morning for spitting of blood; use it in like manner in Plantane-water for pissing of blood, inject it into the bladder; for bleeding at the Nose, either snuff it up, or anoint your forehead with it mixed with Oyle; for the immoderate flowing of the Ternis, inject it up the Womb with a syringe, but first mix it with Plantane-water; for the Hemorhoids or wounds, apply it to the place bleeding.

On the other hand the 1733 *New English Dispensatory*, dealing with earths carefully defined as *bodies opake, insipid, and when dried friable, or consisting of parts easy to separate, soluble in water; and not disposed to burn, flame or take fire.* It still refers to *Cimolia purpurascens*, which had probably long since ceased to be available, gives a description of properties accurate enough to ensure recognition, and confirms the link between Cimolian and English fuller's earth, and cites the areas where the best quality of this was to be found.

James Alleyne, *A New English Dispensatory*, London, (1733)

Robert James wrote in 1752 that Cimolia purpurascens, Offic. Fuller's-Earth is *seldom or never used internally; but applied, as a Topic, it is drying and astringent.* Interest in pharmaceutical clays was apparently beginning to wane.

R. James, *Pharmacopoeia Universalis, or a New Universal English Dispensatory,* 2nd ed., London, (1752).

The last echo of Cimolia purpurascens may be found in Asterix the Gladiator (1964) by the inimitable Goscinny and Uderzo and translators Anthea Bell and Derek Hockridge, Fig. 51.

Brockhampton Press Ltd., Leicester, paperback (1973) White Knight division.

Fig. 51. Washes even purpler! (see text)

The Greco-Roman earths are no longer recommended by Henkelius (J. Fr. Henkel, 1679–1749), whose early treatise on medicinal earths deals with *Steinmark, terra sigillata from Lemnos, Lemniz and Striegnitz and their method of preparation* (1747).

Henkelius, *Mineralogia Redivivus and Chymia Redivivus.* (1747). Dresden: Joh: Nicol.

The 1746 edition of *Pharmacopoeia Londinensis* refers no longer to the ancient varieties of clay, dealing only with *Bolus Armenia* and *Terra Japonica*. By contrast, the compilers had failed to bring the work up-to-date – they still included the magical *Bezoar Orientalis*.

Pharmacopoeia Londinensis, London: T. Longman, etc., (1746).

Terra Japonica, a plant extract (Gambir) has an earthy fracture.

William Cullen (1710–90), lecturer and later professor of medicine in the universities of Glasgow and Edinburgh, realising the preposterousness of many of these disgusting mixtures, returned to the use of pure and simple remedies advocated by Paracelsus nearly three centuries before. He set out thereupon to reform the pharmacopoeia, and the first rational one to be issued was the Materia Medica Catalogue of 1776, which he edited.

A. G. Clement & R. H. S. Robertson. *Scotland's Scientific Heritage,* (1961), Oliver & Boyd, p. 30.

Examination of the Pharmacopoeias of Edinburgh (1783, 1803, 1805) and Dublin (1850) shows the results of Cullen's teaching, but by then pharmaceutical clays get no recognition; and even in England, when the *British Pharmacopoeia* was issued in 1864, clays had totally disappeared from the medical world, their useful properties forgotten or relegated to folk medicine.

Pharmacopoeia Collegii Regii Medicorum Edinburgensis, Edinburgh: Bell etc., (1783).

The Pharmacopoeia of the King's & Queen's College of Physicians in Ireland. Dublin: Hodges & Smith, (1850).

The first type of clay to creep back into the British Pharmacopoeia was in 1898 when *kaolinum* was readmitted, probably because of its widespread use in hot poultices.

Fuller's earth was not accepted for many years or its true nature and potentialities understood. The B. P. Codex of 1907 stated that "White Fuller's Earth" is a synonym for talcum (talc, Creta Gallica or French chalk). The sad thing is that this misdescription frustrated the search for white calcium montmorillonite for many decades. For B. P. Codices 1911–1923 see under cosmetics, Chapter 18.

British Pharmacopoeia Codex. (1907).

Martindale's *Extra Pharmacopoeia*, a publication consulted by many, almost as an official adjunct to the B. P. Codex, said in 1936: FULLER'S EARTH *(china clay) is a silicate similar to kaolin, containing traces of iron and magnesium.* This work claimed unfairly that fuller's earth could cause tetanus.

Martindale, *The Extra Pharmacopoeia,* London: The Pharmaceutical Press, 21st ed., (1936), Vol. I.

Second edition, 1938.

Hackh's *Chemical Dictionary* (1938) repeats this canard – *an impure kaolin, containing magnesium and iron*, while the *New Dictionary of Chemistry* gave some hair-raisingly incorrect information: fuller's earth is usually considered to have anauxite as one of its main crystalline constituents; it contains quartz; and the bulk of the material appears to be amorphous and unidentifiable. In fact it was known to scientists to be mostly calcium montmorillonite with very little free quartz and amorphous constituents.

Edited by Stephen Miall, London: Longman's, Green & Co. (1940), 217.

T. J. Porter postulated anauxite in U S Geol. Survey, Bull. *315*, (1907), p. 268.

A proper recognition of fuller's earth did not come until the *British Pharmacopoeia* published an adsorption test with quinine bisulphate for *the fuller's earth of commerce.*

Recognition of a range of pharmaceutical clays came only with the *British Pharmacopoeia Codex* of 1949, which reflected the growing interest in these clays and the part played by them in presciptions. At last fuller's earth was seen not as an impure kaolin but as a native aluminium silicate, consisting largely of montmorillonite.

Localities of deposits were given correctly as Surrey and Somerset at that time, but fuller's earth had not been worked in Gloucestershire for 30 years and in Dorset at any time. The kaolin spectre had not been entirely exorcised: *fuller's earth is an adsorbent resembling kaolin in its properties,* which is not true, but in other respects the account given is sound.

In cases of Paraquat poisoning, of which there were about 200 a year in 1981, montmorillonite is considered the most effective antidote, as it would be for any of the bipyridinium compounds.

The pesticide chlordimeform is very strongly adsorbed, also interlamellarly and not easily eluted, as it is from kaolinite, illite or vermiculite.

Although the use of fuller's earth as an antidote to poisons has been known for centuries, and the scientific reasons for its success have been known for decades, it is strange that, in a world where heavy metal solutions, alkaloids, cationic pesticides and detergents could be accidentally ingested, fuller's earth is not yet included in Red Cross or First Aid Boxes, in factories, homes and chemical laboratories.

17ii

Brit. Pharmacop. Appendix I, 653, (1948).

B. P. Codex, London: Pharmaceutical Society, (1949), 897.

M. S. Rose, The search for an effective treatment of Paraquat poisoning, *Chemistry & Industry* (17 May 1975), 413–5.

M. C. Hermosin & J. L. Perez Rodriguez, *Clays & Clay Minerals,* (1981), *29,* 143.

Adsorption of alkaloids

17iii

IN 1912 John Uri Lloyd, a chemist and manufacturer of Cincinnati, Ohio, found that clays adsorbed alkaloids and that Fuller's earth, "crude or refined", was especially effective in precipitating morphine from a solution of its sulphate. About 5g of what must have been, from this description, English fuller's earth completely removed 1g of morphine sulphate from solution.

He announced his discovery two years later. "Lloyd's Reagent" consisted of the finer particles of fuller's earth, separated from the coarser material by elutriation. (Today one would use an air separator). The reagent diminished or abolished the bitter taste from the alkaloid solutions and the active principle could be removed quantitatively from it. He later patented the use of a hydrated magnesium silicate.

U.S. Patent Application No. 27, 378 of 23 Feb. 1912.

J. Amer. Pharm. Assoc. (May 1914), p. 625.

U.S. Pat. 1,250,331, applic. No. 38,317 of 6 July 1915.

17iii

B. Fantus, *J. Amer. Med. Assoc.* (1915), *64*, 1838.

J. Amer. Chem. Soc., (1918), *40*, 312.
See also S. Waldbott, Precipitation of alkaloids by Lloyd's reagent, about 1933.

U.S. Pat. 1 235 450.

J. Pharm. Soc. (Japan). (1924), (no. 505), 166.

P. H. Plesch and R. H. S. Robertson, *Nature* (Lond.), (26 June 1948), 161.

Whereas Lloyd used fuller's earth to extract and purify alkaloids, Bernard Fantus was the first to point out its antidotal value. He showed that fuller's earth, especially when administered with tartaric acid or disodium hydrogen phosphate, was indicated in morphine, cocaine and ipecac poisoning. It had less value in strychnine and aconitine poisoning, though still capable of saving life.

Atherton Seidell continued this study in his *Utilisation of the adsorptive power of fuller's earth for chemical separations*. I plotted the amount of quinine bisulphate adsorbed, as shown in his Table V, against the excess of alkaloid used and found that the shape of the curve was exactly the same as that of methylene blue adsorbed onto English fuller's earth, which Seidell had also used.

About the same time F. R. Eldred, F. E. Bibbins and R. N. Reed took out a patent for the separation of cocaine by adsorption onto fuller's earth, which under prescribed conditions does not adsorb ecgonine or benzoyl ecgonine.

S. Ishikawa and K. Takamatsu made a thorough study of the adsorption of alkaloids onto Japanese Acid Clay, a naturally slightly acid calcium montmorillonite. The coefficient of adsorption increases with time of contact, but only up to 3 hours. Temperature is important, usually the higher the greater the coefficient. Also the greater the concentration of alkaloid the greater is the coefficient, but when an optimal ratio of clay and alkaloid is present, concentration of alkaloid has practically no influence over the coefficient. (As Plesch and Robertson found when the equivalent of cationic chemical is present, the cation exchange capacity can be measured. Physical adsorption also occurs when there is more than this amount of chemical).

The presence of too much water in the clay diminishes the adsorption, the best temperature for drying being about 40–60°C. If heated at 100°C the coefficient is "practically destroyed".

Adsorption is greater from an aqueous solution than from an alcoholic solution, depending however upon the solubility of the alkaloids in alcohol. All alkaloids are adsorbed by the montmorillonite, but the coefficient differs with different alkaloids. (For a more realistic comparison all the adsorption coefficients should be calculated in terms of milliequivalents of alkaloids adsorbed per 100 g of dried clay).

If more than one alkaloid is present, the alkaloid that has the greater coefficient will be adsorbed to the greater extent. The optimal condition for the adsorption of codeine is: for 0.5 g codeine, 20 g of Japanese acid clay, concentration of alkaloid 5%, temperature of the solution 95°C, 20 drops of H_3PO_4, with the clay dried at 65°C and contacted for 3 h, – under these conditions 100 g of the clay adsorbed 21.44 g of codeine, (which equals 67.6 meq. of codeine if I am right in assuming that the alkaloid is monovalent and has an equivalent weight of 317.37).

The alkaloid can be freed from the clay best when heated with lime and boiling water, (i.e., by an exchange reaction, using a great excess of calcium ions). The authors also described the preparation of morphine and natural codeine by means of this adsorbant.

Salmon and Guerrant (1928) compared the adsorption of quinine, an oxalate and glucose onto fuller's earth and the active carbon NORIT. The adsorptive power of fuller's earth for quinine, which was very strong, was slightly reduced by electrodialysis of the earth and halved by digestion with hot mineral acid, though largely restored by subsequent electrodialysis. The optimal adsorption of quinine and glucose by NORIT occurs at the same pH as with fuller's earth. They confirmed that, with fuller's earth, both polar and apolar adsorption occurred, i.e., chemisorption and adsorption by Van der Waals forces.

J. Biol. Chem. (1928), *80*, 67.

This line of research was taken up later by V. Knidseff (1932). The adsorption of caffeine by fuller's earth reaches equilibrium within a few minutes of shaking, is independent of the pH but produces a change in the pH and is greater from water than from alchohol solutions. Calcium ions were released to the filtrate – non-stoïchiometrically.

J. Phys. Chem. (1932), *63*, 1191.

D. P. Grettie and R. J. Williams compared the adsorption of many organic chemicals onto silica and alumina gels, a hydrous iron oxide and fuller's earth, which is thought to have been Surrey Powder sold by Eimer & Amend, and consisting then of a mixture of Copyhold Blue and Chart Yellow in equal parts. Fuller's earth was most active with caffeine, quinoline, nicotine, piperidine and adenine, and was not very different from silica gel in this respect. The sesquioxide gels were good adsorbants for glycine, vicine, asparagine, glutamic acid and aspartic acid.

J. Amer. Chem. Soc., (1928), *50*, 668.

S. B. Hendricks (1941) also showed that the alkaloids brucine and codeine, being strong bases, were intercalated by hydrogen montmorillonite through polar transfer and physical adsorption (Van der Waals) forces. Because of their large size, the "cover-up effect" operates: that is, some of the hydrogen ions are not available for neutralising other molecules. As Theng (1974) surmises, strychnine, atropine, and quinine would probably show the same effect because their molecular weights (~ 300) are comparable. On the other hand, nicotine (mol. wt. = 162) can neutralise stoïchiometric amounts of hydrogen ions held by various clay minerals (1967). Barbara S. Emődi (now Neumann) (1949) showed that while cationic dyes give similar exchange capacities, the retention of calcium ions is greater the greater the number of rings in the structure of the dye molecule.

J. Phys. Chem., (1941), *45*, 65–81.

B. K. G. Theng, *The Chemistry of Clay-Organic Reactions,* Adam Hilger: London, (1974).

S. Khan & J. P. Singhal, *Soil Sci.* (1967), *104*, 427.

Clay Minerals Bulletin, (1949), *1* 176.

For a number of years the British Pharmacopoeia Codex recommended quinine bisulphate as a diagnostic test for distinguishing between clay minerals, and one French pharmaceutical firm is known to have used strychnine sulphate for assessing the quality of adsorbent clays, (1970s). In August 1939 I recommended to Professor Berry of the College of the Pharmaceutical Society the use of methylene blue on the grounds that it was easier to use and safer; with R. M. Ward gave details of its use as a measure of cation exchange capacity (1951); and

B. P. (1948), Appendix I, 653, (an adsorption test with quinine bisulphate).

The assay of pharmaceutical clays, *J. Pharm. Pharmacol.*, (1951), *3*, 27–35.

in 1980 made recommendations to the British Pharmacopoeia Commission for its use as a diagnostic test.

COSMETICS

18

Dusting powder, face packs, creams.

18i

> ***Let's be beautiful – with the aid of a few strawberries. Mash half a dozen with an ounce of fuller's earth and thin down to a creamy consistency with witch hazel and benzoin. Smear on the face, keeping away from the eyes. The mask will harden in 20 minutes. Sponge off carefully. It works wonders.***
>
> **Beautician in *The People* 4 July 1937.**

IN the 17th and 18th centuries there are many references to the use of washes or liquid cosmetics for the complexion. The Vicar of Wakefield (1766) noticed his daughters were cooking something over the fire but as he believed these concoctions spoiled the complexion he contrived to spill the whole composition. He was always irritated when the window sill was cluttered with washes for the face and neck. The composition of these washes, often plant extracts, is seldom stated but it is possible that some of them contained clays to produce a matt and peach-like complexion.

Oliver Goldsmith, *The Vicar of Wakefield*, (1766), chh. vi and x.

Fuller's earth sold over the druggist's counter was originally in dried lump form. The pieces were probably sized between one inch and a quarter inch by screening to make the product easy to handle in the little drawer marked TERR. FULLON. The earth was sold for the treatment of common ailments, for household cleaning and in cosmetics.

It was William Matthews, a retail chemist of 12 Wigmore Street, London, who, grasping the potentialities of a milled supply, was first to put a proprietary brand on the market for toilet and medicinal purposes in the early 1850s. Matthews' Fuller's Earth, supplied packed in drums with the green bird pattern, was at one time a household word. An early wrapper reads:-

G. A. Higgins, of Rouse Bros. (London) Ltd., *personal communication*.

USED IN THE ROYAL NURSERIES.

MATTHEWS'S
PURIFIED
FULLER'S EARTH

Is invaluable for Protecting the Skin and Preserving Complexion from Cold Winds, Redness, Roughness, &c.

OF ALL CHEMISTS, 6d. and 1/-.

ROUSE & CO., 12 WIGMORE ST. LONDON, W.

Matthews Fullers Earth (Rouse and Co., Wigmore St., W.)

Fullers Earth has long been esteemed as a Toilet adjunct and is so well known that little that is new or novel can be said about it. Of its softening and preservation action upon the Skin of the Hands and Face there can be no question, and its application after washing is at all times most pleasant. It certainly tends to keep the flesh clean, inasmuch as it thoroughly dries it and thus prevents the ready cohesion of particles of dirt. It whitens the hands, from much the same reason, and is altogether very worthy of extensive patronage.

The Monthly Magazine.

The Monthly Magazine, quoted by Rouse & Co.

MATTHEWS'S FULLERS EARTH – *for the Face.*
" " " *for the Hands.*
" " " *to prevent Freckles.*
" " " *to cure Sunburn.*
" " " *to preserve the Complexion.*
" " " *to heal Chafings.*
" " " *for Roughness and Redness.*
" " " *Nullifies Prickly Heat.*
" " " *Indispensable in Hot Climates.*
" " " *Is Grateful after Shaving.*
" " " *Of Chemists and Perfumers.*

Purified specially for the Nursery and Toilet.

William's son, Herbert Rouse Matthews, sold the old family retail business which had been founded in the early 18th century and became a wholesale chemist at 61 Charlotte Street, London, W.1, where their plate glass window advertised in large white letters Matthews Fullers Earth Powder until the window was smashed by enemy action in 1941. The company was still run by Rouse descendants in 1959.

On the whole the pharmaceutical and cosmetic market proved a steady one – as one application for fuller's earth declined another would take its place.

In 1880 it was reported that fuller's earth *is applied as a cooling and healing dressing by the poor to inflamed breasts, excoriations, etc.*

Cooley's *Cyclopaedia of Practical Receipts,* (1880).

In the early 1900s fuller's earth powder was beginning to be superseded as a nursery and dusting powder by alternatives more attractive in appearance. A. R. Matthews conceived the idea accordingly of making use of its healing properties in an emollient cream. Matthews' Fuller's Earth Cream has gradually replaced the plain earth and has gained a good reputation as a soothing and healing application for all forms of skin inflammation. The cream was particularly appreciated by millgirls in cotton mills and by actresses whose skins suffered from use of greasepaint.

After World War II sales of fuller's earth powder revived again for use in face packs. In 1949 the same firm put out fifteen tons in four-ounce drums. The full instructions are given below:-

MATTHEWS'
FULLERS EARTH
used as a
FACE PACK

Tones up the skin.
Improves the complexion.
Smoothes out wrinkles and crows feet.
Cannot injure the most tender skin.

DIRECTIONS.

- *Shake out about 2 tablespoonfuls into a saucer and mix with water to a stiff paste.*
- *Wash the face in warm water and apply a warm towel to open the pores.*
- *Apply the Face Pack quickly with a clean shaving brush over the entire face and neck. This ensures the application of a perfect mask whilst the pores are open to the fullest extent.*
- *Allow to remain on until thoroughly dry, at least 20 minutes but the longer the better. You will feel the pack drawing the blood to the face, stimulating the muscles and tightening the skin.*
- *Remove with a very wet, warm face glove and apply a good emollient cream, such as*

18i

Fig. 52. Matthews' Fuller's Earth Cream as it was packed for sale in pharmacies.

spreading it liberally over the face.

If you are going to bed and can leave this on, so much the better. If not, gently pat it in for five minutes, then remove and apply usual make-up.

Repeat the Face Pack on alternate days until the complexion is clear of all blemishes after which a weekly use will ensure a Soft and Clear skin.

ROUSE BROS. (London) LTD.,
Toilet Specialists,
61 Charlotte Street, London, W.1.

After several changes of ownership the company became Matthews Fullers Earth Ltd, *Chem. Age*, (26/2/1966) 352.

The decline in popularity of fuller's earth as a nursery dusting powder, readily explainable in view of the availability of powders of white or flesh colour, pleasantly perfumed, and finer in texture and with more slip, had a serious consequence however in that it caused the dermatological properties of the earth to be overlooked. Kaolin was now sold in packets falsely labelled as (white) fuller's earth by Boots, while Mawson and Sons sold the Atherfield clay put out by the misleadingly named Leeds Fullers Earth Company of Kent.

B. P. Codex (1911) 556.

A worse blow came when the *British Pharmacopoeia Codex* of 1911 lent its support to the belief that fuller's earth used for toilet purposes involved a risk of contracting tetanus – a statement applicable to a supply dug up from a germ-laden outcrop, air-dried and therefore not sterilised, but certainly not to the product as commercially marketed, which, even in the days before it was milled, was dried by precisely the process of "roasting" the *Codex* recommends as a safety measure. (A similar "scare" in 1949 in New Zealand where fatalities suggested that the bacteriological standards for talcum powder were inadequate had more foundation, since talc is not normally sterilised in the methods of production.) The canard was repeated in the Codex of 1923, but omitted in the 1949 edition, though it was years before the older family and hospital nurses forgot it.

B. P. Codex (1923), 588.

B. P. Codex (1949), 897.

Fuller's earth is used externally in modern pharmacy and cosmetics in the guise of both powder and creams, and also in face-packs and mud applications.

In the present century fuller's earth powders have usually been mixtures with other powdered substances, such as talc, magnesium oxide, zinc oxide, which contribute slip, lightness in packing density, antiseptic properties, and other useful qualities. Thus Woolley and Forrester give the following formulas:

E. L. Lederer, *Seifensieder Zeitung*, (1932), *59*, 221.

S. W. Woolley and G. P. Forrester, *Pharmaceutical Formulas*, London: Chemist & Druggist, (1929), 10th ed., *1*, 491; *2*, 166.

Fuller's earth powders and creams

I.	Fuller's earth	9 oz.	II.	Fuller's earth	4 oz.
	Talc	6 oz.		Talc	8 oz.
	Boric acid	1 oz.		Lycopodium	4 oz.
	Oil of lavender	10 minims.		Otto of rose	5 minims.

18i

III.	Fullers' earth	4½ oz.	IV.	Fuller's earth	12 oz.
	Boric acid	¾ oz.		Talc	2 oz.
	Zinc oxide	1½ oz.		Rice starch	2 oz.
	Starch	4½ oz.		Boric acid	4 dr.
	Orris root	¾ oz.			
	Oil of bergamot	2 dr.			

No. III was described as a nursery powder. To-day orris root and starch are usually omitted in these formulations since some skins are sensitive to them. Light oxide or carbonate of magnesium or sepiolite are used to reduce the bulk density of powders.

Although the most familiar use of fuller's earth in the popular mind is as a dusting powder for babies, it has been used by adults for several pharmaceutical purposes. Fuller's earth brings relief to sufferers from weeping eczema. Sufferers from this malady had no cure at hand; but they could stop the intense irritation by applying fuller's earth. Spurious powders did not have the desired effect. It is probable that fuller's earth acted by absorbing fatty as well as irritant substances.

Oral tradition (A. L. Johnson).

Oral tradition (paint expert from ICI plc).

Jockeys and other horsemen after riding rub fuller's earth over their backsides to absorb perspiration and to prevent haemorrhoids. Even very bad ones, botryoidal – "like grapes" – are said to have been cured by holding a large quantity of fuller's earth over the affected part at night by means of a bandage between the legs.

86 Clerkenwell Rd., London, E.C.

See section on Scottish fuller's earths.

Dunning Fuller's Earth sold by Fassett & Johnson Ltd, sole agents, under the name EMOL KELEET, was described as *a natural emollient and absorbent dusting powder. Soft as down and perfectly harmless. Dries and soothes moist, sensitive, inflamed, irritated or chafed skins. Softens hard and rough skin. For eczema and acne.* It was used also as a styptic after shaving.

18ii

Foot care

Also in the *Home Guard Handbook*, (1940).

IN World War II fuller's earth was on sale in NAAFI's as a foot powder, though a mixture of fuller's earth, talc and boric acid would have been preferable. In earlier times sore and inflamed feet were soothed in a dispersion of fuller's earth in water.

The U.S. *Manual of Dermatology* gave an alternative formula for the treatment of excessive sweating of the feet:

Fuller's earth powders etc. (contd.)

Zinc peroxide	10 g	germicide
Tannic acid	5 g	hardener
Boric acid	20 g	germicide and lubricant
Bentonite	10 g	absorbent
Talc	55 g	lubricant

Drug & Cosmetic Industry, (Nov. 1941).

and J. Kalisch gave a formula containing absorbents, a mild astringent and an antiseptic:

Colloidal clay	60	parts
Fuller's earth	34	parts
Tannic acid	5	parts
Oxyquinoline sulphate	1	part

18ii

External applications

18iii

RESEARCH during the war established that Lewisite poison gas blisters healed rapidly when bandages containing fuller's earth powder were applied. These bandages made by T. J. Smith and Nephew contained an adsorbent (including fuller's earth), an adhesive (gums), and plaster of paris. A rather similar composition for making casts and splints was protected by an earlier patent of E. Griffiths Hughes Ltd.

Brit. Pat. 559, 693 (28 Aug. 1942).

Brit. Pat. 545, 539 (19 Feb. 1941).

To remove stains or spots of DNOC (dinitroorthocresol), the agricultural spraying poison, fuller's earth should be applied as a paste, allowed to dry and then brushed off, according to official advice.

Min. of Agric. & Fisheries Advisory Leaflet No. 315, (1945).

Another way in which mud is applied to the skin is as a hot plaster for drawing pus. Kaolin is usually used in these "antiphlogistine" preparations, because of its white colour, but Kingzett (1928) mentions bentonite for this purpose; and the Silica Products Company (1930) claims that bentonite is an antiphlogistic rubificient, that is a material which draws pus and at the same time reduces inflammation. One would expect calcium or sodium smectites to have more "drawing power" than kaolinite.

C. T. Kingzett, *Chemical Encyclopedia*, London: The Gresham Publishing Co. Ltd., (1928), *1*.
Bentonite, Bulletin 107, Silica Products Co. (1930).

Fuller's earth was sold until about 1938 by Keltie Products Ltd of Glasgow as a styptic; used in preventing dermatitis in dye houses by dusting onto the face after being given a base coating of cream.

See Scotland.

Fuller's earth in veterinary practice is usually applied as a poultice in the cure of sore places on horses.

Since Matthews' invention of fuller's earth cream various recipes have been published, among them:

Fuller's earth powders etc. (contd.)

Spermaceti	32 gr	Fuller's earth	500 g
White wax	32 gr	Glycerine	20 ml
Almond oil	4½ dr	Tincture of benzoin	70 ml
Lanolin	1 dr	Perfume	a sufficiency
Distilled water	2 dr	Starch paste, thin	400 g, or a sufficiency
Prepared fuller's earth	3 dr		

Woolley and Forrester, *op. cit.*, *2*, 87.

Bentonite because of its great capacity for absorbing water makes a useful salve or ointment base, where oils, fats and waxes must be avoided, in combinations such as:

(a) water	85	or	(b) bentonite	
bentonite	10		zinc or mercuric oxide.	
sulphonamide	5			

a. R. W. Howard, *New Eng. J. Med.*, (1945), *232*, 698.
b. H. Griffin, *J. Pharm. Chim.* (1938), *27*, 159–165.

18iii

A. Axon, *Pharm. J.*, (1947), *159*, 135.

op. cit., *2*, 174.

Bentonite is also used in sterilisable creams exemplified by bentonite, Lanette wax SX, paraffin and water; these are stable, homogeneous, smooth creams compatible with, for example, penicillin. Bentonite has also been used in tooth-paste and in laxatives.

Woolley and Forrester say that clay face packs are generally fuller's earth, kaolin or kieselguhr (though the last sounds too abrasive), made up with starch, paste, mucilage of tragacanth or a non-greasy cream. Some contain magnesium sulphate, nearly all have some borax, and magnesium carbonate.

The method of application is to bathe the face with very warm water to open the pores, and then to apply the paste with a brush and allow it to dry thoroughly. When dry the paste casing is peeled off and the face massaged with a good cream, the final touch being given by dusting on face powder.

Active chemicals, they say, may be incorporated, such as hydrogen peroxide (the oxygen from which would soon be lost by the catalytic action of the clay), lemon juice, and sometimes rock ammonia. All the same these authors scarcely breathed enthusiasm when they wrote:

op. cit., p. 898.

> *The healing and beautifying effect of plastic clay on the skin was known in ancient times, and it is still appreciated to-day, though mainly by middle-aged ladies who are willing to devote ample time to its use.*

A decade later facepacks were much more widely used.

18iv

More on face packs

From 'Last Word on Mud Research'.

> ***Mud the Beautifier is the only type to which a woman will passively submit. This cosmetic mud comes in all colours and can provide traumatic shocks for babies, dogs and returning husbands confronted without warning by a female zombie with a limestone face, speaking in neanderthal ugs lest lip-movement cracks the plaster and the magic escapes.***
>
> **ANON., New Scientist, (21/May/1964), 469.**
>
> ***Face Packs: here is one last quick mud pack for cleaning and tightening up your skin just before you go out to a party. Mix a handful of fuller's earth with enough rose water to make a pack. (If your skin is greasy, add a few drops of peroxide.) Spread the paste evenly over your face and leave it on while you are having your bath and doing your nails.***
>
> **ANON., Daily Express, (London), (22 Feb. 1937).**

At head of Chapter 18i.

IN the summer that year a strawberry face pack was recommended. On the other hand an oatmeal recipe is less seasonal:

Helen Temple's Beauty Parlour, *Woman*, (15/2/1941), (specially formulated for Scotswomen?)

Face pack. *4 oz. fuller's earth*
2 oz. powdered oatmeal
Rose water to moisten when used.

This simple recipe makes a wonderful bracing and bleaching face mask. Ordinary unrefined fuller's earth will do and should be much easier to get, as the prepared fuller's earth comes under the new cosmetic ban.[a] *Powdered oatmeal comes from the grocer and is made in three textures – you want the finest.*

Simply mix the two together and pack away in a carton – say, one of your empty powder boxes. Then when you want to use it, pour a little into a saucer and moisten it with rose water until you have a consistency like Devonshire cream.

Apply and leave to dry, then wash away with warm water and film a little cream on the skin if it is dry.

18iv

a) In World War II.

The week before, this beautician exhorted her readers thus:

on any day when you're looking a good deal less than your best, pop round to the chemist and get a few pennyworth of fuller's earth. Mix it to a paste with milk, leave it on your face for 20 minutes and wash it off with lukewarm water.

Another pleasant variant: if you have a very dry skin, do not have one oftener than once a week. Mix the earth with almond oil and orange flower water; for oily skin, use witch hazel.

The Evening News, (25/3/1938).

Twentieth-century beauticians are clearly well aware of a fuller's earth paste's capacity to draw oil or fatty matter into itself as it dries out. And the public are reminded of an advertisement in the London underground:

Deep pore cleanser draws out hidden oils and grime. MUDD super cleansing treatment. Natural rare clay treatment. A glowing complexion in just 15 minutes! Just use MUDD twice a week. The deep cleansing mask with the rare clay formula. Keeps oily skin dry. Helps stop you getting spots and blemishes. For a clear, spotless skin and a glowing complexion get MUDD. Ask your chemist.

Transcribed *via* my chequebook, 1982.

Other external applications

18v

Mud, mud, glorious mud,
Nothing quite like it for cooling the blood.
Sung by the late Michael Flanders and Donald Swan.

APPLICATION of fuller's earth pastes need not, of course, be confined to the face and neck, though in the modern western world fuller's earth is not bought in the market place for washing the hair, as in Asia and Africa.

Washing the hair.

Fuller's earth, it has been found in America, can be compounded with ointment so that it does not soil underclothes. This is reminiscent of the once well-known use of it in treatment of bedsores. Fuller's earth was

Treatment of bedsores.

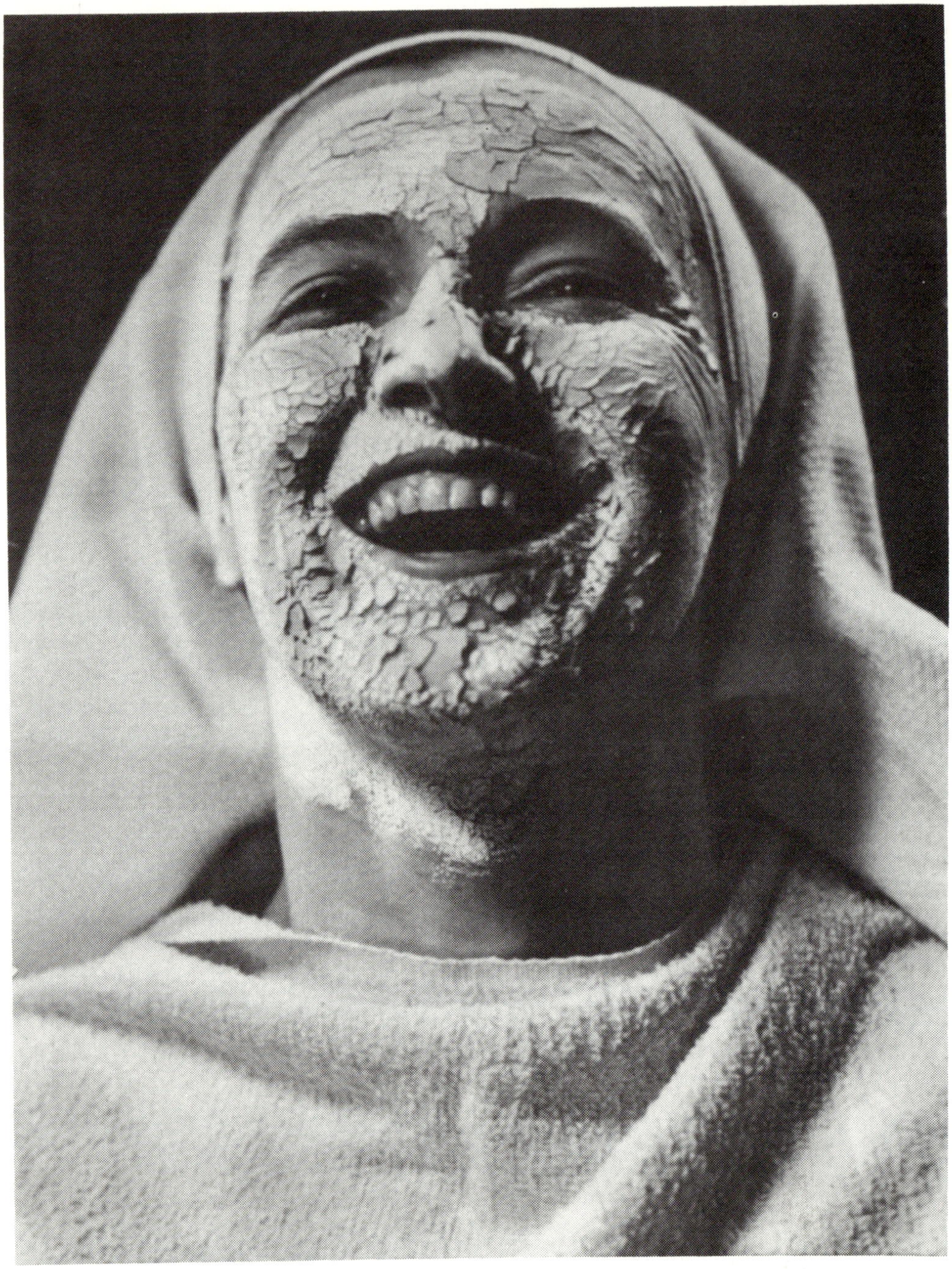

Fig. 53. What price beauty? The Face Mask.

18v

mixed with linseed oil and heated over a steamer. *It is taken up with a bone ladle and spread very smoothly over the patient's back, the consistency being right if the mixture readily comes away from the ladle. No grease is left behind when the patient is washed after the poultice is removed.* The treatment was said to be very refreshing. Doctors sometimes used the mixture to make their hands smooth.

Oral tradition: Mrs Masters, retired nurse.

At Bath, Avon, fuller's earth was mixed to a mud-like paste with the radioactive water of the spa for the treatment of rheumatic joints. At one time the mud bath was a feature of certain spas, such as Nisiro in Rhodes, at Carlsbad until it became Karlovy Vary and at Bath; but the treatment was usually abandoned because it induced micturition.

Balneology, Fig. 54.

Until recently the mud pack has been applied also to the whole body by artistic photographers seeing the female nude in a new way. In 1982 a promoter of Female Mud Wrestling in Los Angeles, California, was enquiring for supplies of 200 mesh fuller's earth at planned venues in Japan, North Africa, and Europe. For this application lubricating properties are no doubt required.

Nude, Fig. 55.

Wrestling, Fig. 56.

Industrial Minerals, (Oct. 1982), (181), 70.

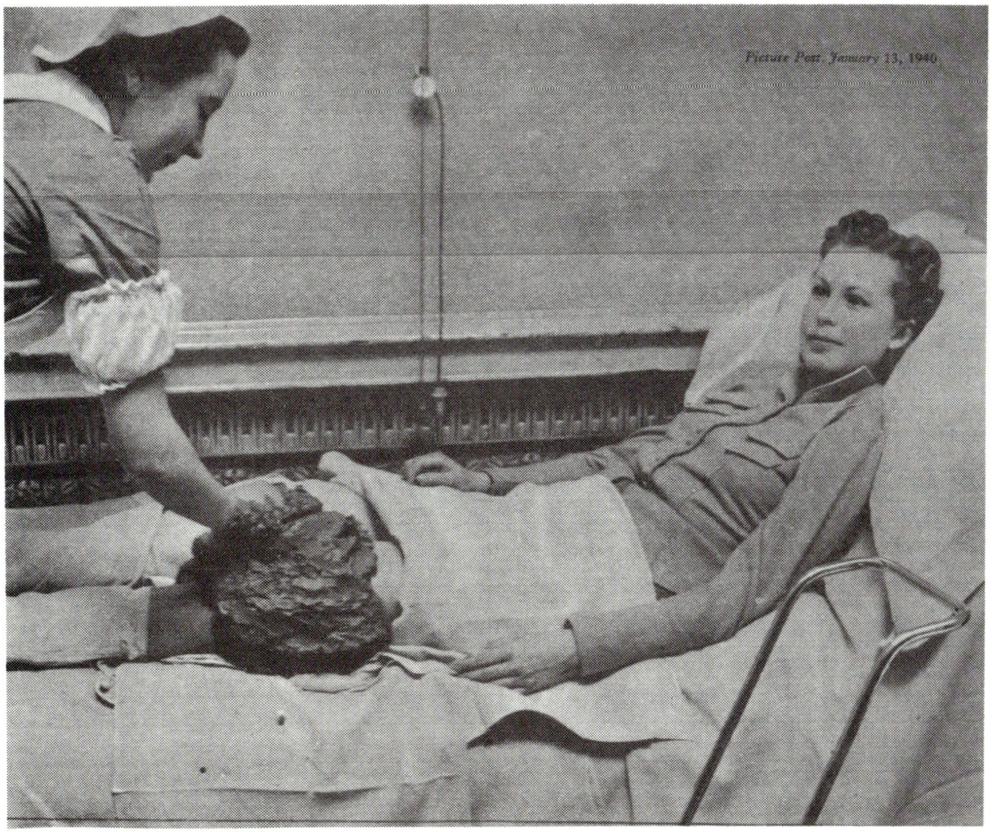

Fig. 54. Mud treatment for a weak knee. The treatment is combined with sprays, douches and immersion baths.

From Picture Post

Fig. 55. The Mud Bath.

Banel, Paris

Fig. 56. Mud wrestling in Teddington

From the Richmond Informer

19 HOUSEHOLD CLEANING

I'm exceedingly annoyed about this stain. If only I knew how to get it out. These women usually have some fuller's earth for removing stains.

Carlo Goldoni, MINE HOSTESS, (1753), Act 3, Scene 9, translated by Clifford Bax, (1944).

Fuller's earth not included in a recent revival.

19i

FULLER'S earth must of course have served from very early times in England for household cleaning as in Pompeii.

The books of household management of the latter half of the 18th century list countless recipes to satisfy the multifarious needs engendered in that age of gracious living.

Mrs. Beeton drew upon these when she produced her famous work which became the gospel of Victorian ladies and their skivvies. *The Housewife's ABC* of 1921, and contemporary periodicals, are clearly still aware of the value of the earth's sorptive properties; for stain-removal earth is often preferable to solvents, which leave a ring.

Mrs. Isabella Beeton, (1836–1865) *Book of household management*, first published 1860 by her husband.

W. S. Fales and J. Hunter, *The Housewife's ABC*, London: The Modern Publishing Co., (1921).

The following are extracts from publications of the 18th, 19th and present centuries showing use of fuller's earth over the period for cleaning dirtied materials, whether by the dry method, in which it is rubbed on as a dry powder, or the wet, in which it is applied in a miniature fulling operation.

The headings of the first four recipes reflect a changing way of life.

To extract lamp-oil out of stone or marble halls etc.

Mix well together a pint of strong soap lees, some fuller's earth well dried, and a little pipe-clay, pounded fine; and lay it on the part which is oiled; then put a hot iron upon it till dry. If all the oil should not come out the first time, do it again; and in putting it on, let it be well rubbed into the stone. By doing it two or three times in this way, it will come out.

Colin Mackenzie, *Five Thousand Receipts in all the useful and domestic arts*. 5th ed., London: G. B. Whittaker, (1825), 788

(British Museum copy is (1823)).

To remove stains from marble.

Make a paste of powdered pipe-clay and fuller's earth; mix with strong soap lye; lay a thick coating of this paste on the marble, and pass lightly over it a moderately warm flat-iron until it is dry. Leave it for a short time, and then wash it off with clean water. If the marble be not entirely free from grease repeat the process till every stain disappears. Discolorisation by smoke may be removed in the same manner.

Mrs. Beeton, *The book of household management*. Edition (c 1900), London: Ward, Lock.

To clean stone.

To remove grease from a stone hearth or stone steps, pour on a strong solution

Fales & Hunter, *op. cit.*, 164.

of washing soda in boiling water. Then cover the stain thickly with a paste of fullers earth and hot water and let remain over night. Repeat if necessary.

To remove milk-stains from door steps.

Doorsteps which have been stained with milk or other liquid can be cleaned by covering the stain with a mixture of fullers earth and water. Leave this on overnight, then wash the steps in the usual way.

Daily Mail, (5/2/1931).

Nancy Spain does not suggest any sources for her great aunt's household recipes in her book *Mrs. Beeton and her husband*; it seems clear though that Colin Mackenzie was at least one of them.

London: Collins, (1948).

Removing grease with fuller's earth 19ii

To get oil out of boards.

Mix together fuller's earth and soap lees, and rub it into the boards. Let it dry and then scour it off with some strong soft soap and sand, or use lees to scour it with. It should be put on hot, which may easily be done, by heating the lees.

Colin Mackenzie, *op. cit.*, 788.

To remove stains from boards.

Take quarter of a lb. of fullers earth and a quarter of a lb. of pearlash; make them into a paste with about a quart of boiling water; spread a thick coating of this over the grease-stains and leave it for 10 to 12 hours; then wash it off with clean water, using sand, if necessary. If the grease-stains are very numerous and the floor very dirty, a coating may be spread all over the floor and left for 24 hours before it is washed off. In washing boards never rub crosswards; but always up and down with the grain.

Mrs. Beeton.

To remove grease-spots from wallpaper.

Grease-spots that are freshly made may be removed by a warm iron and blotting paper. With many of the new pastel shades, however, heat has a bleaching effect. A perfectly harmless remedy is a paste of fullers earth. This should be mixed to a very thick paste with cold water. Spread it well over the stains and leave for 24 hours. With a soft brush lightly dust off the paste. If the stain is an old one several applications may be necessary.

Daily Mail, (27/8/1930).

Cleaning more hard surfaces 19iii

To clean china and glass

The best material for cleaning either porcelain or glassware is fullers earth, but it must be beaten into a fine powder, and carefully cleared from all rough or hard particles, which might endanger the polish of the brilliant surface.

Colin Mackenzie, 782.

Window cleaner.

Kieselguhr, 2 oz. Precipitated chalk, 4 oz. and Fullers earth, 4 oz.

Woolley & Forrester.

Aluminium polish.

Stearic acid, 10 oz. fullers earth, 10 oz. Tripoli, 60 oz.

Id.

Dry cleaning with fuller's earth

Yes, fuller's earth will make them exquisitely white,
F. Ansley, VICE VERSA, London: John Murray, (*1882*), *1929* *ed.*, *257*.

Colin Mackenzie, 184.

To cleanse gloves without wetting.
Lay the gloves upon a clean board, make a mixture of dried fulling-earth and powdered alum, and pass them over on each side with a common stiff brush; then sweep it off and sprinkle them well with dry bran and whiting, and dust them well; this, if they be not exceedingly greasy, will render them quite clean; but if they are much soiled take out the grease with crumbs of toasted bread and powder of burnt lime; then pass them over with a woollen cloth dipped in fulling earth or alum powder; and in this manner they can be cleaned without wetting, which frequently shrinks and spoils them.

Mrs. Beeton.

To take spots and stains from dresses.
To remove grease-spots from cotton and woollen materials absorbent pastes, purified bullock's blood, and even common soap, are used, applied to the spot when dry. When the colours are not fast, use fuller's earth or pulverised potter's clay, laid in a layer over the spot, and press it with a very hot iron.

Fales & Hunter, 164.

To clean straw hats.
Panama hats may be cleaned with hydrogen peroxide applied with a flannel repeating if necessary; or with fullers earth or powdered magnesia well rubbed in with a piece of velveteen, allowed to remain several days and then brushed off.

Id.

To clean kid.
Remove the surface dirt with a soft indiarubber, then with a powder made of equal parts of pipe-clay, powdered alum and fullers earth.

Id.

To clean ermine or fox.
Any white or light furs may be cleaned with benzine or by a dry process, with dry magnesia or fullers earth, rubbed well into the skin, then either brushed or shaken to eliminate the powder. Dark furs are cleaned with magnesia or hot bran.

Dry-cleaning eiderdowns.

Radio Times, (3/10/1936).

If you prefer to dry-clean your eiderdown, do it this way. To begin with you must brush all the loose dust out of it. Then lay it down on a table and fill a bowl with Fullers Earth. Brush the eiderdown briskly in every direction with a clean clothes brush dipped frequently in the Fullers Earth, and turn the eiderdown over and clean the other side in the same way. When quite clean, shake the eiderdown well.

Removing marks from cloth.

Radio Times, (10/7/1937).

Woven fabrics can be cleansed of marks by rubbing with fullers earth enclosed in a muslin bag. The suction cleaner or a clean brush removes the powder afterwards. Powdered magnesia or hot bran is useful on some light coloured, delicate fabrics, while individual marks, shown up by the sunlight, can be scrubbed with hot rich soapsud made with soap flakes. The soap must be removed with a sponge or flannel and the cleaning finished with fullers earth or magnesia.

Feather boas can be cleaned by putting them in a bag with fullers earth and gently kneading the bag once a day for three days. The earth is then shaken out.

Lipstick or rouge can easily be removed from cloth by smearing fullers earth on to the offending spot and then patting out again. Repeat several times if necessary.

Soiled chair backs and coat collars can be cleaned in a similar way.

Clothes balls

19v

AN early recipe for making wash balls is given by Dr. William Salmon in 1710.

> 15. WASH BALLS *Perfum'd:* Take white Soap four Pounds, Orrice ⁱ fix ounces, Macaleb three ounces, Cloves two ounces, all in fine Pouder; mix the Soap with a little Oil of Spike, Rhodium, or the like, of which make Balls.

William Salmon, M.D., *The Family Dictionary, or Household Companion*, London: H. Rhodes (1710).

Orris; mahaleb (fragrant Prunus kernels); *Lavendula spica*; rose wood oil, oil of rhodium, from the scented wood of a Canary convolvulus.

Detergents were made in this form because they were easy to transport and to handle in a shop or, as we can see below, by a pedlar – Fig. 57.

One can see in this picture that the wash balls were about 20 to 35 mm in diameter.

One of the most effective of the old-fashioned ways of cleaning stains from textiles and leather was to use a "clothes ball".

The first patent for the making of detergent cakes (possibly illitic) by bonding with an animal or vegetable mucilage and pressing was taken out by John Vancouver of Tachbrook House, near Warwick (1802). The clay might have been Rhaetic, (see also Ch. 22).

English Patent No. 2636 of 1802. Thanks to Hugh Torrens for this reference.

Clothes ball.

Mix 2 pounds of pipe-clay, 4 ounces of Fuller's earth, 4 ounces of whiting, and a quarter of a pint of ox galls.

Colin Mackenzie, 181.

To make scouring balls.

Portable balls for removing spots from clothes may be thus prepared. Fuller's earth perfectly dried (so that it crumbles into a powder) is to be moistened with the clear juice of lemons, and a small quantity of pure pearl-ashes is to be added. Knead the whole carefully together, till it acquires the con-

Id., 182.

Fig. 57. Eighteenth-century pedlar offering toilet articles. In his left hand he holds soap-balls from Boulogne, in his box he carries soap-balls from Naples.

Engraving attributed to Nicolas Larmessian (1684–1755)

From G. Schaefer, *CIBA Review*, (Soap), (April 1947), *56*, 2047.

sistence of a thick elastic paste; form it into convenient small balls, and dry them in the sun. To be used first moisten the spot on the clothes with water, then rub it with the ball, and let the spot dry in the sun; after having washed it with pure water, the spot will entirely disappear.

Book without title page rescued by pulp merchant in WW II; probably much earlier than 1884.

Clothes Ball

Pipe-clay 2 lb., fullers earth 4 oz., whiting 4 oz., white paper 2 oz., ox gall, 4 oz. Make into a stiff paste with the aid of a little water. Light magnesium carbonate may also be used in this composition.

Id., No. 116.

Clothes Ball

Pipe-clay 4 lb., fullers earth 2 lb., whiting 1 lb., white pepper 6 oz., mix with water.

Clothes Ball.

Fullers earth 1 lb., curd soap 1/2 lb., ox galls, sufficient to make a stiff dough with which form balls.

Id., No. 118.

Scouring Ball, to remove Grease etc., from Cloth.

Soft soap and fullers earth of each half a pound, beat them well together in a mortar, and form into cakes. The spot, first moistened with water, is rubbed with a cake, and allowed to dry, when it is well rubbed with a little warm water, and afterwards rinsed or rubbed off clean.

Enquire within upon everything, 110th ed., London: Simpkins.

Balls for cleaning clothes.

Take some fuller's earth, dried till it crumbles to powder; moisten it with the juice of lemon, add a small quantity of pearlash, work and knead care-

fully together till it forms a thick paste; make into balls, and dry them in the sun. Moisten the spot on clothes with water, then rub it with the ball. Wash out the spot with clean water.

19v

Marshall, Hamilton, Kent & Co., [not later than 1916].

Item 1300.
Pearlash = potassium carbonate.

Cleaning Carpets

19vi

To clean carpets.
Grease-spots can be removed from carpets by means of a paste made of boiling water poured on equal quantities of magnesia and fullers earth.

Mrs. Beeton.

To remove blacklead from carpets.
With a stiff brush brush off any powder there may be present, then cover the stain quickly with a paste of fullers earth and magnesia mixed with a little liquid ammonia. Use this preparation only on light-coloured carpets. Remove the paste by scraping lightly then brushing; repeat the process if necessary.

Fales & Hunter, 16.

To remove soot from carpets.
Cover lightly with dry starch, salt, bran or fullers earth. Remove by flicking soot, etc., into dustpan with a brush. A paste of calcined magnesia or fullers earth mixed with ammonia will remove the mark if allowed to work into the pile before being brushed off.

Magnesia prevents the fuller's earth from sticking to the fibres.

To remove blacklead.
Blacklead accidentally dropped on to wood, or on to light material, should be lightly flicked off and then brushed hard, working in the direction of the grain of the wood, or of the weave in the material. If possible turn the carpet or rug on to a piece of blanket or on dry grass and beat with a cane from the back to extract more of the dried blacklead. If any stain remains and the article is light in colour, apply a thick paste of fullers earth, leave to dry and then brush off.

Daily Mail, (27/8/1930).

Restoring scorched linen

19vii

Restoring scorched linen.
Ingredients – ½ pint of vinegar, 2 oz. of fullers earth, 1 oz. of dried fowl's dung, 1/2 oz. of soap, the juice of two large onions. Mode – boil all these ingredients together to the consistency of paste; spread the composition thickly over the damaged part and if the threads be not actually consumed, after it has been allowed to dry on, and the place has subsequently been washed once or twice, every trace of scorching will disappear.

Mrs. Beeton.

To renew scorched or browned linen.
Add to one quart of vinegar the juice of 6 large onions, one oz. of soap rasped down, a quarter of a pound of fullers earth, one oz. of lime, and one oz. of pearlash, or any other strong alkali. Boil the whole until it is pretty thick,

Book without title page, item No. 166.

and lay some of it on the scorched part suffering it to dry. It will be found that, on repeating the process for one or two washings, the scorch will be completely removed from the linen without additional damage provided its texture has not been absolutely injured as well as discoloured.

19vii

Fales & Hunter, 153.

Restoring scorched linen.

Cottons and linens, if only slightly scorched, can be restored by covering the spot with a paste made with pipe clay and water, and laying in the sun; repeating as often as necessary if after washing the scorch does not disappear. For more severe cases, extract the juice from 2 large onions, mix with 4 tablespoonfuls of fullers earth, ½ cupful of soap jelly, and a cupful of vinegar. Bring to the boil, cool and spread on the fabric. Let dry, wash and lay in the sun. Delicate fabrics may be treated with glycerine and borax. Alcohol will remove the glycerine.

19viii Miscellaneous fabric-cleaning

> ***Out damned spot! out, I say!***
> **W. SHAKESPEARE, *Macbeth*, Act. 5, Scene 1.**

Published in London.

THE first book on removing spots and stains was by L. Mascall (1583) *A profitable boke declaring dyvers approved remedies to taking out spottes and staines, in silkes, velvets, linnens, & woolen cloths.* Some much later examples follow:

Book without title page, item No. 164.

To remove grease-spots from silk.

Take some fullers earth and mix into a paste with water. Apply some to the spot, let it dry and then brush out; after which remove the marks with a little sulphuric ether which must again be washed out with spirit of wine.

Fales & Hunter, 38.

To clean delicate-coloured fabrics.

Mix fullers earth or powdered starch to a paste with glycerine. Apply thickly to the spot and let remain until dry. Remove by brushing with a whisk. The fullers earth should be omitted for dark colours. Repeat, if necessary or, if the stain has nearly disappeared, sprinkle with dry fullers earth and let stand 24 hours.

Id., 42.

To clean corsets.

Rub dry salt or magnesia over the corset, inside and outside, and cover any stained parts with a paste made of fullers earth, talcum powder and magnesia mixed with ammonia. Roll the corset up in a cloth and leave for 48 hours. Brush off the powder.

Id., 105.

To clean mattresses.

Stains on mattresses can be removed in most cases by covering with a thick paste of laundry starch or fullers earth and soap jelly. Allow to remain until dry and brush off with a whisk. Repeat if necessary.

To clean white kid.

A paste made with equal parts of fullers earth, powdered alum and cream of tartar, if spread over soiled or grease-stained white kid, will generally clean when other applications fail.

Id., 186. **19viii**

Removing butter stains.

Butter stains on light, unwashable fabrics can be removed by using a thick paste made of french chalk or fullers earth moistened with some fat solvent such as benzene or petrol. Spread on the stains, allow to dry, brush off and renew the application if necessary. It is safest to do this operation out of doors.

Daily Mail, (17/8/1932).

Removal of ink stains, tar, grease spots, etc.

Ink-stains are vexatious and stubborn, and as inks are composed of many different ingredients it is not possible to give a general ink-stain remover. Writing-ink spots *are best treated with one of the absorbents – cornflour, chalk, fullers earth, talcum powder or magnesia, worked well in with a smooth blunt instrument. Should the stain have dried make the absorbent into a paste with water. Rinse or brush clean. Stains caused by* tar, pitch, axle grease, etc., *can be removed by treatment with eucalyptus oil, then rubbing with fullers earth and completed by (a) washing with hot soapy water, (b) good brushing, or application of the vacuum cleaner.*

Radio Times, (30/4/1937).

Removal of wine stains.

If washable material, use boiling water method, but place salt on stain first. Apply Fuller's Earth to non-washables.

Woman, (18/12/1954).

Removal of ink stains.

Powder a damp stain with Fuller's Earth, then cover with a paste made of water and the absorbent. Brush off when dry. Stains from ball-point pens can be removed from woollens by gently dabbing with methylated spirit.

Id.

More cloth cleaning

19ix

William Tucker, one of the very few scourers to divulge the secrets of his trade, showed how thorough the process of scouring was at the beginning of the nineteenth century.
CAROLINE DAVIDSON, A Woman's Work is Never Done.

A Woman's Work is Never Done, Caroline Davidson, Chatto & Windus (1982), p. 160.

THE truth is more mundane for he copied a recipe in Colin Mackenzie:

Colin Mackenzie, 180.

To scour clothes, coats, pelisses, etc.

If a black, blue or brown coat, dry 2 ounces of fullers earth, and pour on it sufficient boiling water to dissolve it and plaster with it the spots of grease; take a penny-worth of bullock's gall, mix with it half a pint of stale wine[1]*; and a little boiling water; with a hard brush dipped in this liquor brush*

1) probably urine.

19ix

William Tucker's book was *The Family Dyer and Scourer, being a complete treatise in the whole art of cleaning and dying* ..., Sherwood, Neely and Jones, London, (1817), pp. 25–5.

spotted places. Then dip the coat in a bucket of cold spring water. When nearly dry, lay the nap right and pass a drop of oil of olives over the brush to finish it.

If grey, drab, fawns or maroons, cut yellow soap into thin slices and pour water upon it to moisten it. Rub the greasy and dirty spots of the coat. Let it dry a little and then brush it with warm water, repeating if necessary as at first, and use water a little hotter, rinse several times in warm water, and finish as before.

Mrs. Beeton.

To clean cloth.

Ingredients – dry fullers earth moistened with lemon-juice and a small quantity of pulverised pearlash. Mode – mix the fullers earth and pearlash into balls with sufficient lemon-juice to moisten. Scour the cloth with the balls.

Mrs. J. Hathaway, *private communication*, (1941).

To remove blood-stains from linen.

Take lump fullers earth and break down to a thick paste in warm water. Spread it on stains and allow to dry. Wash off.

Colin Mackenzie, 184.

Fullers' purifier for woollen cloths.

Dry, pulverize and sift the following ingredients:

6 lbs. of fullers earth
1 lb. of pipe clay
4 oz. of french chalk

Make a paste of the above with the following:

1 oz. of rectified oil of turpentine
2 oz. of spirit of wine.
1½ lbs. of melted oil soap.

Make up the compound into six-penny or shilling cakes for sale. These cakes are to be kept in water, or in small wooden boxes.

Stain removal by absorbents

Muriel Clark, *Make Do and Mend*, Collins: Glasgow & London, (1978), Ch. 7, p. 99.

MOP up stains as quickly as possible. Towels, rags and tissues are all useful when a stain is freshly made. Absorbent materials like starch and fuller's earth (the white kind) are also useful, and talcum powder will do in emergencies. Fuller's earth and talcum powder share the great merit that they do not leave any mark and cannot harm fabric – especially useful on non-washable articles. If possible, damp the stained area slightly, then sprinkle the powder on. Leave it to dry and brush off. On fabrics where a water-mark would show, e.g., satin, crêpe, velvet and plain-coloured fabrics, steam the stained area with a steam iron held above the fabric but not touching it, or in front of the kettle spout. Then apply the powder quickly.

Note: Almost any stain on any fabric may be treated this way, and in my experience the results have been satisfactory. On dark coloured fabrics, it is best to sprinkle powder on the wrong side.

White fuller's earth comes from only a few places, such as Greece, Italy and Turkey.

Sepiolite has the advantage of sticking less to hairy cloth.

A good trade-name for a household cleansing powder would be FULLO, the cleanser.

Suede and leather garments

19ix

Suede bags and gloves are more difficult to maintain. Shine can be removed with fine emery or a rubber suede-brush. Special suede cleaning products are available and are good value for money, but it is important that things should not be allowed to become too heavily soiled before they are used. Fuller's earth is effective in removing greasy stains – if they are very severe, treat with a paste of fuller's earth and a solvent cleaner. Allow this to dry, then brush off gently. Test the effect of the solvent cleaner before using on coloured suede.

And jackets.
Id., p. 114.

Fuller's earth soap

19x

THE Housewife's A.B.C. contains also a recipe for "carpet soap."

Fullers earth 4 oz., oil of turpentine 1 oz., potassium carbonate 8 oz., soft soap, a sufficiency. Mix the first three ingredients and make into a stiff paste with soft soap.

Fales & Hunter, 514.

As Alexander Watt put it: *Of all the solid matters which have been mechanically combined with soap, the mineral substance known as fuller's earth is undoubtedly the best. Moreover, being in itself a detergent, its combination with soap partakes less of the character of a mere adulterant than other argillaceous or clayey substances.*

The art of soap-making, London: Crosby, Lockwood & Son, (1907), 109.

Watt introduced his Fuller's Earth Soap about 1885, but there can be little doubt that these materials would have been mixed long before. Watt's process was to dry the fuller's earth in an oven, sprinkle the lumps on a slab till slaked and dry again very gently. The earth could then be reduced to a powder very easily and lumps sieved out. The powder was then spread over the surface of the melted soap, preferably resin soap, a little at a time, and crutched in, and finally, perfume added for a toilet soap. In this way at least a third of the mixture could be fuller's earth and two thirds soap; as a skin soap it was said to be *most agreeable, and . . ., when not perfumed, specially serviceable as a nursery soap.*

In "Mr. Davis's process" of blending clays with fuller's earth, pearlash or soda was first calcined and ground up with the fuller's earth or clay until intimately mixed: to every 126 lb. of melted soap, 50 lb. of fuller's earth, "slaked or dried", was added, as well as 56 lb. of dried pipe-clay, and 112 lb. of the alkali. This soap was said to be useful for general washing purposes at sea, and for washing white linen in salt water.

The composition of a modern salt-water soap is given by Woolley and Forrester as: *fuller's earth 40 parts, calcined soda ash 40 parts, coconut oil soap 80 parts.*

Woolley & Forrester, *2*, 189.
sic. Proportions are of course 1 : 1 : 2.

In salt water and carpet soaps, fuller's earth had come to stay, but Watt's toilet soap was not popular. Only in wartime have large quan-

19x

tities of fuller's earth found their way into toilet and household soaps; in peacetime they are quickly dropped because the resultant soap is ugly in colour, cracks and does not lather well, even though it is moderately good as a detergent. Bentonite gives a velvety feel to toilet soaps, and kaolin a satiny whiteness.

19xi

Living, furry animals

Players Cigarette Card. Series of 24, No. 11, (1936).

Sarah Jenkins, *Daily Mail*, (9 Oct. 1959).

WHITE cats, according to the advice given on a cigarette card, are best cleaned before being shown at exhibitions with fuller's earth, which is well rubbed in and then brushed out with a long-fibred brush.

However, even I have to admit that a chinchilla bathing in a baking-tin full of fuller's earth is amusing. Over and over they fling themselves, rolling ecstatically.

19xii

Miscellaneous

The manufacturer will say his polish isn't fairly represented by the television screen. But quite obviously no polish on God's earth is going to give you a black-glass mirror effect when you put it on wood. Artificial dust? Oh, yes; you cover the surface with fuller's earth or something, then you take a cloth and say, "In one moment – wheet! What a wonderful shine!"

From The Quickness of Admen, by PAUL FERRIS,
***The Observer*, 3 January 1960.**

The powder used for drying ink was called 'pounce' (pumice), though powdered gum sandarac and cuttle shell were also used.

IF a fisherman wishes to change from dry fly to wet fly fishing he can do so by drawing the line through a pinch of fuller's earth held tightly between his fingers. The water is then able to wet the line.

Fuller's earth held in a porous bag may be used by draftsmen for removing grease from their paper. White bentonite should be used or else sepiolite which is easier to brush off.

Saddlery. Fuller's earth is also very effective for removing grease stains from saddlery.

KENT

A certain man went to the sea, minding to have gone into Kent for fulling earth.
JOHN FOXE, Acts and monuments of these latter and perilous days (= The book of martyrs), (1563), 1684 ed., III, 591.

FOR many centuries, until about 1820, the main source of fuller's earth in the east of England was in the neighbourhood of Maidstone in Kent.

Though the earliest record of a fulling mill in the area is between 1509 and 1547, and the earliest reference to fuller's earth occurs only in 1530, quarrying had almost certainly been carried on from time to time since Roman times. Two of the pits have had Roman coins found in them.

When the Domesday survey was made in 1086, Maidstone (Meddestone) was an important centre surrounded by the villages of Penenden (Pinneden), Hollingbourne (Hoilingeborde), Langley (Languelei), Otham (Oteham), Barming (Bermelinge), and others. Maidstone was held by the Archbishop of Canterbury, and its chief points of interest were the church, five mills, four and a half fisheries producing four hundred and fifty eels and two salt-pans – from which may be gathered the fact that the sea then came in to this point, but nothing about fulling.

S. Henshall and J. Wilkinson, *Domesday 1086*, (1799).

Douglas Allport, *Maidstone, its history traditions and associations*. Maidstone: Hall, (1842).

It is known however that the Kentish mills of olden times frequently served the double purpose of grinding corn and fulling cloth.

The Maidstone mills leased by Henry VIII (1509–1547) to William Sheldon were detailed as two wheat, one malt and two fulling mills.

W. Page (editor), *A history of Kent*, the Victoria History of the Counties of England, (1932) *3*, 375.

Id., 376.

Power-driven fulling mills, though there are earlier records of these elsewhere, were probably introduced into Kent by the Flemish weavers brought over in 1131 and subsequently.

C. C. R. Pile in his essay on Cranbrooke Broadcloth and the Cloth Makers states that when in the 14th century it became the government's policy to concentrate the woollen industry in selected towns and counties, a different style of cloth being made in each district, Cranbrook was chosen as a centre in Kent because it stood in countryside watered by many small brooks which could be easily dammed to provide water-power, because the mighty oak trees of the Forest of Andred

Cranbrook broadcloth and the cloth makers, (1951).

Andred, the pre-Celtic name of the Weald; Latin Anderida.

could furnish timber for the mills, and because there was a plentiful supply of marl as a cleansing agent in the neighbourhood.

Although local marl is said to have been used around Cranbrook in the early days as a substitute for fuller's earth, it is not improbable that fuller's earth from near Maidstone was carried south along the Roman road (A229) in the summer months when the road was passable. Marlpit Wood, The Marlpit Field and The Marlpit are common field names in Cranbrook, Biddenden and other villages, but it is not known which of these names derives from the use of marl for fulling and which from the later use as an agricultural dressing.

Thomas Philipot, *Villare Cantianum,* (1667), 360, *v. infra.* Raphael Holinshead or Hollingshead, *Chronicles of England, Scotland and Ireland,* 2 Vols., (1578); 2nd ed., enlarged, (1587); 3rd ed., (1808).

Philipot quotes Raphael Holinshead, who died about 1580, as saying that the River Len was *mighty beneficial to the Clothiers, for lying so near the Fuller's Earth; they used to bring their Cloaths from Ten Miles around about to be fulled here. But now I think there are no Fulling Mills hereabouts.*

John Harris, *The history of Kent*, London, (1719), *1*, 51.

In 1345 Roger Vinter, who owned one of the most famous fuller's earth estates, was made a Conservator of the Peace for the County of Kent. His son John sold the Estate in 1409 to Mr. Fremingham, whose daughter sold it to Isley of Sandrich, whose descendant, Sir Henry Isley, was unhappily concerned in Sir Thomas Wyatt's rebellion in 1553.

Owners also given by Thomas Philipot, *Villare Cantianum*, or *Kent surveyed and illustrated.* Lynn: Whittingham etc., (1776), 2nd ed., 90.

(Wyatt of Allington Castle, in the first year of Queen Mary, with other gentlemen of note in the county, raised a rebellion out of disgust at the Queen's marriage with King Philip of Spain; he was taken prisoner, and committed to the Tower, found guilty on his trial and executed on April 11th of that year.)

E. Hasted, *History and topographical survey of Kent*, 1778–99, (1782), *2*, 96.

The estate was forfeited to the Crown; Queen Mary granted it to a Mr. Cutts, who later sold it to Sir Cavaliere Mayeot. He in turn sold it to Covert, and he to Sir William Tufton, whose second son, Sir Charles Tufton, was the possessor in Philipot's time (1667).

C. Greenwood, *An epitome of county history*, (1838), *1*, Kent, 183.

Sir Charles disposed of it between 1649 and 1685 to Daniel Whyte, whose son, also Daniel, sold it between 1702 and 1714 to Sir Samuel Ongley, one of the South Sea Directors, and of his descendants it was purchased in 1783 by James Whatman, of the parish of Boxley, who rebuilt the mansion.

Id.

Turkey Mills on the River Len south of Vinters were used for many years as fulling mills, but on the decay of the woollen trade in these parts were converted by Mr. Gill, then the proprietor, into paper mills. These were subsequently purchased by James Whatman, who in 1739 rebuilt them on a much more extensive scale, and his son, James Whatman, sold them in 1794 to Mr. Hollingworth and retired to Vinters.

E. Hasted, *op.cit.*, *2*, 130.

The Grove, which marches to the east of Vinters, was conveyed between 1399 and 1413 by Isabel de Wavering to Thomas Burbige, whose descendant, also Thomas, owned, besides the Grove, a dovehouse and other lands adjoining it in Henry VII's reign (1485–1509); but thereafter a great part of the land was sold from time to time; what remained with the house descended to Thomas Burbige, who, dying without issue and being the last heir male, bequeathed it to his sister, and she sold it in 1702 to John Watts, who afterwards became famous for his management of the fuller's earth pits. He died "seised of it", and his heirs about 1757 sold it to General William Belfor, Colonel Commandant of the First Battalion of the Royal Regiment of Artillery, who was still the owner in 1782.

"in legal possession of it".

Owling or smuggling was rife in Kent because of the excellent shipping facilities on the River Medway. The subject has been touched on in Chapter 13. It was not only forbidden to export fuller's earth but even to store it near the coast, "lest the temptation to smuggle should arise".

John Ray, a merchant of London, was sentenced in 1620 in the Star Chamber to a fine of £2,000, imprisonment during the King's pleasure, and to be set in the pillory with a paper on his forehead, declaring the nature of the offence, for transporting a quantity of earth clandestinely to Holland; the earth had been purchased from Richard Rods of Maidstone for six shillings per load.

Rushworth, *2*, 348; quoted by Hasted, *2*, 130, and *V.C.H.*, *3*, 39; and *CIBA Review*, (Oct. 1937), *1*, 64.

(or Roodes or Rhodes).

Licensed shipping took place from Rochester. In 1621 there were thirty sailings between 4th April and 2nd November. In the whole year there were shipments of 1,200 chauldrons (about 1,500 tons) from Rochester to Hull, probably fuller's earth clod for the use of the Yorkshire clothiers.

Nevertheless the heavy penalties imposed did not deter the smugglers, for example, in 1622 penalties were inflicted on Stephen Danton of Worth, who dug fuller's earth from his own ground and sold thirty or forty loads to a Dutchman. In the same year Richard Fyle of Sandwich was reported to the authorities for having sold earth suspected to be fuller's from his garden to a Dutchman, who professed to "want it for making pots". This earth, however, proved on examination not to be "the precious treasure", though it might well be used for "scouring baize", according to the Victoria County History of Kent from which these law cases are taken.

V.C.H., *3*, 396 Wortham or Wrotham? Camden gives both spellings.

Cal. S.P. Dom., (1622–23), 377.

Id., 38.

The Crown did not always prove the offence, as when on May 11th of 1662 Richard Fyle was acquitted of smuggling thirty "loads" of fuller's earth to Holland. In the same year, others charged with smuggling fuller's earth to the Continent included a Mr (or Mrs?) Brewer and a Mr. Richard Rhodes. They were also acquitted, on September 23rd, having been charged with illegal export of 450 and 75 "loads" respectively. Edmund Warner's acquittal has been mentioned in Chapter 13 vii.

Fuller's earth was "fetched out" of the county by John Hutchinson, of Plymouth, in 1632; it was stated to have been obtained from "about Gillingham". It was frequently transported out of the rivers Thames and Medway by stealth at this date.

Cal. S.P. Dom., **1631–33**, 429, 533.

In 1640 Mrs. Brewer, of Boxley, *sent her fuller's earth to the waterside, and had three houses where she stored it till the tide served to bring up boats in which it was carried to Rochester and then in greater vessels beyond seas.* Similar offences were recorded at the same date against Mr. Love, of Boxley, and Richard Roodes, of Maidstone – presumably the same man who had offended twenty years before.

Cal. S. P. Dom., (1640–41), 280.

Id.

> *As for the fuller's earth carried out of the country* writes a pamphleteer of the day, *that must certainly be transported out of the river of Medway, for we have none in England but what is about Maidstone on the said river of Medway that I could hear of, except at Wooburn in Bedfordshire.*

V.C.H., *3*, 396, quoting Travers, *New Year's Gift to England*, (1630), 15.

Since sea transport was the only feasible means of supplying fuller's earth to the other woollen centres of England, and the difficulties of engaging in the coastwise trade were imperilled by the suspicion of illegal export, one must assume that the sales of fuller's earth from the Maidstone district were irregular. However, in 1639 the clothiers of Ipswich and Suffolk in general were praying that the prohibition against taking fuller's earth from Rochester by sea should be removed. It was further prayed that such earth as was brought from the ports and lay ready at the waterside in divers parts of Kent in great quantities might be removed further from the coast to avoid secret conveyance.

Id.

In 1641, by royal permission, one hundred and twenty chauldrons (*ca* one hundred and fifty tons) of wet fuller's earth were permitted to be transported to Hull for the use of the Yorkshire clothiers, upon an understanding being given that it should not be transferred at that port to any foreign vessel for shipment to the Continent. Under similar conditions, in the same year, permission was granted to Edmund Adams and other fullers of Lincoln to carry fifty or sixty chauldrons annually by sea out of Kent.

Cal. S.P. Dom., (1641–43), 48.

Id., 60.

Possibly the most prosperous period of the Kent fuller's earth industry was when John Watts owned and managed the pits at The Grove, between 1702 and 1757. In the latter part of the 18th century, though coastwise shipping to other parts of England was made easier, and though Hasted in 1792 records that the fulling and paper mills sent their goods to London by sea and that a great quantity of fuller's earth was sent from home by sea for the use of the clothiers in the distant counties, the industry seems to have declined very rapidly. The quarry faces appear to have been abandoned when fully developed.

Hasted, *2*, 130.

The local markets had long since gone, the manufacture of Cranbrook broadcloth, for instance, had virtually ceased by 1700. Vinters quarries were probably finally closed when James Whatman rebuilt the house and converted the estate into a gentleman's property in 1783. The Grove similarly became a purely residential estate when General Belford bought it in 1782. Certainly *fuller's earth, vulgarly called Walkers earth,* used in Leeds in 1725, came from Kent.

Ralph Thoresby, *Ducatus Leodinensis*, or the *Topography of Leeds*, London (1725), 471.

A few quarries were in use for some years after these main pits ceased to be worked. And there was a final burst of activity around 1801 to 1805 when a new quarry opened up near Maidstone adversely

Malcolm, *Comp. husbandry*, (1805), 1.

affected for a while the sales of Surrey fuller's earth. Since then only very small quantities have been raised until the 1950s.

Memories of an old industry quickly die, and all that remained were the fallen-in pits, a few place-names, and scattered references in the literature.

There are no descriptions of the fuller's earth pits until 1719. John Harris then wrote: *There is a famous Mine of* Fuller's Earth *in this Parish (Oxley), which is wrought by the ingenious Mr.* John Watts *of Groves. The Vein lies about 30 Foot deep, and is about 7 Foot thick. There are two Sorts of it, the* Blue, *and the Dark* Gray, *the latter of which lying under the former, is the best. Over it is a Sand, and neither in that nor in the* Fuller's-Earth, *are any Marine Remains, as Shells, &c., found, though very usual in other places.*

The history of Kent, (1719), *1*, 51.

This description does not tally with John Woodward's in 1729: *shells occurring in the fuller's earth are generally much rotten and decay'd.*

An attempt towards a natural history of the fossils of England, (1727), *1*, 4.

The precise location where the earth was wrought is seldom stated in the older works, the parishes of Boxley and Debtling being sometimes mentioned. The pits of Vinters and The Grove are known to have been worked in Roman times as we have said. A pit used to be worked on the Debtling Road near the Old Turnpike gate; this was said to be filled in and planted over before 1839. The site of this pit has not been identified; but it may have been near the Chiltern Hundreds public house. A four-foot seam of fuller's earth was known on Penenden heath in 1862, and considered at that time to be of insufficient quantity to be workable. A small quantity of fuller's earth was dug in the park of the Moat House at Leeds Castle, or near the cricket ground, up to as late as 1880 to 1890, and some was left unworked. The Kent Fuller's Earth Co. Ltd. worked the Atherfield clay between the two World Wars, north of Leeds Village, but this clay is not the same as the historical fuller's earth of the Sandgate Beds. Fuller's earth has also been recorded south of Aylesford, between Vinters and occurrences later noted at Barming.

J. Smith, *Topography of Maidstone and its environs*, (1839).

W. H. Benstead, The geology of Maidstone and its neighbourhood, *Maidstone Journal*, (1 April 1862); but E. Hunt, Mining Records: Mineral Statistics . . . Part II for 1858, *Mem. Geol. Soc.* (1860), 377, states that fuller's earth at Penenden Heath has not been worked lately.

A fifteenth century Fulling Mill, on sale in 1985, was once in the Leeds Castle Estate.

It is not known when the Barming quarries were worked.

The deeper quarries on the Vinters and Groves estates are easily recognised; other deep quarries are to be seen south of the Chiltern Hundreds and of Ware Street; also at Barming. Shallow quarries are more difficult to see, but careful inspection usually shows some unnatural features in the contouring, especially where the cut was deepest, down the dipslope. The signs of working are too numerous to mention, but have been recorded on field maps. A carefully filled-in quarry however may be impossible to detect on the ground; aerial photographs have been of some assistance for detecting virgin ground.

Fig. 58

It has also been assumed that the Old Turnpike roads passed over virgin ground in Milgate Park. There is a record of the oak trees planted near Milgate House growing very slowly for many years; it is assumed that the roots of the trees suffered from lack of aeration in a seam of fuller's earth. Full rate of growth was restored when they penetrated to the sand beneath. (Even as early as 1598 it was noticed

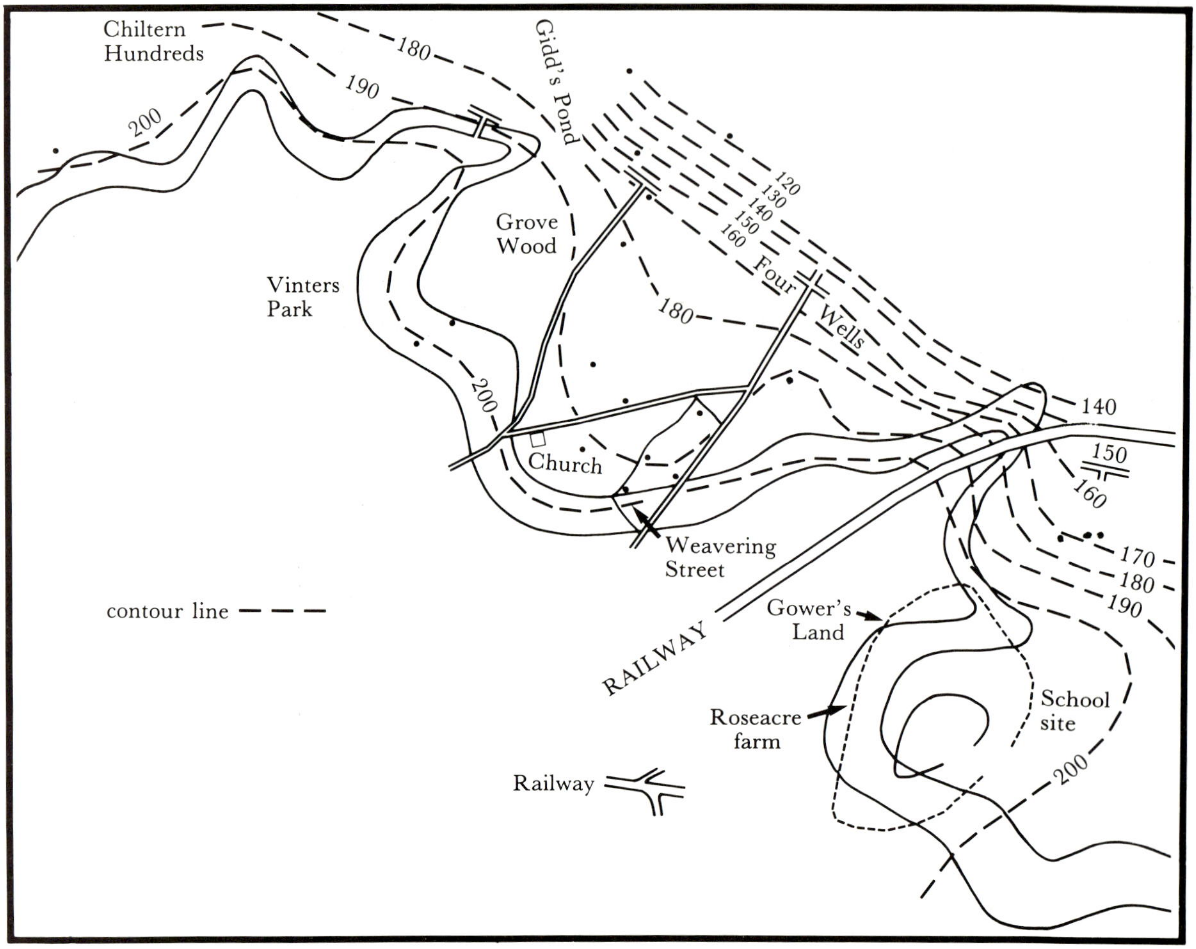

Fig. 58. Strike lines indicating top of the Fuller's Earth from Chiltern Hundreds to Roseacre, Bearsted, Kent.

that fuller's earth had a bad effect on tree growth, the quotation from *Hackluyt's Voyage* being given in Chapter 41 (USSR).

The Geological Survey soon after World War II resurveyed the district and noted fuller's earth in a number of temporary exposures: Oakenwood, Barming, south of the Maidstone-Ashford road (south of Roseacre) in a new housing estate; Mil gate Park, in an air-raid shelter; and in fields south of Foley House near Maidstone.

Also in Fullingpits Wood, Barming Heath.

In an old quarry near Upper Preston, south of Aylesford, the fuller's earth in the Sandgate Beds is only two or three feet thick. Topley also said that at Sevenoaks the Sandgate Beds consist of only about six feet of greenish clay containing some fuller's earth.

W. Topley, *Geology of the Weald*, Mem. Geol. Surv., London: Longmans, (1875), 129.

James Sowerby in his *British Mineralogy* (1804–1817) reproduces a hand-coloured drawing of Kentish fuller's earth (*3*, Tab. 231 and 232), and states that *the earth found at present near Bersted, but not in abundance, like that of Nutfield, is not above three to six feet from the surface, under mixed sand; and differs also from that of Nutfield in being found with much water, and not being in large pieces, or above half a pound, as well as in requiring careful picking and washing; it is then laid out to dry and shedded for carting to the Medway. It is rather softer to the touch than the kind from Surrey, but generally dries darker and has a more ochraceous crust.*

Bearsted.

This is the last and most accurate description of Kentish fuller's earth we have until recent times. Manning and Bray (1809) simply said that it was found at Maidstone.

(1809) *2*.

The workings ceased in Kent partly because the quarries were reaching the depth below which they would have been uneconomic at that time and partly because gravel extraction and farming were more attractive economically. Some of the earth-bearing land was made unavailable by the owners of the gentlemen's estates in the interests of gracious living.

In 1954, The Fullers' Earth Union acquired the Vinters Estate as a reserve. At the beginning of 1958, it bought 10,000 tons a year of selected fuller's earth from the Roseacre estate, operated by F. W. Berk & Co nearby. The FEU, by that time part of Laporte, re-developed its own workings in the 1970s as a supplementary source of feed for the Redhill Works. In 1980/81 the company sold its land in Kent. By then, most of the fuller's earth had been extracted and mineral rights were retained in respect of the remainder. The Roseacre site was later restored and sold as building land.

21 SURREY

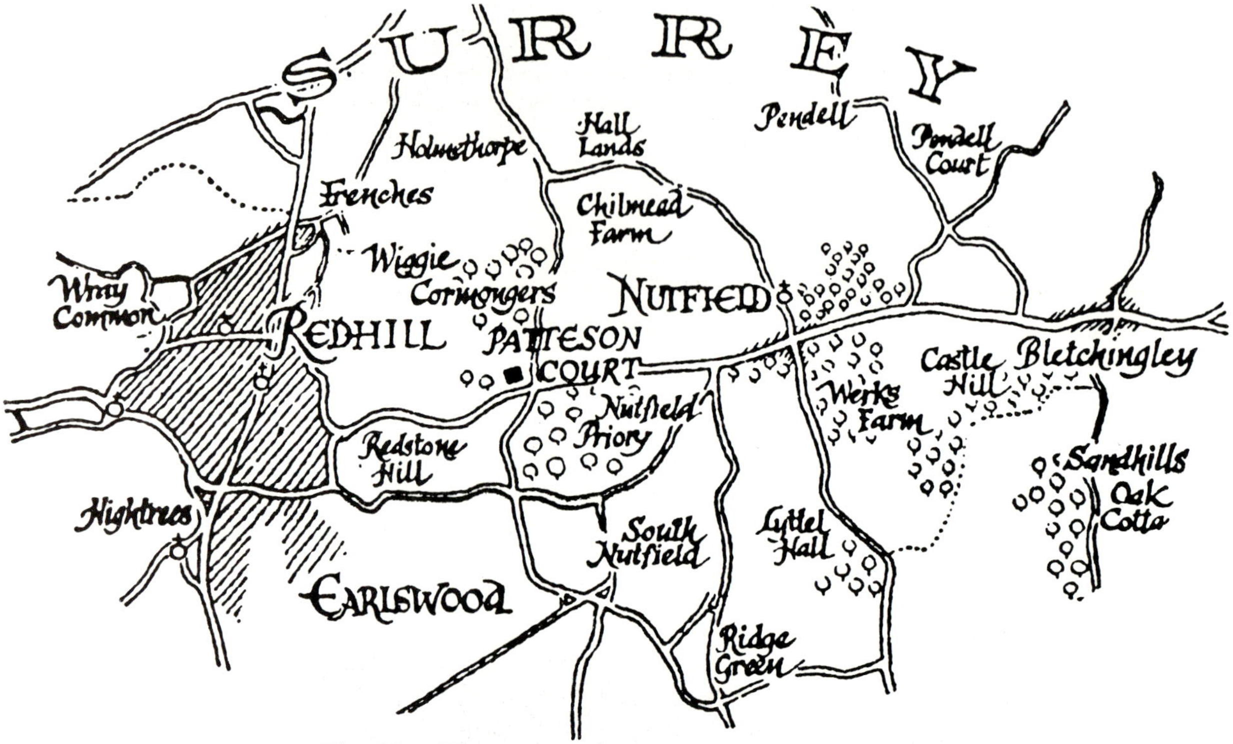

Fig. 59. The area around Nutfield.

The chief commodities of Surrey are Corn, Box, Fuller's Earth and Walnuts

ENGLAND'S Remarques, (1682), p. 185.

Fuller's earth comes from Reigate

COPYBOOK used as late as 1929.

THE Roman villa at Titsey was only a mile north of Limpsfield on the Lower Greensand. The Roman road did not continue much further south into the Weald. Only 5½ and 8 km to the west along the crest of the ridge lay the fuller's earth deposits of Godstone and Blechingley. Titsey itself was on a very ancient track, the Harroway, running from the Port of Rutupiae (Richborough) to the West of England, a vital route in the Conquest of 43 AD. Supplies of fuller's earth could have been delivered by pack horse to military launderies and to the large

Later the Pilgrim's Way.

number of villas along this road, the course of which is conjectural in parts.[1]

We have already suggested[2] that the place-names Wolcnestede (Walkhampstead, part of Godstone) and Blachingelei (Blechingley) in the Domesday Book are linked with ancient fuller's earth workings. Moreover Blachedfeld (now Blackheath, south east of Guildford) and Witley[3], south south west of Guildford) are both on the Lower Cretaceous at a geological horizon which might not be very far from that of the elusive fuller's earth of Headley (Chapter 47).

In 950 Byhrtic and Aelfswith his wife made a will, leaving Wolcnesstede to one Wulfstan Ucca and a large parcel of land to the minister of Wolcnestede. One of the witnesses was Wulfsic se blaca, a name which Lambert thinks should be translated Wulfsic the fuller.

Another place-name in the parish of Blechingley *more than possibly fixes the site of an early fulling mill.* That is Stocklands, which lies below Warwick Wood, with its southern end on the Black Brook, which flows from Pendell to Merstham, on its way to the Mole.

> *The name is found in Blechingley as early as 1336 when Robert atte Stokke was a tenant owing service, and two centuries later, when the land was called "Stokland", it is carefully described as "formerly called le Stok". The Stock was therefore a clearly defined spot, though the name of Stocklands now spreads over some fifty acres; and as there is a brook there, though now much diminished in flow, the evidence that Robert held, in the 14th century, the site of a probably disused stockmill seems fairly strong.*

The woollen industry developed also at Godstone, because of the running water of a tributary of the Eden.

Wool came in plenty from the extensive sheep walks of the North Downs, and early provided the raw material for an important cloth industry in Guildford and south-west Surrey. The woollen industry developed there also because of the ample supply of running water, the Wey giving power to many fulling mills.

Lambert gives evidence that the cloth industry at Godstone was beginning to decline before 1400.

> *In 1391 upon the complaint of the commons of Guildford and other places in Surrey, Sussex and Hampshire, it was enacted that – because that of old times divers cloths were made in the town of Guildford and other places within the counties of Surrey, Sussex and Southampton called Cloths of Guildford, which were of good making and of good value, and did bear a great name; and now because the Fullers and others of the same county do use to buy the cloths of the said counties before that they be fulled and performed, and in making for covetousness to have the said cloths of greater measure over the common assise that late was used, do draw the cloths longer and more large than they were wont or ought to be, to the great impairing of the said cloths and great deceit of the people; For to eschew such damages and deceits in time to come, It is agreed and assented that from henceforth no Fuller nor other person, whatsoever he be, shall buy within the said towns*

1 Later the Romans made paved roads from Canterbury to London, then along Staines Street to Silchester, Speen and Bath.

2 Following Uvedale Lambert, see below.

3 The conventional derivation is from Witta's Wood.

Uvedale Lambert, *Blechingley*, Mitchell, Hughes & Clarke: London (1921), vol. 1, pp 2–17.

Black Brook was known to the inhabitants of Merstham as the "sheep wash".

The first use of the word is tree-stump, and the second, also Anglo-Saxon, is a log of wood. Stocklands would certainly mean tree-stump land, but le Stok does suggest a Stockmill. RHSR.

V.C.H., Surrey, 342.

Rot. Parl. iii, 294a.

Stat. 15 Ric. II, cap. 10.

and counties any cloth before the same cloth be fulled and fully performed in his nature, and also sealed under the seal thereto ordained, upon pain of forfeiture of the same.

Such excessive tentering was part of a losing battle: poor transport and relatively low hydraulic heads on the streams were other causes of the decline.

J. J. Greenwood, *Fullers earth in Surrey 1500–1900*: an economic history (1983), Deering Press, 50 Reigate Road, Reigate, whose work should be consulted for references. I thank Mr Greenwood for his help in preparing chapters 21 and 36.

When the local wool industry had gone, Surrey lacked Kent's compensatory advantage of easy shipment to the new cloth manufacturing centres, and for a period the Surrey earth deposits were worked intermittently only. Greenwood (1982) has traced the early days of the Surrey industry in one of the few scholarly works on the subject. Apparently Alice Gilmyn of an armigerous family of Reigate was granted a lease to dig 'loam' in 1577 by John Mitchell owner of Redstone manor at £4/yr. This was at Blue pit field. Another, Fuller's earth field, was not included in the lease. *Although Norden only mentions "fuller's earth" at Sutton a survey of Reigate manor in 1622 mentions the fuller's earth pits of Richard Heath of Ford (later Copyhold) farm, and pits to the west of this, on the opposite side of Cumbers Lane, the ancient droveway south from Merstham.*

Greenwood continues: *Heath was the only person who can be identified as being involved on more than a casual basis during the 17th c.* The latter half of this century seems to have been a relative boom time as not only are there frequent references to fuller's earth pits in Reigate and Nutfield in Quarter Sessions' presentments but also national antiquarians like Thomas Fuller enthused over the product. His statement in 1662 about the natural commodities of Surrey was that

T. Fuller, *A history of the worthies of England*, (1662), *2*, p. 352.

fuller's earth – the most and best of this kind in England (not to say Europe) is digged up nigh Ryegate, worth 4d. a bushel at the pit and 16d at the wharf in London; 3 s. at Newbury.

It is clear that from the latter half of the 17th c. at least fuller's earth was an article of commerce in this county, as we have seen in the reference to England's *Remarques* (1682). (Box Hill is named from the box trees growing there and Nutfield (Notfelle in the Domesday Book) means the field where the nuts grew).

London: L. Curtis, (1682), 185.

J. Aubrey, *The natural history and antiquities of Surrey*, London: E. Curll, (1718), 214, 237, (written before 1697).

E. Chamberlayne, *Anglia Notitia*. 1667 (–1707).
E. Bowen, *Britannia Depicta*. (1720).

These comments were echoed a few years later by John Aubrey who, unlike Fuller, actually visited the area, and who stated *in this parish (Reigate) adjacent to Nutfield is plenty of fuller's earth which they sell at 4d. a sack* and he made similar comments about Nutfield. Much the same was said by Edward Chamberlayne (1695) and E. Bowen (1720). However, by 1718, when transport problems were presumably somewhat eased, several open pits were at work in Nutfield, and also near what is now Redhill in the parish of Reigate – *the place is very narrow and small, from whence this Earth is dug* suggesting that the method of extraction there was by underground mining. The price of Surrey fuller's earth is given, under 'Production and Price Statistics', for 1693, 1730, 1744, 1759–1839 and thereafter; and the history of mining

Paul W. Sowan, *Fuller's Earth Mine discovered at Nutfield. Historical notes.*

as opposed to quarrying under "Quarrying" below, (Chapters 33 and 36).

Cave Research and Exploration, (12 April 1981), 1. See also Chapter 36.

The Manor of Redstone when sold in Reigate in 1786 was said to have fuller's earth in great plenty on the estate.

Manning & Bray, (1804), *1*, 310, note.

This estate possessed *with a Court Baron, quitrents, reliefs, and heriots, an elegant mansion, garden, orchard, and 109 acres, 3 roods, 1 perch of land, of which 12 acres were copyhold, held of the Manor of Reygate.* Court Baron and the other feudal customs were all connected with copyhold tenure. Copyhold Farm, at the top of Redstone Hill, was situated almost entirely on fuller's-earth-bearing land. The farmhouse became "The Copyhold", home of Wilfred Raikes. It was demolished in 1967 but the name is preserved in the Laporte "Copyhold Works" nearby.

(1809), *2*, 266.

Manning and Bray maintained in 1809 that the fuller's earth *so essentially necessary in our clothing manufacture, is found in greater quantity, and of a better quality in this County, and particularly in this Parish of Nutfield, than in any other part of England.*

This became Hallands House, now demolished.

The principal proprietor was Mr. Edward Russell of Holland House, Nutfield, but there were two other pits in the parish working at the time and another recently opened by Mr. Morris in Copyhold Farm in the parish of Reigate, near to other pits there which were not being worked at the time. The most valuable earth came from Mr. Thomas Dann's pit, one of the smaller ones, in Nutfield. *The Dann family, prominent Quakers, later became lime merchants in Reigate, and provided its first mayor.*

Greenwood, *op. cit.*

A typical quarry section showed top soil and loose material 3–5 feet; sandstone 5 or 6 feet; small fuller's earth and dirt, unusable, 2 feet; sandstone (some used for foundations and inner walls of buildings, some for roads), 6–8 feet; fuller's earth, 4 to 12 or 14 feet (some yellow, some reddish brown and blue).

The sections differed however so that no one pit exactly corresponded with another, and variation was observed in individual quarries.

The earth was taken out in lumps from a pound to a hundredweight in mass. Rain falling on the surface of the earth causes it to crumble off and what so fell was regarded then as of no value. The yellow form of the earth is sometimes impregnated with what they called rust, which they carefully removed. The brown or blue was chiefly sent into Yorkshire, the yellow into Wiltshire and the west. It was sometimes sent in waggons, but was oftener shipped as ballast; but in the latter case bonds were required that it should not be carried to foreign countries.

In 1805.

In the four pits then open only ten to twelve men were employed and their output was between 2,000 and 3,000 tons a year. With this small labour force only wet clod could have been produced at that time, or perhaps only a small quantity of dried clod.

In the fuller's earth bright yellow stones, weighing from fifty to one

(1809), *2*

hundred pounds each, were sometimes found. Manning and Bray (1809) said that an attempt had been made to polish them, but without success on account of their brittleness. Thirty years later Andrew Ure rightly recorded that these pieces of stone were barytes encrusted with quartz crystals. The workmen's superstition was that they *often spoil many tons which lie about them*; and they carefully removed them. Mantell, 1841, more precisely states that they weighed from a few ounces up to 130 or 140 pounds, and described their colour as like sugar-candy. The quartz is sometimes amethystine.

Barytes geodes have also contained sphalerite and the very rare acicular form of pyrite.

Manning and Bray, 1814, *3*, lxxix.

In 1814 the Lower Greensand was called the *fuller's-earth sand or Reigate sand.* Gideon Mantell recognised in 1841 the following strata under the chalk – firestone, galt, Shanklin sand, and Weald clay; the fuller's earth occurred near the top of the lowermost division of the Shanklin sand. The stratigraphy of this district very early attracted the notice of William Smith, who was more familiar with the fuller's earth near Bath. He drew a section of Mr. Edward Russell's pit at Nutfield and coloured it. A similar section was presented to the Geological Society in 1809 by the Reverend E. I. Burrow; this copy may still exist among the William Smith documents at Oxford.

Size 12¼″ × 8¾″ undated.

Trans. Geol. Soc., Ser. 1, *1*, 406.

In 1816 the mineral productions of Surrey were iron-ore, fuller's earth, fire-stone, limestone, and chalk.

J. Nightingale, *English Topography*, London, (1816).

Webster reported that four pits "open to the day" were being worked between Nutfield village and what is now Redhill in 1819.

T. Webster, On the Geognostical Situation of the Reygate Stone, and of the Fuller's Earth at Nutfield, *Trans. Geol. Soc.*, (1821), *5(18)*; 353–357 & pl. 38.

In 1826 Gideon Mantell was fortunate enough to obtain specimens from these quarries of Ammonites Nutfieldensis[1], of great rarity and beauty, preserved in his collection in the British Museum.

(1) Now *Parahoplitoides Nutfieldensis.*

At Godalming about this time there was a certain amount of stocking weaving and wool combing, and some manufacture of worsteds, blankets, sails and collar-cloths; but apart from this there can have been very little sale of earth for fulling within the county.

In places on the Lower Greensand ridge the fuller's earth *lies so near the surface as frequently to be turned up by the wheels of the waggons*. Only thin seams had been found north of the Redhill-Nutfield road in 1839, the main seam being worked at Nutfield on the swell of the hill to the south of the road; the oldest pit then in use was opened some time between 1779 and 1789, and was then fast wearing out. The blue earth in one pit was nearly sixteen feet thick.

Andrew Ure, *Dictionary of Arts, Manufacturers and Names*, London: Longmans, (1839), 529–530.

By 1841 the main pits had been opened on the north side of the road from Redhill to Nutfield. Mantell gives a section at "Colmonger's Farm"[2], where the blue earth was sixteen feet thick. Several small seams of fuller's earth, later known as "bunkum", were noted above the main seam. The output at that time was said by a Mr. Constable of Horley to be about four thousand tons of yellow and two thousand tons of blue earth a year. He gave further information about the places to which fuller's earth was sent then. *The manufacturers of fine cloth make use of the blue only, and that variety therefore is sent chiefly to Leeds and other*

Edward W. Brayley, *A topographical history of Surrey*, London, Tilt & Bogue, (1841), *1*, 145.

(2) Latterly Cormongers House, demolished in 1968. A large date palm tree was shifted to the Ivy House nearby.

parts of Yorkshire, where that cloth is made. The yellow earth has a much wider distribution; being employed in the manufacture of every fabric of coarse woollen goods; it is not only in request over the West Riding of Yorkshire, but also in Lancashire, Cumberland, and Westmorland. Some is sent into Scotland, through the ports of Leith and Fisherrow; and a considerable quantity into North Wales at Welshpool, Montgomery, Newton, Llandilloe and other places where it is employed for the cleaning of flannel. Norwich also receives a supply for its stuff manufacturies. In some kinds of coarse goods a portion of the earth is left in the fabric to give it a substance and the yellow is said to be best adapted for this purpose. The whole of the earth exported from this district is sent from London by sea, except a small quantity retained for the use of the dyers and scourers.

Llandilloe (sic). But note that Llanidloes was famous for flannels

In the library of the Geological Society of London there is a description in manuscript of the fuller's earth pits, by Messrs. Greenough, R. Phillips and Horner, unfortunately with no date. It is worth quoting for its quaint terminology.

Section of Fuller's Earth Pits at Nutfield.

		Ft.	*In.*
A.	Soil	1	0
B.	Lower Mulatto-stone, with Bastard Caps	4	6
C.	Cool	3	0
D.	Mulatto-stone, with cap-stones ("an argillaceous rock with greenish specks.")	7	0
E.	Quoin-stone	3	0
F.	Fuller's earth	10–12	0
G.	Sand	0	4
H.	Hard rock		

Glossary

Cool. *Called also the rubbish of the fuller's earth, from its being in smaller pieces.*

Quoin-stone. *Of the same nature with the other rock masses, but containing no cap-stones. Less indurated than they, but more so than the rock lying upon it. It is used for building.*

Cap-stones. *Rounded masses imbedded in the rock, and much more indurated than the rock itself, though apparently of the same nature with it, sometimes, however, differing in colour.*

Bastard-caps. *The same as the above, but not so much indurated.*

Cool, an old name for bunkum.

Topaz-coloured barytes masses are found in the fuller's earth; ammonites and some other shells are sometimes found in the quoinstone.

About this time William Lambert, a miller, was occupied in the fuller's earth trade and had several men in his employ. In his younger days he was *one of the most successful cricketers that has ever yet appeared.*

He was a very powerful man, weighing fifteen stones, and had huge hands. He died at Nutfield on April 19th, 1851, aged 72, and was buried at his birthplace of Burstow.

Frederick Lillywhite's *Cricket scores and biographies of celebrated cricketers from 1746 to 1826.* from *Cricket* by John Arnott, London: Burke, (1853), 38–39.

21

1851 *Exhibition (London) Catalogue*, Index, *1*, 1, 19, 47, 48 and 55.
W. Topley, see below.
E. Hunt, *Mining Records: Mineral Statistics* . . . Part II for 1858, *Mem. Geol. Surv.* (1860), 377.
J. Cumming, On an Excursion made by the Geologists' Association to Reigate, *Proc. Geol. Assoc.*, (1861), *1(7)* 149.

In 1851 Surrey fuller's earth received an award at the Great Exhibition, at which one of the two British exhibitors was Cawley of Nutfield, the other being Hynam & Gawkroger, merchants.

In 1858 the quantity of fuller's earth sent from the Nutfield deposits had risen to about twelve thousand tons a year. Hunt's *Mineral Statistics* for 1858 record the Reigate Fuller's Earth of Rogers & Hynam and J. Cawley's at Nutfield.

Cumming wrote in 1861 that *so little is the demand at the present time for fuller's earth, that the proprietor merely works the material out from beneath the stone, and thus long galleries are formed.*

(One gallery under the A25 caused a subsidence in 1962. Others were discovered in FEU's workings later.)

Paul W. Sowan, *loc. cit.*, p. 2.

> *A number of candles were procured, and several members* [*of the Geologists' Association*] *inspected the cave, many points in which proved very instructive.* Sowan (1981) commented that *this sounds very like the working lately discovered;* he continues *Hunt's statistics for 1861 and 1862 (1862, 1863) quote output figures for the neighbourhood of Reigate as 8000 tons per annum,* and that *In producing this there is also raised about 17,000 tons of stone.*

W. Topley, *Geology of the Weald*, London: Longmans, (1875), 394.

(1) (Now Cormongers).

William Topley (1875) describes quarry sections in Pimlico Wood (main seam eight feet exposed), on Colmonger's farm[1] (nine feet), and near Copyhold Farm (fifteen feet). He also states that there is fuller's earth at Redhill station and in the cutting at the curve south-west of it. At Redhill Common seven feet six inches of fuller's earth were found by the south of Linkfield Street, where the roads meet, and fuller's earth was seen in the road south of Reigate Church. Gossling describes sections containing fuller's earth in Chart Lane, Reigate, and Grove Hill Road, Redhill. Fuller's earth is seldom seen farther west, but there is some at the corner of the heath a mile west of Milford House and traces of clay near Wheeler Street.

Tracing the Sandgate Beds eastwards, Topley states that in the neighbourhood of Tandridge, there are ten feet or more of fuller's earth, which is dug in some places. The thickness is, we suspect, exaggerated.

The Visitor's Guide to Reigate, Redhill & their Vicinity. Printed and Published by H. Sutton, 3 Station Road, Redhill (1877).

By 1877 there were only two pits at work, one belonging to Mr. J. Cawley and the other to Mr. Matthew Hedley. *Several thousand tons of the fuller's earth are annually raised from the Nutfield quarries, and sent to the manufacturing districts.*

After the discovery in America that fuller's earth effectively removed colour and other impurities from cottonseed oil, it was soon felt that access to the American market was difficult when there were so many producers. There was also the need to introduce quality standards and to maintain uniform products; variation had probably always been a bugbear, as may be inferred from a complaint received by a Paisley firm of drysalters which expresses quaintly but succinctly the fears of the cloth-finisher of a faulty finish and of extra costs:

Stewarton 13 June 1884

Mefrs. Henderson, Hogg & Co.
Der Sirs forward on at wonce thre or four Bags of Fullers Earth as usualy goot thise to hand is of no use thare is no pour to rais Greas out of the goods we had a tryel of it before this, but we gaive a dubel qwantity to thise miled this day and they have gone the sim way and has cased a wonderful extery labour we will set thise a side to ce what is to be done with them
I am yours truly
James Gibson.

100: Henderson, Hogg & Co., Ltd., by R. A. Peel, published by the Company, Stonefield Works, Thornley Park Road, Paisley, Scotland, 25 March 1970.

According to *Mineral Statistics* (Home Office, 1886) C. W. Cawley was mining fuller's earth at the Cockley pits as well as at an openwork quarry at The Park; Thomas Wood Cox had an openwork quarry at Copyhold; and Mrs John Hynam had two open quarries, at Chartfield and Shambles 8,000 tons or so, 'undressed', valued at £9,000.

Home Office, Mines & Minerals, Mining & Mineral Statistics . . . for 1885, (1886) pp. 24 and 128.

The first company to work fuller's earth in Surrey was The Surrey Fullers' Earth Company. Cawley & Company was founded by James Cawley and expectations of a large export trade to America led his son Claude Henry Cawley to issue a prospectus for a Limited Company in April 1886, primarily *to purchase the Park and Cockley Fuller's Earth Works and the business there carried out by Claude William Cawley.* The purchase by this company of property at Wellow and Odd Down near Bath was confirmed in 1887. Even with these acquisitions the company were unable to achieve adequate profits. The dried lump earth then sold at £2 2s. a ton, but when mills were installed prices of from £4 10s. to £14 a ton were obtained. In 1887–8 an interim dividend of seven and a half per cent was paid, but after that competition proved too fierce and amalgamation with the other producers became essential. But before we describe the union – or really cartel – that was formed in March 1890, we should say something about the *Dramatis Personae.*

Cawley father and son described themselves as fuller's earth merchants and both lived at the Tower, Nutfield, so called because a tower had been erected, probably by James in about 1856, in the garden. This folly features in several books about such extravagances. On the death of Claude William the house and estate were sold (1920).

James Cawley (1822–1882).
The house is now called Well House, and the tower may still be seen.

Chart pit, south of the A25, was operated by John William Grece, who built Chart Lodge in 1780. He died in 1820. His son Henry Grece, who died in 1852, sold the pit to Rogers & Hynam. In 1890 the property was owned by John Hynam.

C. W. Cawley was assisted in his plan to achieve a workable union of interests by William Blewitt, an Essex solicitor and director of the London & General Bank. He (Blewitt) was a shareholder in The Fullers' Earth Union Ltd, of which he was one of the first Directors, in Cawley & Co. Ltd. and the Minerals and Fullers' Earth Syndicate, Ltd, a company which was incorporated in 1890, with its registered office in Bath. Among the shareholders was A. G. D. Moger, the first Chairman of The FEU Ltd. The Syndicate continued in existence until 1911.

The Fullers' Earth Union Limited, formed in March 1890, acquired or controlled by agreement the following works in Surrey (the Somerset acquisitions are given in Chapter 22).

Fuller's earth works in Surrey and owners 1890

Name of works	Present owners	Tenure	Remarks
Cockley Works, Nutfield	Cawley & Co. Ltd.	Freehold	Purchased
Park Works, Nutfield	Cawley & Co. Ltd.	Freehold	Purchased
Copyhold Works, Nutfield	Thos. Wood Cox	Copyhold	*
Chartfield Works, Redhill	John Hynam	Leasehold	**
Shambles Works, on Redstone Hill Estate, Redhill	John Hynam	Leasehold	**

* Contract for taking over whole output of company.

** Contract for selling at prices to be fixed monthly by company and giving company exclusive right of grinding at profitable price.

Claude W. Cawley became Managing Director of the new company. (His name was commemorated in the company's telegraphic address: YELWAC, REDHILL). Other original Directors were Wm Blewitt; the Right Hon. Earl Kilmorey; S. Evan Butler; and Wm Wainwright.

George Dames, who sold his leasehold of beds of fuller's earth at Midford to The FEU, was briefly secretary whilst the company was being formed and was succeeded by W. S. Metzner.

Spain Brothers, appointed in 1890, continued as auditors for three quarters of a century.

More formal references are given in the *Surrey Mirror*, 22.3.1890 and *The Times* 24.3.1890.

The Financial News, 18.3.1890.

Bath Journal, 22.3.1890.

Bath and Cheltenham Gazette, 19.3.1890.

Bath Herald 22.3.1890.

Most financial commentators wrote favourably of the proposed Union but the *Daily Oracle*, 24th March 1890, was scathing in its comments about the new company.

> *How pleasant it must be for traders who have been devoting their energies to cut-throat competition amongst themselves to arrange together to invite the public to provide the money to carry on the businesses in future, so that the exhausted combatants may not have to pay the full penalty of their folly, but may be enabled to lead peaceable and quiet, and it may be prosperous, lives ever after. What lovely prospectuses these commercial gladiators can draft! How promising the outlook is when they have ceased to struggle one against the other! What money can be made. What untold wealth is waiting to be picked up.*
>
> *Yet somehow they have not done it in the past, and do not in the future propose to keep the little gold mine to themselves. Having run the trade to the very verge of ruin at their own expense, they magnanimously invite the public to come in and share the profits now that the time of anxiety, worry, squabble and loss is past and there is nothing before them but prosperity, harmony, monopoly and big dividends. We may be very uncharitable and*

unduly sceptical, but we do confess that whenever undertakings of this kind come before us, we think of the weaknesses of human nature, and take a very big discount off the statements and anticipations of the prospectus of the union, syndicate, or whatever it may be which is placing the delectable company before the public in order that outsiders may have an opportunity of sharing in a good thing.

The peppery critic continues:

In the first place, the trade has been brought into a most depressed and ruin-spelling condition by what the prospectus mildly calls "useless rivalries". When that document, which, of course, puts the case as favourably as it can, says that the ill-judged competition "has not only reduced the price of the Earth to an unduly low level, but has also brought out inequalities of charge", outsiders may form an opinion of the horrible plight into which the business has got.

Yet the directors would have us believe that the demand has been greater than the supply! They tell us that the demands of the American trade have been so heavy that orders have had to be not only delayed but refused. Do traders cut prices down to a profitless point, and ruin their own businesses and those of their rivals when orders that it would pay to execute are actually being turned away from their doors? The suggestion is preposterous.

In the section given to Answers to Correspondents, we read: *The Fullers' Earth Union Ltd.: G. DAW – No, we do not advise you to invest in this concern. In our opinion, £95,000 is a preposterous price to pay for the property.*

At first it would seem that the critic's fears were not wholly unfounded. George Dames left within a year to form and run The Fuller's Earth Mining Company near Woburn, and there was a major crisis in the management.

The original prospectus for The FEU Ltd stated that the Founder Shareholders would receive no Dividend in any year until the Ordinary Shares had received a Dividend of £8 per cent for that year.

The intentions were good, the return attractive and confidence firm. The First Report of the Directors in June 1891 described sales as "satisfactory", declared a Dividend of 7 per cent for the year and looked forward to even better trading. The Second Report of the Directors in June 1892 includes the paragraphs:

The Directors regret that the Balance Sheet submitted with this report shows a heavy loss and The Directors discovered that the full information as to the working of the company, to which they were entitled, was being withheld from them, and the position of the company during the course of the year had been misrepresented to them by the Managing Director and, on their finding this, Mr. Cawley was at once removed from his office as Managing Director.

Mr. Cawley was also removed from the Board, the formal reason given being non-attendance at Meetings. Although Cawley & Co. Ltd. had properly passed a special resolution on 2 December 1890 to be

Cawley later ran a hearthstone mine at the base of Juniper Hill, Reigate, until his death in 1908.

wound up voluntarily, this was not completed until February 1892, about the time when Cawley was threatened with litigation.

Also in The FEU's Second Report it was recorded that *Since the change in Management* over the last four months of the financial year, *the business has progressed very favourably*, though all that could be done at that stage was to reduce the loss. No dividend was declared for that year. Indeed The FEU had not acquired all the productive works in Surrey. The Surrey Fullers Earth Co. Ltd., incorporated 27 June 1892, still had Beechfield Works, which was not sold to The FEU until 1919 by the sole Director, Herbert Law Dyson of Kings Villa, Huddersfield. He became a shareholder of The FEU in 1898, however, being introduced by A. G. Moger.

The formation of The Fullers' Earth Union had created some public interest. Thus Conan Doyle made fullers' earth a central theme in *The Engineer's Thumb* (ca 1892).

> *'Then the matter stands thus. You are probably aware that fuller's earth is a valuable product, and that it is only found in one or two places in England?'*
>
> *'I have heard so.'*
>
> *'Some little time ago I bought a small place – a very small place – within ten miles of Reading. I was fortunate enough to discover that there was a deposit of fuller's earth in one of my fields. On examining it, however, I found that this deposit was a comparatively small one, and that it formed a link between two very much larger ones upon the right and the left – both of them, however, in the grounds of my neighbours. These good people were absolutely ignorant that their land contained that which was quite as valuable as a gold mine. Naturally, it was to my interest to buy their land before they discovered its true value; but, unfortunately, I had no capital by which I could do this. I took a few of my friends into the secret, however, and they suggested that we should quietly and secretly work our own little deposit, and that in this way we should earn the money which would enable us to buy the neighbouring fields.'*

In their 5th Report, for the half-year, 15 December 1893, when S. Stanley Brown had become Chairman, the Directors reported *the action against Mr. C. W. Cawley has been settled. The Company has obtained a perpetual injunction restraining him from trading as a Fuller's Earth merchant, and the surrender by him of 1,500 fully paid Vendor shares.*

Eighth Report of Directors, May 1895.

Three years later, the management was still struggling to redress the financial position, especially as competition was still intense and volumes and prices had fallen. The early Debenture-holders had in large part been repaid and a Dividend on ordinary shares of 1 per cent was declared for that year.

The Ordinary Dividend was declared at 1½ per cent in 1896 and 2 per cent in 1897. In 1898 and 1899 the Ordinary Dividend was 3 per cent. It did not exceed 5 per cent until 1904 when it rose to 7 per cent. In 1906 it was 11 per cent. This was maintained in 1907 when it was

THE FULLERS' EARTH UNION, LTD.

COPYHOLD WORKS, REDSTONE HILL, REDHILL, SURREY.

MINERS & MANUFACTURERS
of
BLEACHING EARTHS & CLAYS
including
NATURAL ENGLISH FULLERS' EARTH
ACTIVATED ENGLISH FULLERS' EARTH
CONTINENTAL BLEACHING EARTH
"OSMO-KAOLIN" and other
COLLOIDAL CLAYS

Prepared for
OIL REFINERS
SOAP MANUFACTURERS
THE WOOLLEN TRADE
PAINT, PAPER and PENCIL
MANUFACTURERS
all over the world

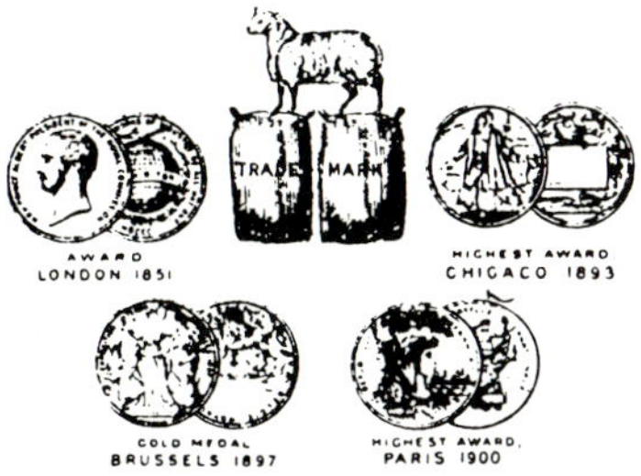

Telephone No.: REDHILL 781/2
Telegrams: "FULLION, PHONE, REDHILL."

Codes:
WESTERN UNION FIVE-LETTER
BENTLEY'S.

19

EXHIBITIONS

F.E.U.

1851	award
1862	International Exhibition
1867	Paris Exhibition
1885	International Inventions Exhibition, London.
1893	Chicago, Highest award.
1897	Brussels. Gold medal.
1900	Paris Highest award.
1904	St. Louis.
1924	British Empire Exhibition.

Fig. 60. Notepaper in use until the early 1930s.

considered that the condition set out in the Prospectus had been met and the first payment to Founder Shareholders was made.

The sales and weights of earth sold by The FEU in years ending March 1894 to March 1920 are given in Chapter 33 i. The *Mineral Statistics* for 1893–1900 are difficult to reconcile with Table 7, but they do distinguish between mined and opencast workings in Surrey. In 1893 250 tons of earth came from mines (value £560) and 7,494 tons from openworks (value £12,197). The one mine then was Park Mine of The FEU, whose agent was W. F. Keevil. He had come from the Bath Works and eventually gave 51 years of valuable service before he died on 24 February 1934. He was notable in building up processing technology, in advising on land purchase and in developing business relationships. An interesting example of his standing is that, in 1911, the South Eastern Railway sought his advice on the driving of the tunnel through fuller's earth in Redhill.

Table 7. C. LeNeve Foster, Mines [and Quarries] Report . . . for the North Wales, &c, and Isle of Man District (No. 9) . . . for the Year 1893 [-1900], (1894-1901).

His son, Owen Keevil, also worked at Redhill, except when he was managing a mine in Arkansas 1911–1920.

See Chapter 53i.

At Park Mine one man was employed underground and one above. Openworks, at Chart Field, Church Hill, and Copyhold employed 81 men. Copyhold was worked by Keevil for The FEU; Church Hill by W. Giles for the Surrey Fuller's Earth Co. Ltd. and Chart Field was still worked by John Hynam on his own account. Nevertheless, The Fuller's Earth Union Ltd. obtained the highest award at the Exhibition in Chicago in 1893.

Park and Church Hill are in Nutfield.

Paul W. Sowan is thanked for extracting these records.

In 1894 Park Mine seems to have been closed, and only 6065 tons of fuller's earth (worth £7,848) were raised from openworks. Further attempts at mining in 1895 and 1897 are referred to in the chapter on quarrying, but in the latter year openworks were at Chartfield and Copyhold at Redhill; and Cockley and Park at Nutfield. The FEU won a Gold Medal at Brussels in 1897 as well as Silver and Bronze Certificates.

Presumably the value ex quarry not the value ex works.
Chapter 36.

In 1898 Beechfield mine produced 4,817 tons of fuller's earth (value £2,168) by two men underground and 35 above, producing fuller's earth and sandstone. In 1899 and 1900 Beechfield was worked opencast along with Chartfield, Cockley, Copyhold, and Shambles. The FEU obtained the highest award in Paris in 1900, and they also exhibited at St. Louis in 1904 and at the British Empire Exhibition in 1924.

An old miner was still alive in Bletchingley in 1970.

The notepaper used in the '20s and '30s showed these awards.

Even though The FEU did not acquire Beechfield Quarry until 1919, the years before World War I were prosperous for the Company. Mr. William Stone, who lived to the age of 102, told me his shares had made him 'a young fortune'. However sales to the USA dropped dramatically from 1914 to 1933 and the Directors decided to make new products, which are described in the Chapter on "The Twentieth Century – England."

The oldest FEU shareholder; he owned No. 1 Albany, Piccadilly, London.

Chapter 38.

SOMERSET (AVON) 22

It had the wool, just as it has to-day; it had water power and plenty of soft water for dyeing; it had a good local supply of fuller's earth in the hillsides; and it was not far from the port of Bristol.

J. B. PRIESTLEY, 'To the Cotswolds', in English Journey (1934).

THE wealth yielded by wool from the 14th to the 17th centuries is commemorated in the magnificent Perpendicular churches of the Cotswolds, such as that at Chipping Campden and the West of England in general.

The fulling industry is also recalled by the large number of surnames such as Tucker and Tuckfield, Walker and Fuller, which occur in the West of England.

Lionel F.J. Walrond, 2 March 1979, "Indeed one gets a fair idea of the distribution of Tuckers *versus* Fullers from Smyth's *Men and Armour 1608*, which is a census of all able-bodied men in the county. The Walker/Tucker boundary just north of Bath is also a cultural division of speech and architecture.

The Somersetshire form of the name is usually Touker, as in Alex. le Toukere (co. Soms. I Edw. III; Kirby's Quest, p. 218); Matilda Toukere, (co. Soms., 1 Edw. III; *id.*, p. 252); (M.D.B. (co. Soms.)1; New York, 24); and John Touker, Mss. Dn and Ch Wells, 1474–5, but in Gloucestershire we find Peter le Walkar, (co. Glouc., 20 Edw. 1, R.); Emery Tuckfield, rector of Trent (J. Collinson).

John Collinson, *The History and Antiquities of the County of Somerset*, Bath: R. Cruttwell, (1791), *2*, 383.

Among the place names which record this ancient industry are:

Fulling Mill, Bath
Tucker Street, Wells
Tuckerton
Tucking Mill, Midford
Tuckingmills (two miles south-west of Keynsham)
Tucker's Arms (two miles south of Bridgwater).
Walkham (near Wells).

Id., B 33, and *3*, 72 and 375.

Of the cloth-weavers and fullers and dyers encouraged to come to England by Edwards I and II, not a few settled in Somerset; they were well established in 1389.

Jarman, *History of Bridgwater*, (1889).

When Leland journeyed in these parts in 1535, West Wiltshire, Somerset and Gloucestershire were still very prosperous with cloth – Dursley stream, for example had four miles of continuous tucking mills. Leland writes: *Then about two myles from Frome market I came to a botome (Wheel Brook). Here dwell certain good clothiars havynge fayre howys and tukkynge myles.* At Pensford *There cummith down a streame that servith divers*

John Leland, *Itinerary*, Part X, *Somersetshire*, (1535).

howys = house (singular).

tukking mills; though of the industry at Thornbury *now much idleness lapsing reigneth there.*

J. Collinson, *op.cit. 3*, 333–334.

The most accessible part of the Somerset fuller's earth, already mentioned in the Roman connection, was in the parish of Combe Hay, one mile north of Wellow and only three miles south-west from Bath.

1.6/4.2 km.

The Domesday Book records that Odo, Bishop of Bayeux, held Come as the place was then called. He was half-brother to William the Conqueror and reputed to be the wisest and wealthiest man in the realm; but he became inordinately ambitious and was banished twice and finally dispossessed. The land here soon came into the ownership of the Hay (or Hawey) family, who gave their name to the parish.

Itinerary, 2, 67.

The existence of fuller's earth so near Bath was of course of particular advantage to the wool industry of that city itself. Leland (1535) says that a little before his time there were three capital clothiers of the names of Style, Kent and Chapman, *by whom the town of Bath then flourished*; and Collinson said that at the time of the Restoration (1660) there were no less than sixty broad looms employed in the single parish of St. Michael. Hence arose a company of artificers called the "Weavers Company", but this had long been extinct (by 1791). A fulling-mill, belonging to the monks of St. Peter and St. Paul in the eastern part of the old city became known as Monk's-mill.

Collinson, B. 28.

Fuller's Earth was still an important production in "Somersetshire" in 1787. John Cary's *English Atlas*, 1 Sept. 1787.

In Taunton the first woollen "manufactury" was established about 1336, but the industry declined towards the end of the 18th century. The loss was partly made good by the rise of a woollen industry in Wellington nearby, but in 1791 Collinson says that the industry had died out in Yeovil and in Buckland-Dinham. There was still some activity in Minehead (where the arms of the town are a ship under sail and a woolpack), and quite a lot at Wiveliscombe, where blanketings, knap-coatings, kersey and other coarse cloths, strouds, ermine and baize were still made for markets in London, Bristol, Exeter and abroad, especially Spain and Guernsey.

Nap, with raised pile. Strouds were blankets made for barter or sale in trading with the North American Indians or the cloth of which these blankets were made.

But the glory had departed and supplies of fuller's earth were no doubt provided by small-scale operators only.

As Priestley puts it, the western area of the Cotswolds became a depressed area. It once

English Journey, London: William Heineman, (1934), Ch. 3, Section 2.

> *. . . made broadcloth, which our grandfathers recognised as the best of all clothing fabrics. The other districts had their own specialties. To this day the best blankets come from Witney, just off the eastern edge of the Cotswolds. Even as late as 1801 a map of England showing the density of population includes this region in its most heavily shaded areas, with an average density of more than a hundred to the square mile. A similar map of modern England shows the bigger part of the region in its most lightly shaded division, with an average density of less than 128 persons to the square mile.*

In the interim, of course, the population of the country had increased threefold.

From the great period of this woollen industry no records of fuller's earth have been preserved; it had been obtained from numerous localities south of Bath before the first records become available in the late 18th century. The earth was dug from shallow exposures or from landslipped masses at Combe Hay, English Combe, Odd Down, Claverton Down, South Stoke, Midcombe, Widcombe, Duncorn Hill, Dunkerton and Wellow, and in other places, such as Combe Monkton and Lyncombe.

W. J. Arkell, *The Jurassic System in Great Britain*, Oxford (1933), 280.

The Reverend John Skinner wrote in his Journal of the soil at Wellow being constantly turned up by the clothiers in search of the earth. It is doubtful whether there were any regular *commercial* suppliers of fuller's earth in the Bath area until the 18th century. The first written record of fuller's earth in Somerset is in 1791, when the Reverend John Collinson writes that in the parish of Mells there are, besides several coal-works, clay equal to the Stourbridge product in the manufacture of crucibles, lead, manganese, pipe-clay, and fuller's earth. But he does *not* discuss fuller's earth in his section on minerals, though he deals with ruddle, talc, ochre, coal, stone, secomia, gaiophragmium, and others. Nor does he notice fuller's earth in his description of Combe Hay (*3*, 333) and Inglishcombe (*3*, 339).

Add. MS 33,688 fol. 29 (*V.C.H*).

1791, *2*, 462.

id., *1*, xvi

Neither secomia nor gaiophragmium are in the O.E.D.

Soon after this time there had appeared in the district one who made the greatest contributions to geology, William Smith. Smith, who as a self-taught land surveyor, aged 18, became assistant to Edward Webb, surveyor, of Stow-on-the-Wold in 1787, travelled over the Cotswold Oolites, the Lias clays, and the red marls to the north; and studied the varieties of strata and soils. In 1791 he became aware of the regularity of strata. In 1793 during the levelling and excavation of the Somerset Coal Canal of which he was the surveyor, he saw more clearly how the strata above the coal were inclined to the east so that they outcropped successively at the surface.

In 1798 Smith moved from Bath to Midford, having purchased Tucking Mill house and the adjacent land. Its grounds were tastefully laid out under his supervision. Here he coloured the first geological map in the world – Five Miles round the City of Bath – which he later presented to the Geological Society of London.

In the following year as Samuel Smiles records it:

Self-help, London: J. Murray, (1872), 144–150.

> *When looking over the cabinet collection of fossils belonging to the Reverend Samuel Richardson, at Bath, Smith astonished his friend by suddenly disarranging his classification, and re-arranging the fossils in their stratigraphical order, saying – "These came from the blue lias, these from the over-lying sand and freestone, these from the fuller's earth, and these from the Bath building stone." His knowledge of the strata in the neighbourhood of Bath was so accurate, that one evening, when dining at the house of the Reverend Joseph Townsend, he dictated to Mr. Richardson the different strata according to their order of succession in descending order, twenty-three in number, commencing with the chalk and descending in continuous series down to the coal . . . to this was added a list of the more remarkable fossils which had been gathered in the several layers of rock.*

A commemorative plaque was unveiled in July 1926 at his house, 29 Pulteney Street, by Dr. F. A. Bather, F.R.S., the author of *William Smith, 'The Father of English Geology'*.

Special publication of the Bath Royal Literary and Scientific Society, (1926).

Thus it was that in 1799 Smith gave the name "The Fuller's Earth" to the whole series of clays and impure limestones, passing further afield into sands, above the Inferior Oolite and below the Great Oolite; the whole amounting in places to a hundred and fifty feet in thickness. The economic fuller's earth is only a few feet thick; not much more than seven feet at the most, but commonly one and a half to four feet.

Tucking Mill was his favourite retreat, although after 1800 he could visit it only intermittently. It is believed that his nephew, John Phillips, also a famous geologist and his biographer, spent part of his early infancy here. In 1799 Smith severed his connection with the Somerset Coal Canal Company and undertook a large amount of surveying work all over the country. In 1800 he dated his prospectus from Midford of the large geological map which was eventually published in 1815 as the *Geological Map of England and Wales, with part of Scotland*, in fifteen sheets.

Memoirs of William Smith, LL.D., London, (1844).

Indeed he was dismissed on 5 June 1799 after the failure of a caisson, which he knew was sited in the unstable Lower Fuller's Earth Clay – against his advice, (Eyles, see below).

"Strata" Smith was frequently in financial difficulties through his single-minded efforts to complete the map of Britain and his unceasing travelling; hence the delay in publication. His friends however helped to bring his work to the notice of a wider public. Thus the Reverend Richard Warner, in 1801, refers the reader for a more scientific and particular account of the mineralogy of the environs of Bath *to a work written expressly on the subject by the very ingenious Mr. Smith of Midford near Bath which we understand will shortly be given to the world*. He remarks himself that the *Freestone, or Oolite, overlies a bed of blue clay, containing rhomboidal selenites; another of yellow clay down to the fuller's earth, which, although but a thin stratum, is visible every where round Bath, by the slips on the declivities of the hills, occasionally by the springs of water which flow out upon it. This is followed by a very deep bed of imperfect or bastard fuller's earth, which abounds with casts or impressions of high-waved and striated* Cardia, Mytili, Nuclei *of the* Nautilus, *and various species of* Anomiae.

Richard Warner, *The history of Bath*, Bath: R. Cruttwell, (1801).

Anomia, the saddle Oyster.

L. R. Cox, G. A. Kellaway, A. Templeman and Major A. Gorham, Eastern Field Meeting, 1940, at Bath, *Proc. Geol. Assoc.* (1941), *52*, 16–35.

In 1811 Smith sold part of his land and invested the proceeds in a stone quarry on the high ground (Combe Down) to the north. The stone was conveyed to the mill by a specially constructed railroad and there was sawn into blocks and loaded on to barges on the canal. Unfortunately the venture proved a failure, and Smith, a ruined man, was obliged, in 1819, to sell up all his property, including Tucking Mill itself. He had already in 1816 been forced to sell his collection to the British Museum. He was able to retain his maps, and continued to improve and extend them. In 1831 he received a life-pension of £100 a year from the Government. He died in Northampton on August 28th, 1839, on the way to attend the British Association from his home at Hackness in Yorkshire.

On a cottage, built by one of Smith's tenants in 1808, a plaque was re-erected in 1932 by the Geological Society of London and the Bath Literary and Scientific Institution. It reads, "Here lived William Smith Father of English Geology", born 23.3.1769, died 28.8.1839. Eyles (1974) has pointed out that the plaque should be placed in front of the much more imposing Tucking Mill House in the neighbouring parish of Monkton Combe, and Palmer (1981), who was born in the house and became a Master Builder in 1949, showed that the eastern part of this enlarged house is older than the western, and therefore the house that Smith lived in – this, not only by the bonding of the stonework at the join but also by the internal arrangements of the house.

Joan M. Eyles, *J. Soc. Biblphy nat. Hist.*, (1974), *7(1)*, 29–34.

Alec. S. Palmer, *Letter* (15 May 1981).

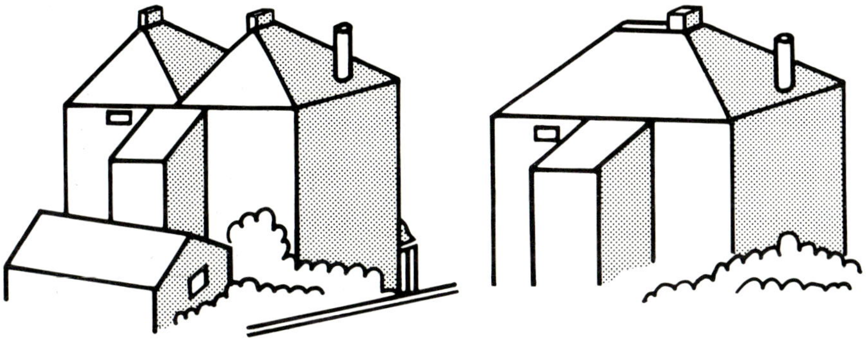

1910 1981

Fig. 61. Tuckingmill House from the North-West.

From photographs.

A photograph of 1910 shows the house before the roof over Smith's house and the extension were unified, Fig. 61.

The Fuller's Earth, named by Smith the "Fullonian" is subdivided thus:-

c. 40-50 ft. the Upper Fuller's Earth Clay
c. 10 ft. the Fuller's Earth Rock – an earthy limestone, with brachiopods.
c. 40 ft the Lower Fuller's Earth Clay.

Commercial fuller's earth occurs only in the Upper Fuller's Earth Clay.

Although in recent years there have been refinements in stratigraphical nomenclature, Arkell was able to say in 1952 that *the best account of the fuller's earth in the South Cotswolds is still that of a reconnaissance made in 1803 by William Smith.* That was true until Penn, Merriman and Wyatt described The Bathonian strata of the Bathstone area (1978), including, of course, the Fuller's Earth.

W.J. Arkell and D.T. Donovan, *Quart. J. geol. Soc.*, (1952), *107*, 227–252.

Inst. Geol. Sci., Pap. 78/22.

In the building of the Somerset Coal Canal the Lower Fuller's Earth clay was probably used as a puddling clay rather than the commercial fuller's earth itself in the neighbourhood of Midford. Hugh Torrens locates the "clay hole" or "borrow pit" above locks 3 and 4 beside the manager's house and still easily discernible. In more recent times Bradford clay from the 'Canal Pit' was used for puddling its western end.

Hugh S. Torrens, The Somerset Coal Canal Caisson Lock, *J. Bristol Industrial Archaeological Society*, (1975), *8*, 4–10.

Fuller's earth may not have been worked in the whole area at that time – at least on a large scale: it was unknown to Manning and Bray (1809) and to Andrew Ure (1839), who wrote that yellow fuller's earth was sent from Surrey to Wiltshire and the West of England woollen mills. However, the fuller's earth of Somerset was noticed by Nightingale in 1816 and in 1822 we hear that the fuller's earth of Combe Hay was much inquired for by foreign chemists.

J. Nightingale, *English Topography*, London (1816) Sutcliffe, *Geology of the Avon* (1822) p. 25.

Much later, fuller's earth was encountered in the railway cuttings between Midford and Wellow, and a fuller's earth section at Combe Hay – in a railway cutting – was described by Richardson (1909).

Railway Magazine, (18/2/1950).

L. Richardson, *Quart J. geol. Soc.*, (1909–10) *21*, 425.

Woodward remarked that the fuller's earth was rarely if ever worked in 1876. It was formerly worked in quarries and in "earth wells". Near South Stoke, at the junction of the Fuller's Earth with the Great Oolite, a shaft penetrated about twenty feet of clay. At the depth of sixteen or seventeen feet a vein of yellowish or buff clay was met with, which was followed in all directions from the central shaft. Woodward recorded that *the blue fuller's earth is frequently of as good a quality as the yellow for particular purposes, as in fulling of coarse cloths*. The seam was usually eighteen inches to three feet in thickness; below eighteen inches it could not be worked at a profit. Sometimes the seam stops suddenly, sometimes thins out gradually.

H. B. Woodward, East Somerset and the Bristol Coalfields, *Mem. Geol. Surv.* (1876), 125.

The ninth edition of the Encyclopaedia Britannica (1870–1903) states that the fuller's earth was *in past times* largely mined in the Downs, south of Bath, for use in the cloth mills of Bradford-on-Avon, Frome, and Gloucestershire; but the Waverley Encyclopaedia (after 1880) says on the other hand that it has long been excavated at Midford for use in the mills at Bradford-on-Avon and Troubridge.

Now Trowbridge.

At Emborough (Embrough) near Radstock the *Ceradotus* Bone-bed (Penarth Group – Rhaetic) occurs as a five-inch bed resting directly upon the Tea Green Marls (late Trias) and was at one time exposed at the base of the Fuller's Earth workings there – not subsequently traced.

L. Richardson, *Quart J. geol. Soc.*, (1911), *67*, 68. John Vancouver may have used Rhaetic illitic clay in his substitute for soap: Chapter 19 (v). Re dating of the Rhaetic here, see *Nature*, (7.11.1985), *318*, 81–2

(The industry was noticed in Rutley's *Mineralogy* (1897) but not in Chambers's *Encyclopaedia*, 1895).

In 1883 a pit beside Hassage Hill, south of Wellow was marked on the 6″ Ordance Survey map and, by 1902, the second edition shows both this and a Fuller's Earth Works at Bath Hill House, NW of Wellow, as disused.

On the 23rd May 1883, Charles Richard Dames was granted a

22

Richard DAMES = Elizabeth
b. 1750 | b. 1751
d. 2.6.1821 | d. 20.3.1828
sugar cooper & refiner, Whitechapel.

Charles Richard DAMES = Elizabeth WALTON
b. 1793 | b. 19. 5.1793
d. 21.11.1861 | d. 30.10.1856
sugar refiner, Whitechapel.

Richard DAMES
bap. 31.8.1820
d. 5.3.1882
sugar refiner, married but no issue.

Charles DAMES = Ann GILBERTSON
bap. 12.6.1823 | b.
d. 12.6.1877 | d. 2.7.1893
tea dealer, Whitechapel d. at Chigwell, Essex. | father, a grocer in Hertford.

George DAMES = Elizabeth Fanny BOWERBANK
bap. 8.12.1826 | b. 1831
d. 15. 2.1861 | d. 7.3.1902
sugar refiner | father, a distiller at Islington.

Charles Richard DAMES = Catherine Eliza WHITE
b. 1849 at West Ham | b.
d. 2.2.1933 Tenby | d. 5.12.1925
Farmer, in 1878 auctioneer, valuer, cattle spice manufacturer, 1883 brewery manager, then fuller's earth purifier.

♀ ♀

George DAMES
b.
d. "before 1899" unmarried,

Baldwin Henry DAMES = Ethel Mary Harrison SLAUNTON
b. 1862 at Epping
d. 28.10.37 at Felpham, insurance

Richard DAMES =
b. 24.12.1859
d. 13. 9.1941
Army Captain

Mary Catherine Ann DAMES = Cuthbert Prideaux SELBY
b. 1901 | b. 1898
d. 1983 in Liss Hants | d. 1948

Ethyl Ann Slaunton DAMES
b. –.9.1899

Guy Maurice Fortescue DAMES =
b. 5.3.1888 at Torquay
d. 3.2.1972

Robert Ian Fortescue DAMES
b. 1925

Family tree of the Dames family

By courtesy of R. I. F. Dames

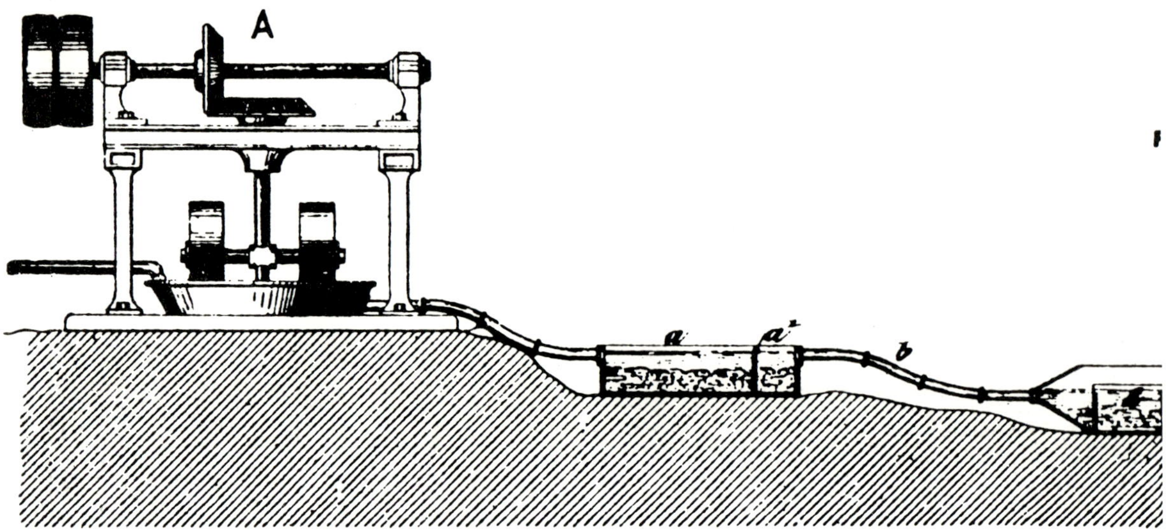

This page and opposite page. Fig. 62A (above) C. R. Dames' edge runner mill and catch pits. Fig. 62B (opposite page, top). A continuation of the same figure showing that launder with baffles. Fig. 63A (below) The concentration tanks. Fig. 63B (opposite, below) The drying shed.

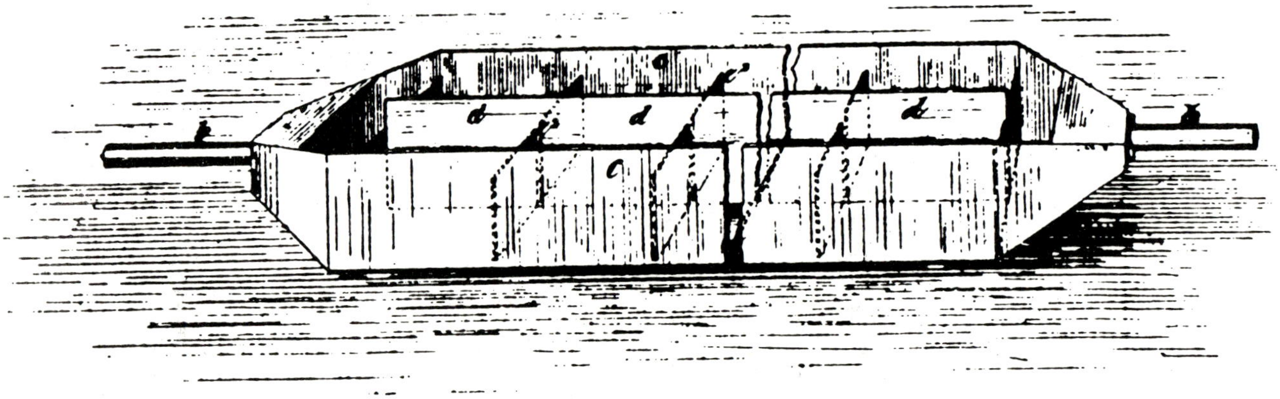

62A

c
b

AD 1883, 23rd May, No. 2582.

Patent for "improvements in the method of and machinery or appliances for the preparation and refining of Fuller's earth", freeing it of all extraneous matters, such as shells, stones, gravel and "insoluble lime".

Meaning undispersible calcite.

Bladud, 16 March 1887.

C. R. Dames of the City of Bath had already registered on 13th March the Midford Fuller's Earth Works Ltd. According to *Bladud*, the Bath Society Paper, *Mr. Dames' patent is the only one in existence by which the separation can be effectually accomplished and the commodity produced in a perfectly pure condition.*

Henry Newson Garrett's notepaper

Charles Richard was joined in the formation of the Company by his brothers George and Baldwin Henry. Adopting a *wet* process for purifying fuller's earth might have occurred to those descended from a long line of sugar refiners, as shown on their family tree, though the extraction of china clay would have given Charles Richard a clue. Anyway he and George went to live at Fern Cottage, Wells Road, Bath, and their names, as well as that of Henry Newson Garrett, Esq., appear first in the 1884–5 Post Office Directory for Bath. Midford Fuller's Earth Works Ltd., appears at the same time.

It is not known if Garrett was Dames' consulting engineer.

The fuller's earth was mined on the South Stoke Manor estate, not far from the Cross Keys Inn on Combe Down. The trucks were lowered towards Horsecombe Brook by self-acting incline. Here the clod was blunged with water in an edge-runner mill to form a slurry. The mill was operated by a steam engine, (though in 1904 another engine had been installed at the adit to the mine).

Called a "pug-mill" in *Bladud* (1887), rather inaccurately.

The slurry flowed from the mill A to a series of little catch-pits *a* close to the engine, so as to remove the largest particles. It then flowed down a sunken earthenware pipe *b* nearly a mile down the valley to the works at Tucking Mill. (Fig. 62A). (This pipe existed still in 1979).

See John Powell, Fullers Earth from Midford, *J. Bristol Industrial Archaeology Society* (1978), *11*, 22–25.

(The 25 in O.S. map of 1904, Somerset Sheet XIV, 13, shows the lay-out).

Here the slurry passed into a long trough called a "maggie" *c* where more grit was collected between the baffles e^2. The trough was also divided longitudinally *d*. Fig. 62B.

"Maggie" is not in O.E.D.

The purified fuller's earth slurry then flowed into one of four large settling tanks provided with a sluice-board and with a perpendicular row of two-inch diameter holes, closable by wooden pegs, *e*. The surface water, ("drinkable and very soft") was thus drawn off. Fig. 63A.

See figs., previous pages.

After about 30 days the concentrated paste was dug out of the tank leaving the finest particles of calcite at the bottom. These were thrown away.

The paste could not be dried on checker-plates as at Redhill because it would have passed through the holes. Dames' ingenious solution was to have a porous floor suspended over a hypocaust. Hot air passed along nine* wide flues, running from a furnace at one end to a tall chimney which stood at some distance from the drying shed, *f*. Fig. 63B.

* A photograph by J. Powell of the ruins shows only seven hot gas channels, 25. Feb. 1978. The chimney was demolished about 1975.

The top of the chimney was 300 ft from the furnace which was 15 ft wide and 8 ft deep and had 3 firedoors. Most of the moisture in the paste was drawn through the plates into the hot gases at low pressure, and the rest by boiling of the slurry itself.

In 1887 the dried earth was sacked and shipped by barge on its way to markets "all over the world". It was still mostly used for scouring and fulling of cloth, but also largely used for refining oil. It was also said to be valuable in dressing wounds.

See also *Bath (Midford) Refining Works*, Bath Society Paper, 16 March 1887.

In 1884 disaster struck, one of the tanks gave way at the side, the deluge doing damage estimated at about £500. In 1885 the company was wound up and sold to Mr. H. N. Garrett, formerly an Associate Member of the Institute of Civil Engineers, "under whose management the trade has wonderfully increased".

George Dames' future career is described under Surrey and Woburn, but Charles Richard Dames was at 12 Camden Crescent, Bath (1890–91 Directory) and at Middlesex House, Batheaston (1892–93). In 1895 he was secretary of the Northey Stone Company, 4 Old King St. His daughter was born in Bath in 1901. He had met his wife while hunting with the Beaufort and he retired from the commercial world. He died in 1933 at the age of 84 at his daughter's home in Tenby.

I am obliged to Mrs. Mary Selby (née Dames) for information about her father.

There had been another works beside the now derelict canal and railway; this was supplied from mines due north on the way to Hodshill, where Roman remains were found. This mine also belonged to Manor Farm estate, South Stoke.

G. W. Hignett, Manor Farm estate, 28 Feb. 1958.

Between 1883 and 1890 as many as six fuller's earth works were set up in the Bath area, such was the revival in demand. Competition was so acute that the prices were cut almost to an uneconomical level. Of these Cawley & Company owned in 1886 Wellow mine and works as well as Evergreen Works at Odd Down under the management of Sheppard Walter Allen, who became quarry master in 1884/5, and 'proprietor' in 1888/9.

Probably 'manager'.

Union of interests was imperative. When in 1890 The Fullers' Earth Union Limited was formed, it acquired or gained control by agreement of:

Among the Founder Shareholders, Theo Stretch introduced Samuel Butler and R. Taunton Raikes.

Works in Somerset and owners

Name of Works.	*Present Owners*	*Tenure*	*Remarks*
Combe Hay, Upper Works.	S. Butler	Freehold	Purchased
Combe Hay, Lower Works.	Major S. Evan Butler	"	"
Southstoke Works.	W.H. Handley	"	"
Evergreen Works, Odd Down	Cawley and Co. Ltd.	Leasehold	*
Wellow Works.	"	**	Purchased
Beds of Earth at Midford	G. Dames	Leasehold	"

* Purchased, works not now carried on [1890].
** Freehold, with power to mine under adjacent lands.

The Union also acquired quarries of Bath Stone, "valuable for building operations and commercial working", and "plantations admirably adapted for timbering of the Works."

The works were already equipped with modern machinery and had excellent arrangements for canal and railway haulage. The drying and milling capacity were increased to take advantage of the "insatiable" American demand.

When The Fullers' Earth Union had been formed, S.W. Allen left Evergreen, and became manager at Midford, and moved to Cleeve Hill House, Midford.

Bath P.O. Directory 1890–91.

Curiously, H.N. Garrett also worked fuller's earth at Wellow for a while as well as on Combe Down. The Time Book for Wellow begins on 15 December 1892. Some 218 tons of earth were produced in the first 4 months of 1893. The last entry was on 4 January 1894, when (Albert) James Palmer joined Samuel Menear at the Midford Works.

Leased presumably from The F.E.U.

Time Book owned by Alec S Palmer. His father James was appointed foreman 7 Apr. 1893.

Beside Bath Hill House, Wellow, the three cast iron fire doors to the drying shed may still be seen, though the chimney stack was removed in about 1964.

In 1899 Garrett paid £158.9.5½, in rent, royalties, and wayleave on 1646 tons. Bags were hauled to the Midford siding of the S. & D. J. Railway.

In 1895 the 'Babcock' [boiler] House was added to the works at Midford, and in 1898 the Somerset Coal Canal was closed; Midford fuller's earth was thereafter despatched by rail and by road. At the auction of Midford Castle on 26 June 1901, the reserve price for the Fuller's Earth Works was not met, but on 30 Dec. 1905 Garrett finally got full ownership. In 1907 Hy. Newson Garrett described himself as "Fuller's Earth Refiner and Rectifier" and seems to have recovered from the difficult times recorded in 1896.

Letter HNG to James Palmer 1 Jan 1907.

The advice of W. F. Keevil, who had joined one of the Somerset companies in 1883, was sought when The FEU was formed and, in 1904, he was manager of the works at Combe Hay and at Wellow. The final act of Garrett was the formation of H. N. Garrett (Bath) Ltd. to take over certain fuller's earth mines and works near Bath and *lately carried on* and owned by H. N. Garrett. This company was wound up on 9 August 1915, The FEU having taken over on 29 July and become the sole operators in the Bath area. In 1914 to 1915 a new Davey Paxman boiler was installed at Top Works; photographs exist of its erection and of the miners.

Registered 22.3.1913.

I am greatly indebted to Alec S. Palmer and to Dr. John Broome, the present tenant of Tucking Mill Cottage, for a large amount of information, of which this is but a meagre summary.

In 1916–17 milling was transferred to Combe Hay, but as the markets for rather pure fuller's earth could be supplied from Surrey there was less need to make a purified grade at Midford. Moreover one of the largest outlets for Bath earth was for mine dusting in the neighbouring coalfields. The foreman James Palmer was therefore required to mix the sedimented impurity with the purified fuller's earth to meet this 'low-grade' application. Palmer who had taken so much pride in operating Dames' process of purification found it unpalatable to accept these instructions and left in 1919. Nevertheless the wet process was not abandoned for many years though some clod was taken

The Bristol and Forest of Dean Coalfields – to check explosions.

Depreciation on the "Garrett Engine" was £379.4s.0d. annually from 1920; it is thought it was installed in the previous year.

by road from Top Works to Combe Hay. L. R. Cox wrote that the wet process, still in use in March 1940, was much the same as in 1893. The works were finally closed soon afterwards.

L. R. Cox, Easter Field Meeting 21–26 March 1940, *Proc. Geol. Assoc.*, *52*, 16–21.

In World War I the demand for Bath fuller's earth increased, especially for the fulling of woollen blankets and overcoat cloth.

In 1908; 3,757 tons were raised in the county; Gen. Rep. & Statistics Mines and Quarries, (1909), 172.

Production of Bath fuller's earth

Year	1913	1914	1915	1916	1917	1918	1919
Tons	3559	4256	4425	5577	4109	3783	3469

A carefully blended mixture of blue and yellow earths, milled and screened to a 200-mesh powder, "YB20", was popular among pharmaceutical suppliers from the 1890s to the 1960s at least. The yellow earth was worked intermittently just west of Vernham Wood.

In the 1920s and 1930s Combe Hay scarcely made a profit but The FEU kept it going in the interests of a faithful workforce. An opportunity to revive its fortunes arose when the laboratory at Redhill developed a process for making synthetic bentonite. Bath earth responded well to the treatment, so the first plant was located there. Until then some Bath earth was taken to Cockley acid-activation works for use in the preparation of several grades of bleaching earth. After 1942, when the Dunford & Elliott Rotary Dryer was installed, the Bath works had a new lease of life. Another fillip was when cat litter was made there.

25 April 1942 Bath Works damaged by bombs.

The Fullers' Earth Union Ltd. owned the land under their works, a house called The Firs, and a 200-acre triangular patch of land, which provided most of the feed to the works. The rest of the earth-bearing land belongs to the Duchy of Cornwall, but is difficult to work because of faulting. The royalty was one shilling in 1950 and 1s 2d in 1958.

5p and ~6p.

When the Duke of Cornwall visited his lands around Bath in 1970, The FEU General Works Manager, Derek Oliver, was presented to him. The Combe Hay Works were closed down in 1979, because the cost of mining had risen to an uneconomic level.

Bath Chronicle, 26 July 1979.

As Michael Flanders and Donald Swann sang in their song "Greensleeves" **"And royalties went to Royalty."**

23

WOBURN

Gums must be stor'd of Guinea's arid coast;
Mexican woods, and India's brightening salts;
Fruits, herbage, sulphurs, minerals, to stain
The fleece prepar'd, which oil-imbibing earth
of Wooburn blanches, and keen alum waves intenerate.

= soften

JOHN DYER, The Fleece, 1757.

Warren R. Dawson, *Bletchley Gazette*, (3/5/1947) and *Bucks Standard*, (14/7/1951)

Now Honesty End.

FULLER'S earth was worked for centuries in the parish of Wavendon. Wavendon, like many other Buckinghamshire parishes, was divided into districts or end-ships, amongst which were Church End, Duck End, and Hogsty End. Hogsty End is that part of Wavendon Parish which was afterwards detached and incorporated in Woburn Sands.

[No *t* in the name from 12th to 16th century]

Just as the Surrey fuller's earth occurs near the village of Blechingley, the meaning of which may be taken as Bleaching-ley or -field, the Woburn fuller's earth lies scarcely five miles from Bletchley. Indeed the Roman station of Magiovinium may have been sited near Little Brickhill on Watling Street so as to provide a laundry service for the troops.

The Woburn Sands and Aspley fuller's earth deposits are now in the county of Bedfordshire.

The earliest written reference to fuller's earth in Britain is in the Inquisition of the Manor of Aspley, 7th February 1295:

Quoted in *A history of our district*, by the Aspley Heath School Historical Society. Aspley Guise: The Powage Press, (1931), 30. (Based on research by W. F. Cooper).

there are there forty customary (tenants) and two cottagers of which each will mow the lord's meadow for one day with one man, and each mowing is worth 4d	*14s*	*0d*
each will hoe for one day with one man, and a hoeing is worth ½d		*21d*
each of them will cart for one day the lord's hay or (?) fuller's earth at the lord's food, and each work is worth 1d	*3s*	*6d*

It is not until 1539 that the fuller's earth here is mentioned again, in a deed of Henry VIII. The workings were almost certainly at the outcrop at that time, between Aspley Guise and Woburn Sands.

In the parish of Wavendon, the Manor of Hugh de Bolebec was later in the hands of the Earl of Oxford and afterwards the Courtenays, from whom it passed to John Sheppard in 1557. Some of the pits of fuller's earth are mentioned in the grant on the estate.

Owing to its inland situation and the poor state of the roads, the

Woburn fuller's earth was never extensively worked in early days, though a pamphleteer in 1630 wrote that apart from the better known deposits in Kent, the only other one known to him was at Wooburn in Bedfordshire. However the Merchant Adventurers reminded the clothiers of the kingdom in 1639: *as the clothiers still suppose there is no other earth nearer than Kent*, they are *put in mind that Wooburn in the county of Bedfordshire is the best in all men's judgement in England*. The *Philosophical Transactions* of 1684 says *there is fuller's earth of a yellowish colour at Brickhill in Northamptonshire* (sic); the place was correctly given in 1723.

V.C.H., Kent, (1932), *3*, 396, quoting Travers *New Year's Gift to England*, 15.

V.C.H., Kent, (1932), *3*, 396, quoting *Cal. S.P. Dom.*, (1639), 489.

Error for Buckinghamshire.

VI. *An Account of the Pits for* Fullers-Earth *in* Bedfordshire; *in a Letter from the Reverend Mr.* B. Holloway, *F. R. S. to Dr.* Woodward, *Pr. Med.* Gresh. *S. R. & Coll. Med.* Lond. *Soc.*

BEDFORD, 6 *July*, 1723.

Transactions of the Royal Society, (1723), *32*, 419.

Holloway said there were several *Fullers-Earth Pits* then open at Wavendon near Woburn, but men were at work only in one when he visited the place. The seam was approached vertically, first through reddish sands to a stratum of red sand-stone which had to be broken through before dropping a further 7 or 8 yards to the top of the fuller's earth; the following varieties being recognised: the top foot, called cledge, is contaminated with sand and thrown away; then an inch of clay looking like Terra Japonica. The lower part of the seam is not tinged with red, and seems to be the more pure and fitter for fulling. Sand occurs below the seam but the men seldom penetrate as far.

A gum, the inspissated juice of *Mimosa catechu*.

He recognised that the Cretaceous rocks could be traced for many miles and that these occurrences confirmed what Woodward had said about the regular disposition of strata over large tracts of country. He suggested that other occurrences of fuller's earth ought to be found elsewhere along the outcrop. This is an early instance of the application of stratigraphy for economic purposes.

V. A. Eyles, *Toward a History of Geology*, ed. Cecil J. Schneer, MIT Press (1969) 177.

John Woodward, physician and geologist, to whom Holloway's letter was addressed, was the first British author to publish a work solely devoted to mineralogy, 1729. Woodward said "The fuller's earth of England is very various, and it very much exceeds any yet discovered abroad in goodness, which is one great reason why the English surpass all other nations in the woollen manufacture," – an exaggeration, it is now known.

John Woodward, *An Attempt towards a Natural History of the Fossils of England*, (1729), Page 4, Sect. 1, Art, 1, a10 and a12.

The Woodwardian Collection which still exists in Cambridge in splendid cabinets, contains a sample of "*Fullers-Earth*, of a grey colour,

John Woodward, F.R.S. (1665–1728), V. A. Eyles, *Nature*, (29 May 1965), *206*, 868–870 with a Cast of greenish yellow. From *Wooburn* in *Bedfordshire.*" It is interesting because it shows current bedding, evidence of its derivation from volcanic ash. This was illustrated in an article on the origin of fuller's earth by Cowperthwaite *et al.* (1972). The sample preserved at Cambridge was possibly sent to Woodward by Holloway, shortly before the collection was complete.

Sedimentation, petrogenesis, & radioisotopic age of the Cretaceous f.e. of southern England, I. A. Cowperthwaite, F. J. Fitch, J. A. Miller, J. G. Mitchell & R. H. S. Robertson, *Clay Minerals*, (1972), *9*, 309.

W. R. Dawson, *verbatim.*

Thomas Pennant, the well-known naturalist, author and traveller, in his *Journey from Chester to London* (1782), speaks thus of the Wavendon fuller's earth pits:

> *Near the road side are the noted pits of fuller's earth, that invaluable substance which is supposed to give the great superiority to the British cloth (honestly worked) over that of other nations.*

Pennant then makes a précis of Holloway's paper, but adds:-

> *The great use of this earth is cleansing the cloth, or imbibing the tar, grease and tallow, which are so frequently employed by the shepherds, in healing the external diseases which sheep are liable to; neither can the wool be worked, spun or woven, unless it be well greased. All this grease must be gotten out, before the cloths are fit to wear. Other countries either want this species of earth, or have it in less perfection. The British legislature therefore have, from the days of Charles I, guarded against the exportation of it under severe penalties.*

In Pennant's time, the working of fuller's earth at Wavendon was evidently a going concern, but increase in the depth of overburden brought this phase of the industry to a standstill. At what date the "earth wells" were introduced is unknown.

Magna Britannica, being a concise topographical account of the several counties of Great Britain, by the Rev. Daniel Lyson and Samuel Lyson in London: T. Cadell & W. Davies in the Strand, (1806) *1*, 477–478. Under Buckinghamshire.

Nevertheless the Lysons (1806), quoting Pennant, write of the celebrated fuller's earth pits at Wavendon, one of which is only occasionally worked, in a close shaft. On the other hand the pits at Aspley Guise had not been worked for many years. The sale of the earth even from the Wavendon *pits has of late much diminished, the dealers got into a practise of procuring earth of an inferior quality from other parts of the Kingdom, which sell as the produce of this neighbourhood.*

Manning & Bray, *The history of Surrey*, (1809), *2*, London: J. W. Parker.

J. Sowerby, *Mineralogy*, (1809), *3*, 59.

Surrey.

English Topography, London: Baldwin, Craddock & Joy (1816).

Quoted by A. C. G. Cameron below.

The district however remained one of the recognised sources of fuller's earth in England; the Third Edition of the Encyclopaedia Britannica (1797–1810) and Manning and Bray (1809) record it as occurring at Aspley near Woburn in Bedfordshire, J. Sowerby (1809) says that Bedfordshire and Hampshire are the only two places *mentioned by most authors where fuller's earth occurs* but elsewhere give details of the earths of Surrey and Kent. J. Nightingale (1816) explains that the fuller's earth procured at Wavendon on the borders of Woburn is dug in Buckinghamshire not Bedfordshire as other authors state. However, Phillips and Conybeare wrote in 1822 that fuller's earth "was once extensively dug in Bedfordshire" – so operations had presumably ceased.

The Reverend John McArthur, Minister of Kilcalmonell and Kilberry, Argyll, upon enquiry learned that on the Duke of Bedford's property fuller's earth *has been of great value as an article of commerce, and considerable quantities have been exported to foreign countries,* (1845).

New statistical account of Scotland, (1845), 7, 422.

The Reverend J. R. Jackson also records the working of fuller's earth at Woburn in his "Minerals and their uses, in a series of letters to a lady" (1849).

The quality seems to have been good, for the Encyclopaedia Britannica's eighth edition in 1861 states that the best fuller's earth comes from here and Surrey. Des Cloizeaux considered it important enough to include among his list of nineteen occurrences of fuller's earth in the world (1862.)

It was not until 1885 that further information was given by A. C. G. Cameron of the Geological Survey. He maintained that the fuller's earth which is dug from near the base of the Bedfordshire Greensand is of superior quality, and characterises it as a purifier of water. Fuller's earth was at that time worked at Aspley Heath where cylindrical holes, called 'earth wells', were dug deep in the sandy rocks to reach the fuller's earth, which occurs as a nearly horizontal stratum not far above the Oxford clay.

British Association Proceedings (1885), (13).

See Water-softening properties and cation-exchange, chapter 59.

H. B. Woodward (1887) said that the fuller's earth occurs rather below the middle part of the Lower Greensand, as follows:-

sand	130 feet
fuller's earth	12 feet
sand	80 feet.

H. B. Woodward, *The Geology of England and Wales*, (1887).

George Dames, who had been Secretary of The Fullers' Earth Union for about a year, decided, probably early in 1891, to reinforce the efforts of the British Fullers' Earth Co. Ltd., founded by Mr. A. J. G. Swinney, a mining engineer, and Mr. George Edward Dixon, India Merchant of Leeds. These two joined George Dames of Woburn Sands on 23rd July 1891 in forming the Fuller's Earth Mining Co. Ltd. Thomas Riley, a traveller from Leeds, may have introduced the parties, and was supported by Sam K. Riley, Boot & Shoe Engineer. Between them they raised £5000 to buy as a going concern Dixon and Swinney's business, which consisted of one mine on the edge of Aspley Heath and a drying and grinding works beside the railway station at Woburn Sands where a "Niagara" mill had been installed.

Swinney was "removed from office as Director" 23 Jan. 1892 but continued to work the lower mine (as on 25 June 1892, below).

Riley bought fuller's earth also from the Midford Works. *Letter* from H. N. Garrett to J. Palmer 1. Jan. 1907: *Mr Riley has written wishing to be remembered to you for the New Year.*

The Fuller's Earth Mining Company was a more impressive undertaking. The clay was hauled up to the works in little trucks, dried and then ground in a "Cyclone" or pulverising mill. The whole plant was in operation before the end of 1891, according to A. C. G. Cameron, who led a party of the Geologists' Association to both works on 25th June 1892, Fig. 64. The well-timbered mine was lit by many candles to let them see the earth *in situ*.

Both works were on land known as Ireland's Ground (Auctioneer's catalogue of sale of the works on 22 June 1900).

Excursion to Woburn Sands and Sandy, *Proc. Geol. Assoc.*, (1892), *12*, 395–403.

George Dames said that fuller's earth in boiler water prevents the growth of incrustation. *It is well known to every housewife that boiling a lump of fuller's earth softens the hard crust in a kettle so that it can be easily removed*

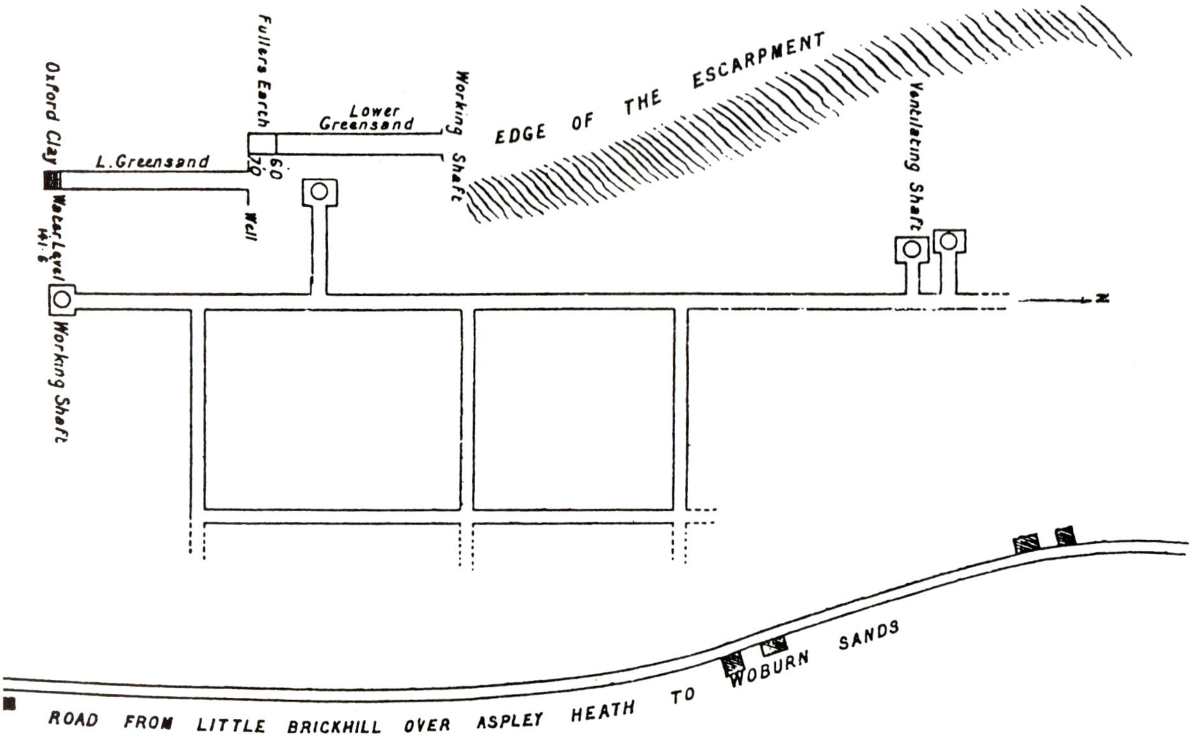

Fig. 64. Sketch plan of underground works, The Fuller's Earth Mining Co. (Limited), Aspley Heath, Woburn Sands.

A.G.C. Cameron (1892)
Scale 70 feet = 1 inch.

merely by rubbing. The textile market had declined but Woburn fuller's earth was already being exported to America to decolorise cottonseed oil. *The artificial preparation of ultramarine, paper making, the extraction of grease spots from wall papers, scouring butter boards, hearthstones and the like, are all facilitated by the use of fuller's earth. Therapeutically, besides being used externally, it is even swallowed or taken as snuff. It could be mixed with blacking in boot polish.*

Cameron returning to the subject in 1893–4 maintained that *a great deal of the fullers' earth of commerce comes at present from Woburn Sands.* Indeed the industry there employed some hundred men. He also referred to a curious feature of the stratum of fuller's earth, though fairly horizontal – it rests on undulating ridges and furrows. These were later revealed in an opencast mine near Aspley Lane, Woburn, and mapped by I. A. Cowperthwaite, Fig. 65.

By 1892 "hundreds of tons of fuller's earth" [a year] were shipped at Liverpool for New York.
"There is now an almost phenomenal demand for it", in reference below.

Trans. Fed. Inst. Min. Eng., (1893–4), *6*, 204–209.

Sedimentation, petrogenesis and radioisotopic age of the Cretaceous fuller's earth of southern England, I. A. Cowperthwaite *et al.*, *Clay Min.*, (1972), *9*, 309.

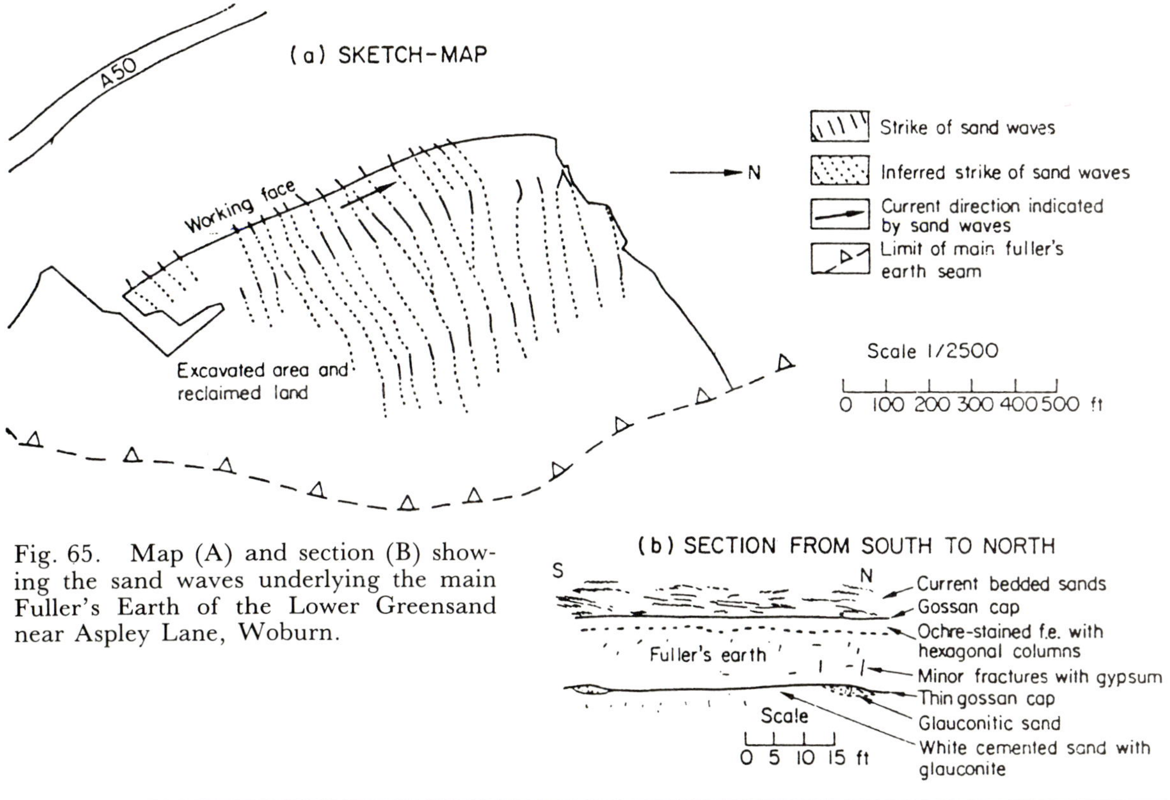

Fig. 65. Map (A) and section (B) showing the sand waves underlying the main Fuller's Earth of the Lower Greensand near Aspley Lane, Woburn.

The Fuller's Earth Mining Co. was wound up in 1896 for reasons which are not known. Can it be that George Dames had fallen ill? All that is known is that he died some time before September 1899 and by order of the High Court of Justice these noted fuller's earth works were offered for sale on the 18th September 1900, *as a going concern, in one lot.*

Mining by the new owner continued fitfully until World War I; at least 590 tons were got in 1915, but all activity had ceased by 1918.

"Uncle George died before I was born" said Miss E. A. S. Dames, born Sept. 1899.

Victoria County History, Bedfordshire (1904). The area sold was 17 acres 3 roods 17 perches.

R. H. Rastall, Mineral composition of the Lower Greensand and strata of Eastern England, *Geol. Mag.*, (1919), 265.

Summary of Progress for 1934, Geological Survey (1935), 48.

James A. Purcell of the *Woburn Reporter* office kindly supplied much information for this chapter.

Berk's £300,000 development at Woburn, *Foundry Trade Journal*, (28 May 1964), 667.

Fullers Earth Lodge is near the entrance to Aspley Lane from B528.

Rastall notes that 6 feet of fuller's earth occurs near the top of the Parish Sand Pit at Aspley Guise (quoting from Cameron); and in 1934 a further exposure was reported by E. E. L. Dixon at Clophill, about 5 miles east of Ampthill. This was acquired by The Fuller's Earth Union Ltd, which had workings in that area in the 1950s.

Fuller's earth workings, started by F.W. Berk and Co. Ltd., were in Tidbery Fields at Aspley Heath, not far from the Iron Age fortifications known as Danesborough. The overburden was about 22 feet. Digging and carting ran from early May 1951 to February 1952, but was resumed in 1956 when their associated company, the St Albans Sand & Gravel Co., extracted over the next few years some 50,000 tons for processing at London Colney. Berks started a much larger operation in Aspley Wood in 1963 and when supplies were running out about 1980 their successors Steetley Minerals Ltd started a spectacular open cast mine on the west side of the Woburn – Woburn Sands Road.

Other points of interest are the discovery of a Roman coin at Aspley Guise, the name *Ful*brook Secondary School at Woburn Sands, Fullers Earth Lodge and Kiln Ground, the last being 1½ miles SW of Woburn Sands on the Little Brickhill road.

GLOUCESTERSHIRE

24

> ***The advantages of its situation are a bed of Fuller's Earth, a copious water supply and plenty of game.***
> **CHEDWORTH ROMAN VILLA, National Trust, (1953), p. 5.**

Discussed in Ch. 8 ii.

BATHONIAN fuller's earth extends at intervals for 50 miles from the Avon Valley at Bath to the Vale of Morton at Stow-on-the-Wold, though there are few good exposures. The clay becomes more calcareous towards the north and the quality deteriorates beyond Sodbury.

A rather gritty fuller's earth could be used in a small mechanical fulling machine, especially if the grit were allowed to settle out. We have already noted that the earliest mill of this sort was set up at Barton in 1175 or so; but it is safe to assume that foot fulling continued in Gloucestershire for a long time after that. The impure fuller's earth was satisfactory when the cloth was not heavily beaten, as in the more recent process.

In the Stroud area, the Minchinhampton Custumal of the 13th century lists one William Bigge, who, for a water-course and fuller's earth delvings and for a certain site between the streams, paid of 16d per annum.

My thanks go to Lionel F. J. Walrond, A.M.A., Curator, Stroud & District Museum, Lansdown, Stroud, GL5 1BB, for all this information. A custumal is a written record of the customs of a place.

Other rentals per term for digging fuller's earth included

Roger Fuller of Woodchester	1d	John de Dodebrigge	1d
Richard Fuller	1d	Robert de Bristol	1d
William Hockvale	1d	Reginald Fuller	3d
Simon Fuller	2d		

As this was the time when surnames were being fixed, it is significant that four out of the eight who paid to dig fuller's earth bore the name of Fuller. The linguistic boundary north of Bath has already been alluded to under Somerset.

Chapter 22.

A record that John Clutterbuck, a name of Flemish origin, in 1514, and Richard Clutterbuck and John Haynes in 1556, were 'digging fuller's earth' at Eastington presents difficulties because this village lies on the Lias Clay. Could it be that they had a licence to dig fuller's earth where it occurred a few miles away? The same comment would apply to William Harmer in 1533 who was digging fuller's earth at King's Stanley, but he was even closer to the mineral outcrop than the Clutterbucks. All the parties were of clothier families and would have known

That is one rich in calcium montmorillonite.

R. Cave, Geology of the Malmesbury district (Sheet 251, New Series), Inst. Geol. Sci., (1978).

J. Woodward, *op. cit.*, Title 3, p. 2, no. 2.6.

See H. Torrens, *op. cit.* quoting G. Household, (1969). *The Thames and Severn Canal.* Newton Abbot: David & Charles.

Railway Magazine, (18 Feb. 1950), reproduced the diagram.

Waverley Encyclopaedia, (no date).

A. B. Searle, *An Introduction to British Clays, Shales and Sands*, (1912), pp. 73 and 190.

"Hewitt's Henbury Decolouring Earth", brochure, C.H. Hewitt, Henbury Hill, Henbury, Bristol, c. 1950.

of the superior quality of a true fuller's earth. It is hard to believe that any cloth worker of that period would have used Lias clay in place of fuller's earth which was locally obtainable from Brimscombe and other places near Minchinhampton. Yet it is a fact that other licences were given to people living in the Severn Plain in the 15th and 16th centuries. It is also known that the Lower Fuller's Earth around Nailsworth, in the same area, was at one time dug for local use for fulling, though the earth is too impure for commercial exploitation today.

A Fullers-Earth; out of a Coal-pit, on Broad-Mow-Green, near Little Dean, Forest of Dean, was probably misidentified by John Woodward.

The Sapperton canal tunnel on the Thames and Severn canal became distorted where it passed throught the plastic Fuller's Earth Series. In 1790 leakages occurred there only a year after the tunnel was opened. We have seen how William Smith had advised against building a caisson in similar clay at Midford (1799).

Two bands of true fuller's earth were found when the Box railway tunnel was cut about 1850. A diagram is given in the report by Isambard Kingdom Brunel, engineer of the Great Western Railway, on the section from Kemble to Stroud.

Fuller's earth was also recorded as occurring in this county about 1900, and Searle (1912) wrote that *in parts of central Gloucestershire "Fullers clays" lie over the solid strata of white limestone.* The only time it was ever worked commercially was for a short while in 1919, when 495 tons were prepared in the Golden Valley. Production soon ceased because requirements dropped after the wartime peak of 1916.

About 1950 Mr. Charles H. Hewitt was drying and milling for decolorising vegetable oils a non-smectitic clay at Wick. Although it consisted of illite and kaolinite and a smaller proportion of calcite, this clay was found to be almost as effective as an acid-activated montmorillonite in bleaching vegetable oils, the recommended dosage being:

0.5% for palm kernel and copra oils.
0.75%–1% for groundnut oil.
1–2% for cottonseed oil.
1–3% for olive oil, all treated at 180–190°F.

Linseed oil had to be contacted with the powder at 200°F. the earth preferentially removed the red colour and caused the yellow colour to be removed on subsequent deodorisation. The oil retained in the dry steamed filter cake was as low as 15–16%. The clay had no effect on palm oil but was good for removing chlorophyll from gut tallow.

BERKSHIRE, now OXFORDSHIRE

25

Baulking baulked; Brett bentonite unbaulked
HEADLINES, Industrial Minerals, Sept. 1975, p. 10 and Oct. 1976.

ON 24 November 1969 the Natural Environment Research Council announced the discovery of fuller's earth of Lower Greensand age near the villages of Baulking and Fernham in Berkshire. The former was described by E. G. Poole, B. Kelk, J. A. Bain, D. J. Morgan and G. N. McKissock, and the latter by the first four authors with H. J. E. Haggard and B. R. Young, in 1970. (A bed of fuller's earth, said to have been found in 1872 in the Reading Beds at Katesgrove Kiln, Reading, has not been confirmed).

Press notice, *Financial Times* and other papers alluded to this, e.g. *Ind. Min.* (Jan. 1970), 37.

Inst. Geol. Sci., Reports No. 71/4 and 71/12.

W. Whitaker, *Mem. Geol. Surv.*, (1872), *4*, 605.

S. G. Henderson tried to persuade the many landowners and tenants to quarry the deposit themselves. Naturally enough, The FEU or, rather, Laporte, was also keenly interested and Derek Oliver, the General Works Manager for fuller's earth, was a frequent visitor to the two sites while drilling was in progress. It seems, however, that an unrealistic impression of the value of the earth had sprung up, a local newspaper, for example, publishing an article saying *this clay may be worth £60 million*. In the end, more sober opinions prevailing, Robert Brett & Sons Ltd. acquired the right to work but not the ownership of the mineral rights and not without a struggle at two planning enquiries, the second being an appeal against a refusal.

These long and expensive procedures were wrily described by Desmond Siberry (1982). With Henderson's help, Brett Bentonite actively developed the deposit. J. A. Bain confirmed in the laboratory that sodium-exchanged Baulking bentonite had a higher position on the Plastic Limit/Plastic Index plot than other treated clays or even than Wyoming bentonite itself. Chemical and thermal analyses were given by Robertson (1980), who found that there was almost no substitution by aluminium in the tetrahedral layer of the calcium montmorillonite.

Industrial Minerals (Aug 1978) 11; (Mar. 1980) 14; (Feb. 1982) 49...

J. A. Bain, *Clay Minerals*, (1971), *9(1)*, 1–17.

R. H. S. Robertson, *Mineralogia Polonica*, (1980), *11(1)*, 6–13.

When the county boundaries were redrawn, the area was allocated to Oxfordshire (1974); and in 1983/4 Laporte acquired a 75 per cent holding in Brett Bentonite Ltd. and BRETTBOND.

26 OXFORDSHIRE

To render the surface completely puddled, to use a term employed in ground works, and thereby retain water equally with any clay.
R. W. DICKSON, Practical Agriculture, (1805), *1*, 157

Partly from *The Oxford Region*, published for the British Association, (1954), p. 115.

TWO of the earliest water-mills adapted for fulling cloth at Barton and at Winchcombe Abbey are on or close to the Great oolite, which even here may have clays rich in montmorillonite. Both mills were at work by 1185–1195. The fulling mill at Bourton-on-the-Water (1206) was also at or close to the same horizon. Brightwell on the other hand (1208) is close to the Lower Greensand, but Witney, which had two such mills by 1223, and Bletchingdon, are on the Forest Marble, in which a calcitic fuller's earth occurs. All of these early mills were located where water power and usable fuller's earth occurred.

Geology of the country around Witney, *Geol. Surv. Explan.* Sheet 236 (1946), 67–69, 135.

W. S. Horton, *Geologist*, (1860), *3*, 249.

At Witney flagstones and building blocks would have been quarried at the same sites as the calcitic fuller's earth. North-east of the cemetery in some large quarries, now largely overgrown, the Wychwood Beds clay *is divided about the centre by a band of slaty sandstone, which, although not equal to the Stonesfield slate . . . is capable of being worked by the hammer into a rough kind of roofing tile.*

The Gibraltar quarry, in the parish of Bletchingdon, in the Kemble Beds – the lower part of the Forest Marble, yielded stone for ornamental chimney-pieces and pavements, as well as tombstones. Another small quarry closer to the Rock of Gibraltar inn, provided puddling clay and rock lining for the construction of the Birmingham-Oxford Canal. The Great Siege of Gibraltar was raised in 1783 causing such feelings of relief and joy that the name was attached to quarries, streets, pubs and so on all over England. The quarry was therefore in use some time after 1783. In another exposure nearby Henderson and I took a sample of the clay which contained about half and half montmorillonite and calcite.

N. G. Gedye, *Encyclopaedia Britannica*, 14th ed., (1929), *4*, 720; and O.E.D.

It was James Brindley who was engineer to the Worsley to Manchester coal canal (1759–1761) and who first used the word puddling in 1762 to mean 'the process of converting clay, etc., into puddle, or of lining or covering something with puddle to make it water-tight'.

Transactions of the Society of Arts, (1796), *14*, 238, 239, 240.

The use of puddling clay in civil engineering was described in some detail in 1796. William Smith would have known in 1799 that the Lower Fuller's Earth Clay was good for puddling. John S. Macaulay referred in 1834–47 to the construction of the interior of a dam with well-tempered clay, called puddling; and as late as 1861 Samuel Smiles refers to "the former practice of working the new layer of puddling stuff as to unite it with the stratum immediately beneath."

John S. Macaulay, *A Treatise on Field Fortification*, (1851) 125.

S. Smiles, *Lives of the Engineers*, (1861), *1*, 35, 353.

Today bentonite is commonly used for making lakes, ponds and irrigation channels watertight and for grouting permeable ground. Perhaps the art of working cheaper smectitic clays into porous soils by puddling could be revived.

See chapter 61.

Cloth-making in Oxfordshire

BY 1954 the Witney blanket industry was one of the two survivors in the region of the once-important Cotswold woollen manufacture, Early's of Witney being established in 1669. In 1954 the factories in the town employed 1200 workers and made about half a million blankets a year, almost a fifth of the British output – before the competition of the duvet.

The Oxford region, published for the British Association, (1954), p. 115.

The making of tweeds, fine flannels, and suitings still flourished at Chipping Norton. Elsewhere the textile trade had died out, including the worsted plush manufacture at Banbury and carpet-making at Abingdon. The last to close, at the beginning of World War II, was the plush-mill at Shutford, near Banbury, a town once famous for its cloths, certainly well fulled and used for livery uniforms in many European courts.

id., p. 143.

The fuller's earth stated by Searle (1912) as occurring "near Oxford" has not been identified.

A. B. Searle, *An introduction to British Clays, Shales and Sands*, (1912).

27

HAMPSHIRE

Isle of Wight

> ***A great variety of soils – marl both shell and stone, chalk, fuller's – and brick – earth, tobacco-pipe clay, stone of different qualities and white sand.***
> **REV. MR. WARNER, General View of the Agriculture of the Isle of Wight, (1794).**

J. Nightingale, *English Topography*, London: Baldwin, Cradock & Joy, (1816).

A WRITER of 1816 states more specifically

on the north side of this stratum (of coal at the foot of Brimbridge (Bembridge) Cliff, and running through the southern parts of the isle, reappearing at Warden Lodge, in Freshwater parish) lies a vein of sand, and another of fuller's earth.

A thin seam of fuller's earth here is said to be the cause of landslides and to be obscured thereby.

Headley

> ***The fuller's earth of Hampshire, England, so much celebrated, is of a greenish-yellow, tolerably firm, crumbles easily in water, receives a polish from the finger nail, and is very powerfully detergent.***
> **BENJAMIN SILLIMAN, Sr, *Amer. J. Sci.*, 1st series, 1, *2*, 217.**

Quoted by D. T. Day.

T. O. Bergman, 9.3.1735–8.7.1784.

Richard Kirwan, *Elements of Mineralogy*, 2nd ed., London: printed by J. Nicols for P. Elmsley, (1794), 184–186.

HAMPSHIRE fuller's earth is famous because it was the first one to be chemically analysed – by Torbern Olaf Bergman no less – in 1782. Details of his analysis are given in a later section (chapter 70 iv).

Bergman gave the best description so far of a fuller's earth, noticing its brown colour, with a scarce perceptible shade of green, and streaked with pale yellowish veins. *The great use of this earth is to deprive the cloths of greasy substances with which they are impregnated, before they are soaped.*

Benjamin Silliman, appointed to the first Chair in chemistry at Yale in 1802, visited Britain, studied pure chemistry under Thomas Hope and acquired practical skills under Friederich Accum in his London laboratory about 1805. It may have been at this time that he obtained his specimen of fuller's earth which he used as a reference when examining a clay, near the town of Kent on the Housatonic River in Connecticut, which he inspected in 1820 and considered very closely resembled the Hampshire fuller's earth.

B. Silliman, (1779–1864), *Amer. J. Sci.*, 1st Series, I, *2*, 217.

John Kidd, M.D., Professor of Chemistry at the University of Oxford, quotes Bergman and adds *fuller's earth is met with . . . in very great abundance in different parts of England, as at Woburn, Ryegate, and Maidstone. A great proportion of the fuller's earth of commerce comes from Hampshire.*

John Kidd, *Outlines of Mineralogy*, printed at Oxford for Longman, Hurst, Rees and Orme, & Messrs Rivington, London (1809) *1*, 176.

James Sowerby admits (also in 1809) that Hampshire and Bedfordshire were the only two places in England mentioned by most authors as a source of fuller's earth, but he comments that *it is remarkable, that of many acquaintances of whom I have inquired there is not one who can find out any place in Hampshire where it is found, and I suspect something wrong in all that quote Bergman's analyses*. Among these acquaintances who influenced his judgement were Dr. Thomson and Mr. Hatchett.

James Sowerby, *British Mineralogy*, London: Richard Taylor & Co., (1809), *3*, 59, 60.

See chapter 35 on Prof. Thomas Thomson.

Admittedly, Nightingale (1816) wrote that there are no manufactures of any consequence in this county. The few that are consist chiefly of shalloons and coarse woollens, though malt was made in pretty large quantities. The *Victoria County History* stated that fuller's earth has been dug at Fuller's Bottom, now more delicately called Fuller's Vale, near Headley, but the industry ceased with the decline of the cloth trade, for which it was much in requisition, though in the 18th century fulling, paper-making, tanning and brewing were carried out extensively in Alton.

J. Nightingale, *English Topography*, (1816), London: printed for Baldwin, Craddock & Joy, Hampshire.
Shalloons, closely woven woollen cloth chiefly used for linings.
V.C.H., Hampshire, London: Constable, (1920), p. 453.

Much earlier there are more numerous references to the fulling industry. Thus in Headley (in Bishop's Sutton Hundred) five mills were mentioned in a perambulation of the parish taken in the reign of Edward VI; these included a fulling mill and a water course held by Thomas Fygg; and a messuage and fulling mill abutting on Lacyes Marsh. *Another waste (marsh) lies at the west of Erthpytlane*: this does rather suggest the topography of Fuller's Bottom. And at Binsted, in 1611 we have mention of a messuage called "Teaselies" or "Theselies."

V.C.H., Hampshire (1908), *3*, 51–52.

Ed. VI (1546–1553) *Eccl. Com. Ct. R., bdle* 136, No. 1.

Add. Chart. 8,503 and 8,504.

According to Andrew Ure (1839) fuller's earth is in greatest abundance in Bedfordshire, Berkshire, Hampshire and Surrey; but we cannot assume that the quarries were still operated at that date. What we do know is that the provenance of Hampshire fuller's earth was completely forgotten for many years.

A. Ure, *Dictionary of Arts, Manufactures & Mines* (1839), 529–530.

Topley (1875), however, referred to fuller's earth at Alton, Milford, and further north at Linchmere and Artington; the railway cutting at Petersfield showed, beneath the Folkestone beds, 2 ft. of yellow clay, but at Headley a greenish clayey sand could be seen then – as now.

F. Drew in W. Topley, *Geology of the Weald*, (1875). 1 mile W of Milford House.

27

T. W. Shore (1890) was "getting warm" when he identified as fuller's earth a sample of clay at 155 m AOD (above sea level) in a well on Greyshott Down. He also quoted records which showed that fuller's earth was formerly worked at Fuller's Bottom. Strangely, neither Shore nor Brierley (1871), who analysed the well sample, described the actual quarries although they did say that the earth lay between 90 and 120m AOD. For the analysis see chapter 35i.

T. W. Shore, *Hampshire Field Clay, Papers & Proceedings* (1890) *1* (4), 33–45.

J. Brierley, *id.*, (1891), *2*, 84–88.

A. C. G. Cameron of the Geological Survey stated that fuller's earth was at one time dug where the Lower Greensand comes to the surface in the eastern point of Hampshire. This may well be true, since old, shallow workings would no longer be recognisable where so many houses have been built in Bordon. Nevertheless the claim of Fuller's Vale is still attractive.

A. C. G. Cameron, *Trans. Fed. Instn. Min. Eng.*

Ian A. Cowperthwaite and I (1981) were impressed by the size of the quarry workings north of Rose Cottage and on later occasions I showed that the quarry runs between 100 and 107 m above sea level, and was probably entered near the cross roads at the head of the valley, 1400 m east of Headley church. The quarry is about 600 m long and still 5 m deep in places. These quarries and others on the other side of the road suggest quite important workings. In a recent remapping for the Haslemere Sheet the British Geological Survey found by repeated drilling no evidence at all of fuller's earth.

Visit (8. Apr. 1981).

Thanks to friends who drove me there.

The I.G.S.

Hampshire fuller's earth, if the provenance were correctly reported, could however have come from Bordon Camp where the seam in the Lower Greensand, now built on, was thick enough to be of local use. All the same an occurrence of altered volcanic ash in the Hythe Beds is not an impossibility since volcanic ash fell on Southern England at various times in the late Jurassic and early Cretaceous. The precise dating of these episodes has not yet been achieved. According to Jeans *et al.* Surrey and Woburn fuller's earths contain ash particles of two periods (123 and 140 m.y. ago) even though the Plinian explosion occurred in the Aptian perhaps 109 m.y or so. The younger of the Surrey and Woburn dates is approximately Hauterivian and the older is late Jurassic – one Callovian glauconite gave a date of 139 m.y.

C. V. Jeans, R. J. Merriman and J. G. Mitchell, origin of Middle Jurassic and Lower Cretaceous Earths in England, *Clay Minerals*, (1977) *12*, 11.

The Phanerozoic Time-scale, ed. W. B. Harland *et al. Q. J. Geol. Soc.* (1964), *120*, 203.

OTHER ENGLISH COUNTIES

28

Sussex

IN Sussex, we have the testimony of three 18th century and early 19th century authors that fuller's earth was "found at Tillington, and consumed in the neighbouring fulling mills", and of a fourth of slightly later date that it used to be dug near Petworth – two miles from Tillington. However the industry seems to have had a new lease of life since Hunt (1860) states that the fuller's earth at Petworth was being worked at that time. The occurrence was mentioned also by W. Topley in *The geology of the Weald* (1875).

Andrew Ure, *op.cit.*, citing W. Marshall, *Agriculture of Sussex*, c. 1786–96, Section V, Minerals III, 15, and A. Young, *General view of the agriculture of the county of Sussex*, (1806) 15; see also W. Marshall, *The rural economy of the southern counties*, London, (1798).

E. Hunt, *Mining Records*: Mineral Statistics... Part III for 1858. *Mem. Geol. Surv.* (1860), 377.

Wm. Topley, *Mem, Geol. Surv. Gr. Br.* p. 125.

In 1971 the area was re-examined by the Institute of Geological Sciences, who found two beds of fuller's earth in the neighbourhood. In two places it was about 4 m thick, but the quality did not appear to be as good as that of Surrey fuller's earth. The same authors found other occurrences at Bognor Common, Pulborough, Washington borehole, Nep Town borehole, Fulking, and Hampden Park borehole, all in the Lower Greensand.

B. Young, D. E. Highley, and D. J. Morgan. Calcium montmorillonite (f.e.) deposits in the Lower Greensand of the Tillington area, West Sussex, I.G.S., open-file Report 1979/1. Same authors, *Trans. Inst. Min. Met. Section B, 87*, B93–96.

Rev. C. P. N. Wilson (1821) was probably the first to notice a thin layer (80–100 mm) of fuller's earth in the Upper Chalk. Calcium montmorillonite is a common constituent of chalk, though usually in disseminated form. His observation was made at Bepton Hill at 122 m, 5.6 km south of Midhurst.

Quart. J. Sci. (1821), 220–221.

Lincolnshire

NORTH of Langton Hall, Langton-by-Spilsby, in Lincolnshire, a large excavation has been made by the roadside, exposing over twenty feet of Carstone, a ferruginous clayey sand. At the mouth of the pit, there is a well which is said to have been dug through fifteen feet of grey and white sand; and water is said to have been found in white sand which rested on a kind of "soapy fuller's earth".

A. J. Jukes-Brown, The geology of part of E. Lincolnshire, *Mem. Geol. Survey Explanation of sheet 84* (Old Series), (1887).

This fuller's earth is in what H. H. Swinnerton calls the Sutterby marl; it is probably of Lower Greensand age.

The rocks below the Red Chalk of Lines and their cephalopod faunas, *Quart. J. Geol. Soc.*, (1935), *91*, 1.

Elsewhere it is stated that two hundred yards north of the house is a well, eighteen feet six inches deep, in which fuller's earth was found about six feet down. Fuller's earth was observed here and in some other places in the neighbourhood by A. C. G. Cameron, when he was mapping the district.

(Stephen Langton, Cardinal and Archbishop of Canterbury, [d. 1228] was the son of Henry, Lord of the manor of Langton-by-Wragby; a family related to Mr. Langton of Langton-by-Spilsby).

The Sutterby marl can be traced in several outliers capped by Carstone and a still smaller cap of chalk as in the hill south of Dalby (road cutting), two ridges west of Brinkhill called Warden Hill and Cloven Hill. A larger outlier is at South Ormsby. The Lincolnshire Wolds are much dissected and the Carstone has a most irregular outcrop.

Whether these Lincolnshire deposits were ever worked for earth for fulling or not is not clear, but working seems unlikely in any case to have been on any scale. The squire, Mr. Langton, had a sample of this clay in the 1930s. However one must point out that a sample of 'fuller's earth' from Langton in the Geological Museum, London, contains only minor amounts of smectite.

Yorkshire

Encyclopaedia Britannica, 9th ed. (1870–1903).

THE fuller's earth said to occur at Renton has not been traced nor indeed has the place.

Nottinghamshire

Rec. Boro. Nott., *2*, 41, in V.C.H. Notts, *2*, 346.

ON 1 August 1405 John Ingham, walker, was complained of by John de Aldsworth for a plea of debt, and claimed amongst other articles a strike of fuller's earth (unum stryk de walkerherth).

V.C.H. Surrey, *2*, 279 F. R. Buckley, *The Listener*, (25 March 1954), 515.

Chambers' Encyclopaedia (1895) lists Nottinghamshire among those counties which have fuller's earth, and the *Victoria County History of Surrey* confirms this. Could this refer to "The Trip to Jerusalem", which claims to be the oldest inn in England? It is built up against the Castle Rock at Nottingham, with four of its rooms hollowed out of the Castle Rock itself; in these large sand-stone rooms fuller's earth oozes from its crevices. Sowerby (1817) gave a coloured picture of Nottingham fuller's earth, which he said was more opaque than soft and waxy in appearance – and therefore not so pure.

James Sowerby, *British Mineralogy*. (1804–1817), *3*, 232. "Women and children strip the Fuller's earth of its coloured covering."

Isle of Man

THE occurrence of a clay said to have the properties of fuller's earth is recorded by the Geological Survey, the locality being Glen Wyllin.

Warwickshire

A. C. G. Cameron, *Trans. Fed. Inst. Min. Eng.*, (1893–4), *6*, 204–209.

A CLAY, "probably of the same nature as fuller's earth, as it was used as a substitute for soap, has been dug in this county".

Wiltshire

The mudd or earth cleanses and scowres incomparably
JOHN AUBREY, *The Natural History of Wiltshire*, written between 1656 and 1691.

Edited by John Britton (1847); reproduced (1969), David & Charles (Publishers), Newton Abbot, Devon, pp. 25 and 35.

JOHN Aubrey was wrong in identifying this mud or earth at Gogges on Minety Common with fuller's earth. The clay was black, drying gray, at least 18 feet thick and obviously Oxford Clay.

He was on surer ground when he wrote *At Heddington Wick, at the foot of the hill, is a kind of white fullers'-earth which the cloth-workers doe use.* Heddington is at the north end of an inlier of Lower Greensand, in which fuller's earth is known to occur, the place names Stockland 1 km and Blackland 2.6 km NNE of Heddington, may be "fullonical". They are near or on LGS inliers. Also *there is at Burton-hill, juxta* Malmesbury, *fuller's earth, as also about* Westport, *and elsewhere there abouts, which the clothiers use.* These are on the Corallian.

Lower Greensand Fuller's Earth occurs between Nash Hill and Nethermore Farms ca. 3 to 2 km NW of Sandy Lane, the Roman VERLVCIO: I. A. Cowperthwaite. The 0.6 m seam may have been a single *in situ* ashfall.

> *At Clarendon-parke is lately discovered (1684) an earth that cleanseth better than Woburne earthe in Bedfordshire; and Mr Cutler, the clothier of Wilton, tells me he now makes only use of it. He formerly bought fuller's earth from Woburn at ten groats a bushell.*

Near Salisbury on the chalk.

Also on the Chalk, at Broade Chalk, "there is a kind of fuller's-earth, which the cloth-workers doe use for their chaines", but the description does not tally with a smectite.

Near Salisbury

chains – longitudinal threads or warp.

29

WALKER'S EARTH OR "WAYBOARDS"

Wales

R. I. Murchison, *The Silurian System*, (1839), *1*, 304 and 434–435.

Lichfield Memoir of the Geological Survey, Explanation of Sheet 154, (1919), 314.

IN the Silurian rocks of the Welsh Marches the great pioneer geologist Murchison noticed a number of bands of yellowish-white clay which were called locally "Walker's Earth" or "Walker's Soap" because it was used for cleansing purposes. The product of one such band at Daw End near Walsall was once offered for sale as fuller's earth.

Herefordshire

R. I. Murchison, *op.cit.*

J. Nightingale, *English Topography*, (1816).

(Stoke Edith is at the North end of the Woolhope (Silurian) inlier).

IN Herefordshire the Walker's Earth occurring between beds of Aymestry or Ludlow Limestone sometimes causes landslips. It was sometimes dug near Stoke in this county as a fuller's earth, and the Woolhope Limestone has a two-foot seam of rather calcareous clay, which was used up to about 1900 for washing clothes, in spite of the fact that it contains more illite than montmorillonite.

Derbyshire.

The Geological Observer, (1853), 2nd ed., London, Derbyshire, 559–560, 681–694.

G. M. Walkden, *Geol. J.* (1972), *8(1)*, 143–160; *Proc. Yorks. Geol. Soc.*, (1977), *41(3)*, 347–366.

SIR Henry de la Beche (1853) was the first to conclude that the similar 'wayboard' clays of the Lower Carboniferous of Derbyshire were decomposed volcanic material. A modern study of these clays assigned them to the category of 'potassium bentonites' (Walkden, 1972, 1977).

Cumberland

J. Wilson, *Chem. Tr. J.* (1934), *95*, 202.

G. M. Walkden, *Yorks. Geol. Soc.*, occasional publication No.3, (1978), p.175.

AN occurrence of bentonite in Cumberland was reported by Wilson in 1934. Walkden wrote to me to say that shale bands can rarely be seen at the surface but he has established the bentonitic characteristics of some of them and regards these as representing non-local, volcanic-ash, atmospheric dusts.

SCOTLAND

"Not good is a web without fulling", one of the aphorisms of Phelim M'Dougall, preserved in the Book of the Dean of Lismore. (Language and Literature of the Scottish Highlands.)
J. S. BLACKIE, Edinburgh: Edmonston & Douglas, (1876) p. 89.

The Dean of Lismore was vicar of Fortingall, and belonged to a sept of the clan MacGregor at the time of the Reformation (1560).

THERE is a tradition that Saint Columba's feet were sore with boils when he reached Dowally and that these were cured by the application of clay – a story which reminds one of the curative properties of Bath fuller's earth. Wool was certainly woven in Scotland in the 8th century. King David I (1124–1153) brought to the principal towns from England wooldressers, dyers, and weavers; at the same time he controlled the importation of wool. King William the Lion (1165–1214) granted to the burgesses of Perth the right of having their merchant guild, though the artisans, fullers and weavers, were excluded. In the reign of Alexander III (1249–1286) wool was exchanged with Flemish merchants for linen, silks, and broadcloth. The Scottish industry had perforce to develop separately from that of England; indeed *no Scotchman is to be taken apprentice without a penulty of 6s 8d by members of the Weavers' Company of Durham, at its foundation in 1450.*

Dowally, north of Dunkeld. The tradition persisted for many centuries, when people came for cures by Uir Coluim Cille, St. Columba's earth.
C. Rogers, *Social Life of Scotland*, Edinburgh: Grampian Club, (1884), *1*, 390–391.

Scotland in the Middle Ages, Cosmo Innes, Edinburgh: Edmonston & Douglas, (1860), 163. *Fullones et telarii.*

28 Henry VI = 1450.

Cloth made in most parts of Scotland was a coarse homespun, presumably fairly well fulled, to keep out the wind and rain. Hodden gray cloth was made from a mixture of undyed black and white wool. In the larger towns and monasteries no doubt there were craftsmen able to make a better type of cloth. Walkers had their own place in the annual procession on Corpus Christi and Candlemas in Aberdeen in 1531. The order of precedence was *Fleschearis, Barbouris, Skinneris & Furroweris, Websteris & Walcaris togidder, Baxteris, Hammermen.*

See *An Economic History of Scotland, 1100–1939*, S. G. E. Lythe & J. Butt, Blackie: Glasgow & London (1975) for a history of the woollen industry: also S. G. E. Lythe, *The Economy of Scotland*, Oliver & Boyd.

A. M. MacKenzie, *Scottish Pageant*, Edinburgh: Oliver & Boyd, (1948), 90.

According to Fynes Moryson, (1598), *The Eastern Scots carry into* France *course cloathes, both linnen and woollen, which be narrow and shrinkle in the wetting. They also carry thether Wooll, Skines of goats, Weathers and of Conies, and divers kindes of Fishes...* it is probable that cloth was later exported to Veere and Bruges to be finished since English fuller's earth would often have arrived there illegally.

Itinerary, from Hume Brown's *Early Travellers in Scotland*, Edinburgh: Douglas (1891), 87.

In 1600 an Act was passed by the Privy Council to import 100 families of strangers skilled in the making of woollen cloth, in order to improve the quality of cloth made in Scotland. They were to have a

The History of Galashiels, Robert Hall, Galashiels: Alex: Walker (1898).

minister or pastoure quha sall teich to thame the Worde of God in their awin language, providing that he aggreis with the religioun presently proffert within this realme.

Sir Robert Sibbald (1684) refers to *Cimolia, quae et smectis dicitur, nostratibus* **Fullers-Earth**. I presume he knew of its existence in Scotland since other earths and minerals mentioned by him were local; for example loadstone – Magnes – came from the cliff of the bay of Dumbarton.

Scotia Illustrata, Edinburgh (1684).

If I translate aright "In Rupe Arcis Britanno-dunensis."

Martin Martin, a native of Skye, was given a grant by the Royal Society to make a report on the Western Isles in 1695. In Arran he was told by the natives *that some places afford fuller's earth*, and in Skye he observed that *the rivulet at Dunvegan Church, and that of Nisbost, have fuller's earth, and that the villages of Borve and Glenmore afford two very fine sorts of earth, the one red, the other white; and they both feel and cut like melted tallow* – a good description of a smectite. Cameron (1893–4) observed that *in the Isle of Skye, the so-called rock-soap* (a dark-coloured clay), *sometimes used for washing cloth, is said to have a chemical composition resembling fuller's earth.* A sample of clay collected by George Henderson and me from the foreshore at Nisbost in October 1949 was found by X-ray diffraction to contain montmorillonite with a small amount of saponite as well as about 20% feldspar and less than 10% illite. A clay found by us in a basalt exposure at Glenmore was found to be almost wholly saponite.

Martin Martin, *A Description of the Western Isles of Scotland*, (1703), [1934 edition, pp. 254, 191,] Stirling: Aeneas Mackay. His journey was 2 years after the famous Warner trial.

A. G. C. Cameron, *Trans. Fed. Inst. Min. Eng.* (1893–4) *6*, 204–209.

Journey 5–10 October 1949.

DTA endotherm at 850°C confirmed saponite, but another at ~550° suggests some montmorillonite also.

The ministers, who reported on their parishes to the Statistical Account of Scotland (from 1792) and to the New Statistical Account (1845), recorded several occurrences of fuller's earth. Indeed there had been a previous 'Questionair' composed by the Society of Antiquaries and addressed to the Gentlemen and Clergy of North Britain before 20 October 1771. Pennant reprinted this in the second edition of his famous Tour. Question XXI was *Is there any marl, fuller's earth, potter's clay, or loam, or other remarkable soils, as ochre, etc, in your parish.* Thus, *beds of fuller's earth and potter's clay were found in different places* near Hamilton, Lanarkshire. At Falkland *marl and fuller's earth have been discovered but in small quantities* and at Sanquhar *in the (coal) mines have also been found some beds of fuller's earth.* In the parish of Uphall are *slate-stone, some coarse fuller's earth, potters clay, brick clay*, and various sorts of calcareous rocks. Robert Forsyth (1808) very interestingly also refers to the *Forth alluvium soil* [*which*] *when first dug from the natural bed is of a bright blue color and of a substance resembling the richest soap and sometimes even serves as a substitute for fuller's earth.* The Carse clays probably consist of a mixture of clay minerals, illite, smectite and a little kaolinite, with quartz of fine silt grade – not a promising composition.

Thomas Pennant, *A Tour in Scotland*, [1769], London: B. White, 2nd ed. (1772).

Sir John Sinclair, Bt (editor).

Statistical Account (1792), *2, 194;* N.S.A., *6*, 260.

S.A. (1792) *4*, 441.

S.A. (1793) *6*, 455.

Robert Forsyth, *The Beauties of Scotland*, Edinburgh: John Brown, (1808), *5*, 130.

id, 3, 397.

In the description of the parish of Aboyne and Glentanner, the Rev. Robert Milne Miller, NSA (1845), *12*, 1051, wrote that *on the left bank of the Aultdinny* [*burn*] *is fuller's earth, which has been successfully tried as a substitute for soap.*

[N.S.A. = New Statistical Account.]

The *great quantity of fuller's earth* recorded by Forsyth at Lamb Hoga in the island of Fetlar, Shetland was probably an impure kaolin, but we are on firmer ground when we read (1794) *there is likewise in this parish* [*of Campbelto(w)n*], *abundance of fuller's earth*. Even more was found in the parish of Southend further south in Kintyre. *There is also, in some vallies here and along the banks of the rivers, what is commonly called Fuller's Earth*, and Forsyth (1808) repeats this. The minister in 1845 says more: *In the neighbourhood of the . . . manse, there is a large bank of Fuller's earth, which is exposed to view by the . . . river during heavy rains, but has never been made use of either for agricultural or commercial purposes*. The minister, the Rev. Angus McVicar, had no knowledge (1949) of this occurrence, but Henderson and I were more fortunate at Lagnacraig. The 1845 account says: *It is found in some of the glens, and particularly at Auchnaclach and Lagnacraig, partly white, but generally a mixture of white and brown. The white however is the best*. The writer said that although it was not sufficiently appreciated at that time, the earth was of the sort which cleans oil out of cloth and tar out of wool. He compared it with the fuller's earth from near Woburn Sands.

id., (1808), *3*, 397
S.A. (1794), *10*, 551

S.A. (1792), *2*, 367

(1808), *5*, 435
N.S.A. (1845), *7*, 2

Nevertheless it does seem probable that in the western Highlands and Islands argillified igneous rocks were known to those who made fine cloths, made of yarn dyed in bright colours, the ancestors of the tartans, which are such a feature of the area.

(These occurrences were reported in a lecture given to the Geological Society of Glasgow, 13 Dec. 1945).

When the island of Canna was last for sale, one of the assets mentioned in the prospectus was an occurrence of fuller's earth. This the proprietor said (1944) is in a cave with a very narrow entrance at a point farthest from the house. The only practicable access is by boat, as there is no road within three miles of it. The occurrence was known to Heddle (1901)[1], who also noted the fuller's earth of Uphall, mentioned above, and of Maxton[2] in Roxburghshire, where it was used for a while in the wool industry around Jedburgh. He states that fuller's earth is found in a small seam below Bridge House bridge over the Lyne, on the east side of the water, at West Linton, Peeblesshire.[3]

Dr. John Lorne Campbell of Canna, letter (3.12.1944).

1 M. F. Heddle, *The Mineralogy of Scotland* (1901), *2*, 149.

2 *Chambers Encyclopaedia* (1895), FULLER'S EARTH.

3 S.A. (1791), *1*, 131.

Cameron (1893–4) says of the Maxton occurrence that nodules of unctuous clay are frequently found in the [old Red] sand-stone. The colours vary from dark blue, almost black, through many shades, including light green. A resident in Maxton said that a hard form used as a slate pencil, when ground down, is of the nature of fuller's earth. Not likely!

A. C. G. Cameron, *op. cit.*

See also *Nichol's Mineralogy*, p. 171.

Mr. P. D. Ridge-Beedle recollected an occurrence of fuller's earth at Crinan, but no confirmation has been received.

Letter 8.7.1948.

According to the 9th edition of the *Encyclopaedia Britannica* (1870–1903), fuller's earth was got from Quarry Wood, about 6 miles from Elgin on the road to Forres for use in local waulk mills. The occurrence is not mentioned in later editions. Apparently it occurs in blobs, larger pockets and lens-shaped masses in many of the 30 stone quarries around there, but no commercial quantities have been seen.

Alex. Christie, of 2 Commerce St., Elgin, 30.11.1947.

Yellow fuller's earth was used in Scotland, Wales & the North of England for fulling coarse woollen goods. Blue fuller's earth was used for fine cloth chiefly in Yorkshire, Cambridge County Series (1909).

Journal of the Society of Chemical Industry, (31.5.1893), *12*, 477.

Used for staunching blood after shaving.

W. F. Cole and J. S. Hosking, chapter X in *The Differential Thermal Analysis of Clays*, (ed. R. C. Mackenzie), Mineralogical Society, (1957), p. 260 & 256. See also Fig. 1 of R. C. Mackenzie in *Bull. Gr. Franç. Argiles* (1957), *9*, 7–15.

R. C. Mackenzie, *Silicates industriels*, (1960), Nos. 1 & 2.
See ch. 47 about Ponza montmorillonite.

J. B. Johnston, *Place Names of Scotland*, Edinburgh: D. Douglas (1892).

There's another near Calder, by Nairn, S.A. *4*, 353.
Also Waukrigg Mill, Galashiels 1890; 3 waulkmills there before 1581.

Indeed from 1790 (Galashiels) and 1800 (Elgin) English fuller's earth was imported and local supplies of inferior earth were used no more.

Fuller's earth was discovered c 1863 on Lord Rollo's estate when Duncrub House was being built. It is a weathered igneous rock, in part white, in part brown and white, similar to the clay at Lagnaha, Kintyre. The clay was found in the stream running quite close to Keltie Castle. Dr. John Clark "late of Hertford Street, London", introduced it to the notice of the late Dr. Percy, the metallurgist, who consulted Sir Alfred Garrod as to its utility.

The Scottish Fuller's Earth Company started operations in 1893; the Hon. Bernard Rollo was manager of the plant for purifying the earth – a pan mill, settling tanks, kilns, boiler house, an engine and a packing-house – all clearly copied from Dames' process at Midford. At Findony Mill in Dunning Village, the clay was finely ground and air-separated. The industry continued until about 1914, supplying Castle Keltie woollen mills, and the pharmaceutical industry. The latter was supplied by another company, Keltie Products Ltd., who finally went out of business before 1938. The styptic grade was sold through Fassett & Johnson Ltd and a soap-filling grade was tried for a while in World War I by Burroughs & Wellcome Ltd. The plant was quite large, the tank room containing five settling tanks being 80 feet long by 9 feet broad, the packing-house was 84 feet long.

Cole and Hosking (1957) published a differential thermal curve for Dunning fuller's earth. Because it had two endotherms, at 600°C and 700°C, it was classified as of 'abnormal type'. The clay as dug, dried and milled has a very low cation exchange capacity (12.5 meq./100g) and the purified styptic powder one of only 40. Even in the laboratory, Mackenzie (1957) obtained a value of only 49.5 meq, so difficult is this clay to purify. However, when he made allowances for the 46 per cent of impurities, he showed that the montmorillonite in the clay had a CEC of 85 meq and was able to calculate its ionic composition. In view of the great difficulty of removing impurities this clay has no commercial value today.

Both at Lagnaha and at Dunning this unusual montmorillonite was formed by the alteration of porphyrite injected into the Old Red Sandstone.

Nothing is known about the fuller's earth said to occur at Woodhill in Strathardle, a dozen miles ESE of Pitlochry.

In Scotland fulling is commemorated in the many place names based on the word wauk or waulk, to full or dress cloth, such as Wauk Mill (Haddington) 1561; the Waulk miln of Partick 1587; waulkmills near Arbroath; Waulkmill 2 miles N of Scone Palace and Waulkmill, a farm near Elgin. Henry Ballantyne set up his mill at Walkerburn in 1855. There must be many more such names.

In the Highlands Ballintore, Easter Ross, and Auchintore, South of Fort William, mean bleaching township and field, respectively, the ending *todhar* originally meaning dung. It is possible that manure was once used for processing linen. Before the days of cheap soap, if fuller's earth could not be found locally, stale urine was the commonest fulling agent. Certainly an English traveller putting up at an inn at Dunbar could scarcely bear the stench of the bed clothes. (See end of this chapter.)

Walkerburn is now the home of the Scottish Museum of Wool Textiles at Galashiels.

There is a Muilean Luaidh, fulling mill, 2 miles from Tain, Ross-shire, as well as an Allt Luadh. Fuller's earth in Gaelic is criadh an fhùcadair. Four miles N of Dufftown there is a place called Maggieknockater, derived inelegantly from *magh an fhùcadair*, the fuller's field. In Highland Perthshire the surname *Mac an fhùcadair* was once quite common, but after 1746, those who left and settled on the East Coast retained their name in the form of McNucator or Nucator; whereas those remaining behind at home discreetly changed their name to Walker. The name Walkinshaw is thought by Bardsley to derive from the word for a fuller, and mentions the curious feminine form Walkster (as in Baxter).

Fulling burn

Criadh really the genitive of cré, clay.

Which stimulated J. M. Oliver to write some humorous verses. . . "So here's a sang o' Bonnie Maggie Nockiter" Edinburgh: Lindsay & Co Ltd, N.D.

W. A. Gillies, *In Famed Breadalbane*, Perth (1938).

C. W. Bardsley, *A Dictionary of English and Welsh surnames*, (1901).

Waulking in the Islands

MARTIN Martin tells a story about fulling in Borera.

M. Martin, *op. cit.*, p. 129.

> *The latter* [*at a harbour on the south side of the island*] *happened to come to a house where he found only ten women, and they were employed (as he supposed) in a strange manner, viz, their arms and legs were bare, being five on a side; and between them lay a board, upon which they had laid a piece of cloth and were thickening of it with their hands and feet, and singing all the while. The Englishman presently concluded it to be a little bedlam, which he did not expect in so remote a corner; and this he told to Mr. John Maclean, who possesses the island. Mr. Maclean answered he never saw any mad people in those islands; but this would not satisfy him till they both went to the place where the women were at work, and then Mr. Maclean having told him that it was their common way of thickening cloth, he was convinced, though surprised at the manner of it.*

The unusual method of waulking the cloth in the Highlands and Islands has been recorded on many occasions. Martin himself watched the women at work in St. Kilda in 1685. *Both sexes wear course flannel shirts which they put off when they go to bed; they thicken their cloaths upon flakes, or mats of hay, twisted and woven together in small ropes; they work hard at this employment, first making use of their hands, and at last of their feet; and when at this work, they commonly sing all the time, one of their number acting the part of prime chantress whom all the rest follow and obey. Often there would be five women on each side of the board, their arms and legs bare.*

M. Martin, *A Voyage to St Kilda*, in 1685, (1702), Stirling: E. Mackay, (1934). 456.

Thomas Pennant describes a grooved board, like a washing board; and 12 or 14 women at a waulking in Skye in 1772.

T. Pennant, *loc. cit.*, (1774 ed.), *2*, 285.

James Headrick (1807) described how the women manufacture most of the woollen cloth that is used in the island of Arran. *There is a dyer and a fuller established near Lamlash, who dresses the cloth, and it is far from being inelegant. I have often admired the brilliancy of the Highland dyes, and ascribe it partly to the purity of their water.*

James Headrick, *View of the Mineralogy, Agriculture, Manufacturies and Fisheries of the Island of Arran*, Edinburgh: Archibald Constable, and London: John Murray, (1807), p. 361.

Alexander Ross wrote in 1885 on Old Highland Industries.

Trans. Gaelic Society Inverness, (1885), *12*, 402–3.

> *In the Highland districts women make use of their feet to produce the same result [as a waulking mill], and a picturesque sight it is to see a dozen or more Highland lassies set round in two rows facing each other. The web of cloth is passed round in a damp state, each one pressing and pitching it with a dash to her neighbour, and so the cloth is handled, pushed, crushed, and welded as to become close and even in texture. The process is slow and tedious, but the ladies know how to beguile the time, and the song is passed round, each one taking up the verse in turn, and all joining in the chorus. The effect is very peculiar and often very pleasing, and the waulking songs are very popular in all the collections.*
>
> *I have on various occasions watched the waulking process, but seldom in recent years. It is often the occasion of a little boisterous merriment and practical joking, for, should a member of the male sex be found prowling near by, he is, if caught, unceremoniously thrust into the centre of the circle and tossed with the web till, bruised with the rough usage and blackened with the dye, he is glad to make his escape from the hands of the furies.*

See also James Hogg, *Highland Tours*, recorded at Ballachulish 13.6.1818.

Neil Gunn gives a fine literary account of waulking the cloth in *Butcher's Broom*.

Chapter III.

Many examples of waulking songs (òran luadhaidh) from Barra, South Uist, Eriskay and Benbecula, dating from the early seventeenth century were recorded by John Lorne Campbell of Canna and the even older tunes were transcribed by Francis Collinson (1977). David MacCormick's collection of 1893 was published by the same Press in 1969.

Hebridean Folksongs II Oxford University Press, (1977), 378 pp.

President, Folklore Institute of Scotland, July 1949

In the old days of superstition the women believed that a faery-sprite known as the *loireag* supervised the waulking of the cloth. If a libation of milk were not set down for her, the *loireag* would place the cows "under enchantment". She had a very sensitive ear and if a woman sang out of tune, she would be tempted to unravel the warp from the web.

A. A. Macgregor, *A Last Voyage to St Kilda*, London Cassels (1931) 95, quoting A. Carmichael, *Carmina Gadelioa*, Edinburgh, (1928).

Tenting in Shetland

AN wan fine day she wis streetchin a hap (shawl). Weel them days the wey 'at they streetched the haps, they used to put the pins i the grund an then streetch the hap on this pins.

Recorded from Henry Hunter, Muness, Unst, Shetland by Alan Bruford.

Tocher 28, 197 (Edinburgh Univ.)

The related art of laundering

SIR William Brereton (1636) observed at Dunbar *the sluttish women washing their clothes in a great tub with their feet, their coats, smocks and all, tucked up to their breech.*

Travels of Sir William Brereton (1636), Chetham Society, (1884); quoted in *Early Travellers in Scotland*, Edinburgh: David Douglas. (1891), p. 135.

By the time he got to Edinburgh he was really suffering from Scotland's lack of fuller's earth.

> *The sluttishness and nastiness of this people is such, that I cannot omit the particularizing thereof. . . . Their pewter pots, wherein they bring wine and water, are furred within, that it would loathe you to touch anything which comes out of them. Their linen is as sluttishly and slothfully washed by women's feet, who, after their linen is put into a great, broad, low tub of water, then (their clothes being tucked up above the knees) they step into the tub and tread it, and trample it with their feet (never vouchsafing a hand to nett or wash it withal) until it be sufficiently cleansed in their apprehensions, and then it looks as nastily as ours does when it is put unto and designed to the washing, as also it doth so strongly taste and smell of lant or other noisome savours, as that when I came to bed I was constrained to hold my nose and mouth together. To come into their kitchen, and to see them dress their meat, and to behold the sink (which is more offensive than any jakes) will be a sufficient supper, and will take off the edge of your stomach.*

Nett, to clean. Fr. *nettoyer*.

Lant, stale urine.

Jakes, privy.

The only good thing that Brereton had to say was that the women near Kirkintilloch were *only neat and handsome about the feet, which comes to pass by their often washing with their feet.*

id., p. 150

Twenty years later Sir Richard Franck hinted that the women's method of washing was not altogether innocent. *Thrice in a week you shall rarely fail to see [in the county of Galloway] their maid-maukins dance corantos in tubs. So on Sunday some as seldom miss to make their appearance on the stool of repentence.*

id., p. 185.

maukins = 'Sluts'–with a hint of lewdness; a coranto was a French dance.

Thomas Kirke (1679) confirmed that in Scotland the *banks and borders of these rivers (especially near their towns) are adorn'd with hardy amazons, though inverted, their value being (chiefly) from the waste (sic) downwards, which parts they readily expose to all the dangers of a naked encounter. From the way they bounced in a buck-tub came the invention of fulling-mills, the women taught the men, and they put in practice.*

id. p. 255.

This unseemly practice has not wholly ceased, Rev. Charles Rogers, *Social Life of Scotland*, Edinburgh: Grampian Club (1884) 252.

An early picture of *Women jumping in tubs at the River* is given in Burt's *Letters from Inverness* (1724–1734).

Letters from a Gentleman in the North of Scotland to his Friend in London, London: S. Birt 1754, page 52.

Perhaps in the East of Scotland gravestones would have served as a surface on which to waulk the cloth. Certainly Dorothy Wordsworth (1803) was horrified to see linen treated in this way in Melrose. *The churchyard was full of graves, and exceedingly slovenly and dirty; one most indecent practice I observed: several women brought their linen to the flat table-tombstones, and, having spread it upon them, began to batter as hard as they could with a wooden roller. . . .*

Dorothy Wordsworth, *Recollections of a Tour in Scotland*, (21.9.1803). Edinburgh: Douglas, (1894), 265.

A more idyllic picture was given by Lady Strachey in *Memoirs of a Highland Lady*. Near the Doune, Rothiemurchus, in 1803, *between*

Fig. 66. "Scotch washing", a print by Isaac Cruikshank, 1810. The Scottish beetle was much larger than the Welsh and Irish types probably because of its additional use in banishing *voyeurs*.

From Caroline Davidson (1982), A Woman's Work is Never Done, p. 140.

London: John Murray 1911, p. 30.

The arm-length club for hammering out the dirt from clothing on a flat rock by a burn was called in Gaelic a *slacan* or *slachdan*.

Guy Mannering, (Feb. 1815) ch. XXIV.

the two large alders hung in gipsy fashion the large cauldron used for the washing; a rude open shed, just sufficient to protect the officiating damsels from the weather; tubs, cogues, lippies, a watering-pot and a beetle – a bit of wood, bottle-shaped, with which the clothes were thumped, Indian and French fashion – lay all about among the yellow broom under the alders and hazels on this happy island. Isaac Cruikshank depicts a similar scene (1810), Fig. 66; and Sir Walter Scott (1815) wrote *bleached on the bonny white gowans and bittled by Nelly and hersell.*

IRELAND

31

Syringo, who much wealth had got
By urinal and chamber-pot
WILLIAM MOFFET, **Hesperi-Neso-Graphia. 1728.**

1791 edition, Dublin, p. 47.

IN Ireland, urine was often collected by a man with two large jars in creels on an ass and cart. Tubs for urine were placed outside the door of "apple-houses": these were ceilidhe-houses and small shops which sold apples. No doubt there were many other centres of collection, because in Ireland commercial deposits of fuller's earth, like adders, did not exist.

This section derives almost wholly from *Cloth Finishing in Ireland* by A. T. Lucas, National Museum of Ireland, to whom I give many thanks. Reprint from *Folk Life Journal, 6*, 67 pp.

True, some clays were thought to be of use in fulling. The manuscript diary of Donald Stewart, an itinerant mineralogist to the Dublin Society at the end of the 18th century, contains frequent references to deposits, such as lithomarge, which would be useful in fulling. In 1812, an earth resembling fuller's earth was described as occurring along the River Laune in Kerry and another 'saponaceous earth' near Old Head, Kinsale, Co. Cork. But neither kind was used commercially, nor were the deposits of good quality, which caused the bureaucratic Cromwell to issue instructions to the Lord Deputy and Council to take steps to prohibit the export of this commodity. *Neither does the earth appear to have been imported for it is not mentioned among the goods traded between Britain and Ireland in the 16th century, although it was extensively exported from England to the Continent in that period.*

Information supplied to Lucas by Dr. J. S. Jackson, his colleague at the museum.

Edward Wakefield, *An Account of Ireland Statistical and Political*, London, (1812), *1*, 113.

Ireland under the Commonwealth, ed. Robert Dunlop, Manchester, (1913), *2*, 585.

A. K. Longfield, *Anglo-Irish Trade in the Sixteenth Century*, London (1929).

The earliest account of fulling in Ireland was by the English traveller Thomas Dineley (1681): *The manner of tucking and thickening cloth, without a mill, is thus: they place the cloth double upon a large wicker or twiggen door, called here a hurle, and work it with their hands and feet until it become thick by rowling, this they dash with stale chamberlye in working. Thus making a virtue of necessity.* This description reminds one of Martin Martin's observations in St. Kilda (1698).

Observations in a Voyage through the Kingdom of Ireland, (1681), (Dublin 1870) pp. 22–3.

The Rev. Horatio Townsend described the method of making "frize" for domestic use on Clear Island, Co. Cork.

Upon a square hurdle, to keep the cloth from the dirt of the ground, eight women take their seats, four opposed to four, at such a distance as that the extended legs of one set just reach the drawn up feet of the other. The frize, placed between, is pushed alternately by each party, with as much force as they can exert, against the feet of the other, until by frequent repetition of this

A Statistical Survey of the County of Cork, Dublin, *1810*, 307–8.
Frieze, a kind of woollen cloth with a nap on one side only; now especially of Irish manufacture. OED, 1933, Compact Edition, (1979).

laborious process the piece is sufficiently tucked. Partly from the necessity of keeping the frize wet, and partly for the convenience of having their limbs at liberty in this laborious exercise, the fair operators find it necessary to disencumber themselves of superfluous clothing. The work is therefore always performed in a state of half nudity, nor does the approach of a curious stranger suspend their labours. Unconscious of any thing extraordinary in a situation too common to excite any surprise among their own tribes, their simplicity very naturally supposes, that it will be viewed with equal indifference by others.

During the nineteenth century, fulling was gradually taken over by men, who could exert a greater force on the cloth and to avoid distraction worked in pairs. Robert Heron (1793) clearly disapproved of the former social aspects of waulking in contemporary Scotland.

Robert Heron, *Journey through part of Scotland*, Edinburgh, (26 May 1783) vol. 1. p. 286.

It is a fact in the hiſtory of the manners of the Highlanders, that they are accuſtomed to ſing at the performance of almoſt every piece of ſocial labour: Rowers in a boat ſing as they ply the oars; reapers ſing as they cut down handful after handful of the corn; and here were waſhers ſinging as they rubbed and rinſed their clothes. This accompaniment of muſic certainly renders the labour more cheerful; but, I ſhould ſuppoſe, that it cannot well fail to retard it, at the ſame time.—No better proof needs be aſked of the low ſtate of the induſtry of any people, than that their labour is often interrupted by ſinging and by converſation.

(Figures below) Cloth finishing in Ireland.

Fig. 67. Drawing to illustrate kicking of cloth in a trough.

Fig. 68. Drawing to illustrate table bound with rope used for fulling in Kilcredaun, Carrigahold, Co. Clare.

Both from A. T. Lucas

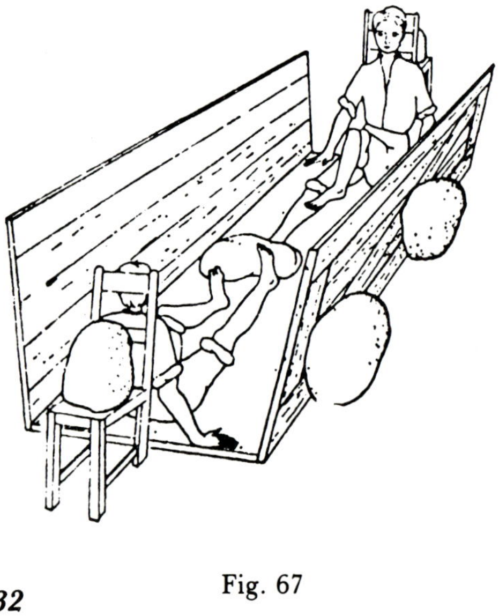

Fig. 67

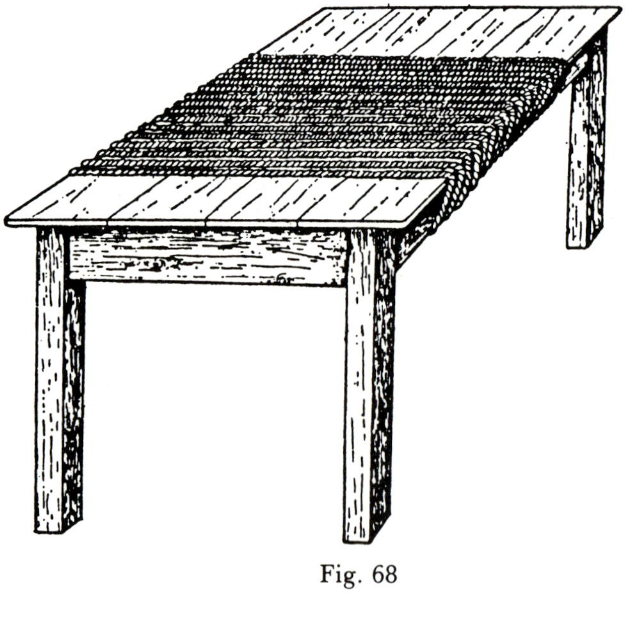

Fig. 68

But Lucas thinks that the change over *to the male operation, which is virtually standard practice in the period covered by the oral accounts, may have taken place as the result of increasingly puritanical ideas about feminine decorum during the 19th century which made the inevitable exposure of the legs involved in the kicking of the cloth seem grossly indelicate.*

Beetling in Ireland in 1841 is illustrated by Mr. & Mrs S. C. Hall in a *somewhat uptight-sounding account of the natives, accompanied by propa-*

Collected by the Irish Folklore Commission.

Lucas, *op. cit.*, 31.

Mr. & Mrs. S. C. Hall, *Ireland: Its Scenery, Character, etc.*, (1841–3)

From *A Woman's Work is Never Done. A history of housework in the British Isles 1650–1950*, by Caroline Davidson, Chatto and Windus 1982.

Fig. 69. Irish peasant women beetling their linen in a river. From Mr and Mrs Samuel Carter Hall, 1841–3.

From Caroline Davidson (1982), p. 141.

31
Comment in *The Observer*, (11 Jan. 1970).

John Millington Synge, *Collected Works*, edited by Robin Skelton, Oxford University Press; London (1966), *2*, 76.

George Seton, *St Kilda past and present.* W. Blackwood, Edinburgh & London, (1878), 243.

ganda for the Temperance movement but relieved by its delicate and romantic engravings which make waifs look like sylphs, Fig. 69.

I would like to end this chapter with two quotations from Caroline Davidson:

> *One of the last people to describe it was the Irish playwright, J. M. Synge, who spent much time on the Aran Islands off the west coast of Ireland in the 1890s and early 1900s. 'As I walk round the edges of the sea', he wrote, 'I often come on a girl with her petticoats tucked up round her, standing in a pool left by the tide and washing her flannels among the sea-anemones and crabs. Their red bodices and white tapering legs make them as beautiful as tropical sea-birds, as they stand in a frame of seaweeds against the brink of the Atlantic'. However, he had one more prosaic comment to add. 'Their habit of using the sea-water for washing causes a good deal of rheumatism on the islands, for the salt lies in the clothes and keeps them continually moist.'*
>
> *Women also washed their clothes in seawater on the island of St. Kilda in the 1830s.*

ENTREPRENEURS

32

IN medieval times fuller's earth deposits were owned by the large landowners or by religious houses. Clothworkers at first obtained licences to work the earth for their own and their colleagues' use.

The first part of the table gives the names of individuals who worked fuller's earth professionally, paying a royalty to the Estates who owned the mineral rights.

In the 19th century gentlemen owning small estates employed workmen to dig the earth and sold it to London merchants, some of whom bought fuller's-earth-bearing land and joined the ranks of the other owners.

When it became necessary to dry and grind the mineral, the capital cost was often too much for an individual to bear. At first ordinary companies were formed, and these were replaced later by limited liability companies, one of them – The Fullers' Earth Union Ltd—being a cartel of various interested parties. The FEU did not acquire all the producers until 1919.

Mid 19th century.

See p. 188

Competition arose again in 1951 (Berks) and in 1978 (Bretts), and the two early companies were themselves merged into much larger companies with great resources, Laporte Industries Ltd (FEU Ltd) and The Steetley Company Ltd (Berks). And Laportes eventually acquired Bretts.

Table 4.

DATE	ENTREPRENEURS HANDLING ENGLISH FULLER'S EARTH
1519	A patent demising the clay pit [at Woburn] to John Sheppard; *The Newport Hundreds.*
1577	Lease of Blue pit field by the Gilmyns of Reigate from John Mitchell owner of Redstone manor at £4/yr. †
1622	Richard Heath owned pits at Ford later Copyhold farm. *Survey of Reigate Manor.* †
1630	Richard Heath, mason; Richard Ace and William Bishop, the last two on neighbouring land. †
1632	R. Heath bought Redstone Manor. †
1620	John Ray (1620); Stephen Danton of Worth (1622); Richard Fyle (1622); Mrs. Brewer of Boxley (1622); Richard Rhodes (1622); the last two in 1640 also; all accused of illegal export.
1693	Hump. Bellomy, Richard Hopkins, Timothy Fowler, and – Shepherd, of Worcester, buyers and sellers of FE.
1702–57	"The ingenious Mr. John Watts", The Grove, Maidenhead (1719).
1804–14	Messrs Branton, Ward & Day, London Merchants.
1804	Messrs Foster, Brunton & Goddard, London Merchants.
1805	Mr. Silva perhaps the largest shipper of FE in London.
1807–45	Messrs William & Thomas Jones, London Merchants.
1809	Mr. Edward Russell, Holland House, Nutfield.
1809	Mr. Morris, Copyhold Farm, Reigate.
1809	Mr. Thomas Dann, Nutfield.
1815	Messrs Ward, Sky & Wood, London Merchants, who succeeded Messrs Branton, Ward & Day.
~1830	Mr. John William Grece, built Chart Lodge, Redhill, in 1780; died 1820; his son, Henry, (died 1852), also worked FE at Chartfield Quarry. Joined by Mr. Matthew Headley, from Manchester, who leased a pit in Nutfield near Copyhold Farm, in which he later leased Shipley field.
~1840	Mr. William Lambert, Nutfield.
~1840	Edward Larmer of Reigate, maltster and miller, also acted as a fuller's earth merchant. †
Late 40s	Mr James Cawley, formerly of Blechingley, started at site of what was later called Cockley Quarry.
1847–71	Messrs James Wishart & Sons, Leith Merchants.
c. 1850	Messrs Hyham & Gawkroger, Merchants of Thames St., London, exhibited at the 1851 Exhibition.
Early 1850s	William Matthews, Retail Chemist, 12 Wigmore St., London, first proprietary FE in toilet and medicine.
1852	Chart sold by H. Grece to Rogers & Hynam.
1871	Messrs Wm Dawson & Co, 6 Morrison's Court, Glasgow, Merchants
1877	Mr. Matthew Headley and Mr. J. Cawley were the only two operators of FE in Surrey in this year.
1880	Mr. Thomas Wood Cox, Copyhold Works, Redhill. *
	Mr John Hynam, Chartfield & Shambles Works, Redhill. *
1883	Mr Charles Richard Dames' patent for purifying Midford (Bath) FE.
1885	Mr. Henry Newson Garrett leased Midford FE Works, but acquired them in 1905.
Before 1890	Mr. S. Butler, Combe Hay Upper Works, near Bath. *
1890	Major S. Evan Butler, Combe Hay Lower Works, Bath. *
1890	Mr. W. H. Handley, South Stoke Works, near Bath. *
1890	Mr. G. Dames owned beds of FE at Midford. *

† From J.J. Greenwood (1982)
* Founders of the Fullers' Earth Union (1890).

Table 5.

	COMPANIES WORKING FULLER'S EARTH
13.3.1883	The Midford Fuller's Earth Works Ltd: C. R. Dames.
16.4.1886	Messrs Cawley & Co. Ltd.: C. W. Cawley.
23.5.1887	London Fuller's Earth Co's properties purchased by Messrs Cawley & Co. Ltd.: C. W. Cawley.
1890	The Fullers' Earth Union Ltd., amalgamation of interests marked* Note 1.
?	Queensborough Fullers Earth Co. liquidated ~1910.
?	The British Fullers' Earth Co., Ltd., Woburn: A. J. G. Swinney.
early 1891	The Fuller's Earth Mining Co., Ltd., Woburn: G. Dames.
1892	Surrey Fullers Earth Co., Ltd., incorporated 27 June, quarry at Nutfield, wound up in 1920, sold to The FEU Ltd.
1893	The Scottish Fuller's Earth Co., Dunning, Perthshire.
22.5.1913	H. N. Garrett (Bath) Ltd., formed with the object of selling to The FEU Ltd in 1915.
1940	Messrs F. W. Berk & Co Ltd, introduced a foundry bonding clay DISTRIBOND.
1954	Laporte Industries Ltd acquired The FEU Ltd.
1971	The Steetley Co. Ltd acquired Berk Ltd.
1978	Brett Bentonite Ltd., Baulking FE, 1983/4 acquired by Laporte.
	Note 1. The Minerals and Fullers' Earth Syndicate Ltd were founder shareholders in The Fullers' Earth Union Ltd.

33i PRICE STATISTICS

What a pity! Fuller's earth isn't here!
Myself, on perusing Thomas Tooke's A History of Prices, London, 1838.

The earliest price known is T Fuller's (1660): 4d a bushel at pit, 16d at wharf in London & 3s at Newbury (ch. 21).

A last, meaning a load, had locally accepted values; it might have been 12 casks in 1693, possibly 2 tons. *A Compendium of Husbandry*, (1805), *1*, 46.

A. Ure, *op. cit.*, 529–530.

John Cary, *English Atlas*, printed (1 Sept. 1787); under "Surrey".

A PITY indeed, because we have discovered so few prices, ex works, ex warehouse or delivered, of fuller's earth before 1850. A few more 17th century prices may yet be found in the lawsuits dealing with the illegal export of fuller's earth. In the case against Edmund Warner for this offence, the price of Kent earth was stated to be 7 Guilders a Last in Rotterdam in 1693, compared with 25 to 30 Guilders a Last for potter's clay.

In 1730, the price of Surrey fuller's earth at the pit was 6d a sack, and 6s a load or ton. Malcolm says the price was 4d a bushel, or 8s a ton in 1744. The sacks must have been strong to contain 186 lb. The bushel held 93 lb, which compares with 70 lb of salt or 56 lb of some cereals. Andrew Ure says also of Surrey fuller's earth that the price "seems to have varied very little these last eighty years", i.e., 1759–1839. Curiously Cary (1787) does not mention fuller's earth as one of the productions of Surrey.

Mr. William P. H. Sinclair of the Scottish Woollen Technical College, Galashiels, kindly sent extracts from the ledger of Richard Lees, of Boldside, Galashiels; they date from about the year 1790:

In Galashiels, Richard Lees sold to John Hogg of Earlston:

4 st. of earth at 7½d a st.	(2/6)
also to Adam Cocran of Galashiels:	
2 st.	(1/3)
to John Hunter of Doveshaugh: 8 st.	(4/6)
to John Lees: 13 st. (8 st.	4/6
(1 st.	7
(4 st.	2/6
to George Mercer, Galashiels, 1 st.	
to Robert Clapperton, Selkirk, 8 st.	4/6
to John Gray, Dryburgh, 4 st. 8 lb.	—
to William Johnston, Galashiels, 2 st.	1/3

His prices were therefore equivalent to £4 10s 0d a ton for hundredweight (45.36 kg) lots and £5 0s 0d a ton for smaller quantites.

By great good fortune a long series of fuller's earth prices have been preserved in the Purchases Day Books and Sales Day Books, dating from 1800, of the firm of James Johnston and Company, of Elgin, Limited, Newmill, Elgin, founded in 1800 by Alexander Johnston. We are greatly indebted to Mr. E. P. Harrison for the most laborious and dusty task of extracting every reference to fuller's earth from so many hundreds of pages of these books.

It was apparently the practice of the Founder of this firm to buy fuller's earth in large quantities, use some himself, and sell the remainder to people in the district. The earth was bought from merchants in London between 1802 to 1845 and from Leith between 1847 to 1871, when the purchases ceased. The earth was brought by sea usually to Burghead, though sometimes to Findhorn, or occasionally, Lossiemouth. At first the earth was brought in casks and hogsheads, but the latter weighing about a ton must have been most inconvenient to handle and we find the firm asking in 1819 for the earth to be shipped in bulk. A 20-ton shipment of 1824 lasted for seven years, but the heaviest traffic was between 1839 and 1848 when 36 tons 11 hundredweights were bought.

A hogshead was (of beer) 63 gallons; 8 bushels (dry weight) were 62 gallons!

From 1847 to 1854 deliveries from Leith were made in sacks, but bulk deliveries were resumed in 1857, the cost of sacks presumably being regarded as too heavy by the merchant. Dundee's first jute power mill was installed in 1854, and there were only four of them in 1872, so we may suppose that jute sacks were in short supply during this period.

Note: 16 power mills in 1882; 39 in 1907; 59 in 1928.

When Alexander Johnston resold the earth, he seems to have been content with a modest 300 per cent; so his enterprise was not unrewarded. But it can be said that the value of fuller's earth was really appreciated in those days.

Although it is never expressly stated that these prices do not include shipping, in every invoice the price per ton is quoted and there are a few extra charges which vary from invoice to invoice and are included under the three items "cartage, wharfage and sufferance" – or demurrage as we should call it now. It would appear at first sight that the prices quoted are the f.o.b. (free on board) prices delivered from the merchants' warehouses, but there is other evidence to show that the prices include sea-freight to Banffshire. There are, for example, no invoices to shippers among the firm's records, only a single note that the sea-freight cost 10/- from London in 1839. In 1809 we have Manning and Bray's statement that the earth was generally sold at the London wharves at 24/- to 26/- a ton. This would indicate that the sea-freight in the period 1807–1811 was about 10/- to 15/- a ton.[1]

The prices of wool, wheat, and alkalis had fallen greatly in the period from 1809 to 1830 and they remained very low for about twenty years. In Fig. 70 the Elgin delivered prices are compared with the price of Spanish Leonesa wool in England. If the enormous price paid during the Peninsular War be omitted, it can be seen that this commodity price rose from 1780 to 1812, fell rapidly till 1820[1] and then more slowly

[1] And especially from 1815. The London merchants Ward, Sky & Wood however did not reduce their price. Very naturally the Elgin firm changed their supplier in 1823, when they noticed they were being cheated.

Table 6 Fuller's Earth Prices (Elgin)

Date after 1800	*Bought from*	*Shipped*	C–Cask H–Hogshead B–Bulk	*Weight cwt.*	*qr.*	*lb.*	*Price per ton*
13.4.02	Branton, Ward and Day of London	Per "Resolution"	1C	4	2	27	40/–
15.4.04	Foster, Brunton and Goddard, London	Per "Banff" to Findhorn	1H	15	–	9	45/–
11.9.05	Branton, Ward and Day	do.	1C	12	3	9	42/–
1.7.07	Wm. and Thos. Jones, London	Per "Macduff"	1H	15	3	–	35/–
5.8.08	Branton, Ward and Day	–	4C	15	–	–	40/–
13.5.09	do.	–	1H	16	–	14	35/–
18.4.10	do.	–	1C	9	2	14	36/8
13.3.11	Thos. Hall	–		13	–	–	30/–
30.12.11	Branton, Ward and Day	–	1C	6	2	–	40/–
7.1.12	Branton, Ward and Day	–	1H	20	–	–	40/–
–.10.12	do.	Sent from Carron Wharf, London	2H	37	2	–	40/–
4.3.14	do.	–	1H	19	2	14	40/–
–.7.14	do.	–	1H	17	3	21	40/–
–.10.14	do.	Shipped to Burghead	2H	40	2	21	40/–
–.9.15	Ward, Sky and Wood, London	–	1H	15	2	21	40/–
	(This firm succeeded Branton, Ward and Day.)						
–.7.16	do.	Shipped to Burghead	1H	17	3	14	40/–
–.10.16	do.	–	1H	15	3	–	40/–
–.6.17	do.	–	1H	18	1	–	40/–
–.1.18	do.	–	2H	27	2	–	40/–
–.9.18	do.	–	1H	7	–	21	40/–
–.11.18	do.	–	2C	5	1	27	40/–
–.8.19	do.	–	16C	44	8	–	40/–
–.	(Asked for in small casks.)						
–.6.20	do.	–	2H	36	–	14	40/–
19.3.21	do.	–	2C	29	–	–	40/–
12.4.21	do.	–	4C	62	2	–	40/–
14.6.22	do.	–	5C	97	–	–	40/–
26.6.23	Wm. and Thos. Jones	–	5H	103	–	–	26/8
2.8.24	do.	–	B	20 tons			27/–
	(Three months' correspondence covering this period missing.)						
7.10.31	do.	–	?	36	1	9	25/–
10.5.32	do.	Shipped to Burghead	B	5 tons			25/–
5.3.34	do.	–	B	5 tons			25/–
1.6.35	Wm. and Thos. Jones	–	B	5 tons			25/–
8.6.37	do.	–	?	35	3	8	25/–
4.1.38	do.	–		15	3	6	25/–
22.3.39	do.	–	B	10 tons			25/–
	(the carriage by sea amounting to 10/–)						
28.4.41	do.	From Dawns Wharf in "Hope"	B	3 tons			30/–

Table 6 *Continued*

Date after 1800	*Bought from*	*Shipped* (C–Cask H–Hogshead B–Bulk)		*Weight* *cwt.*	*qr.*	*lb.*	*Price per ton*
1.6.42	do.	To Burghead	B	6 tons			30/–
5.4.45	do.	–	B	5 tons			21/–
23.1.47	Jas. Wishart and Sons, Leith	Per "Favourite"		120	2	–	25/–)
29.1.48	do.	–		131	–	–	25/–)
28.7.49	do.	Per "Ceres"		21	–	–	40/–)
8.11.52	do.	–		40	3	8	35/–)
10.4.54	do.	–		43	3	4	35/–)
		(The last five items shipped in sacks.)					
11.5.57	do.	Per "Martell"	B	2 tons			40/–
2.4.60	do.	Per "Helen and Mary"	B	3 tons			40/–
30.7.64	do.	–	B	2 tons			46/–
9.6.66	do.	–	B	2 tons			55/–
1.7.70	do.	– (In 24 bags)	B	2 tons			47/6
4.5.71	do.	–	B	2 tons			47/6
20.5.71	do.	– (In 24 bags)*	B	2 tons			54/–

*A pencil note on this entry states: "Dawson's Price to Kynoch 47/6", presumably Kynoch of Keith, with whom James Johnston was very friendly. James Johnston bought dyestuffs from John Dawson and Co., 6 Morrison's Court, Glasgow, but no references to fuller's earth were found in the list of invoices from Dawson's.

till 1855, when there was an upturn. The fuller's earth prices roughly follow these tendencies, but the second rise started in 1845, after being very depressed.

It is not improbable therefore that in 1839 the London price of fuller's earth was only about 15/-, and the ex-quarry price about 4/- per ton. After 1845 prices steadily rose. It is probably safe to assume that Andrew Ure's price of 25/- to 26/- at London wharves was quoted by him, as is the custom of dictionary writers, from earlier sources such as James Malcolm (1805). *op. cit.*, *1*, 45.

The existence from 1803 to 1846 of the Surrey Iron Railway and its extension to Merstham in 1805 is not clearly reflected in the Elgin fuller's earth prices, but the arrival of the London, Brighton and South Coast Railway in 1841 may, at least temporarily, have reduced the freight to London. It certainly increased the competitiveness of the Surrey products in comparison with those of Kent, with their ready access to the Medway.

Gideon Mantell, (1841), writes that some fuller's earth is sent into Scotland through the ports of Leith and Fisherrow. Jas. Wishart of Leith certainly did not take an exorbitant commission for warehousing.

At the time of the formation of The Fullers' Earth Union in 1890, the average price of Surrey and Somerset earths of all grades was

Table 7 The Fullers' Earth Union Ltd
Sales 1894–1920

Sales of earth	*Tons*	*Realising*	*Rate per ton*
Year to 31.3.1894			
Surrey Works	2 563	£6 272– 4– 2	48/11
Somerset Works	1 955	£4 104–10–10	42/-
	4 518	£1 0376–15– 0	45/11
Year to 31.3.1895			
Surrey Works	3 671	£9 000–17– 3	49/2 ½
Somerset Works	2 450	£4 786–18– 6	39/1
	6 121	£13 787–15– 9	45/– ¾
Year to 31.3.1896			
Surrey Works	3 605	£7 950–10– 8	44/1 ¼
Somerset Works	2 937	£5 612–16–10	38/2 ½
	6 542	£13 563– 7– 6	41/5 ½
Year to 31.3.1897			
Surrey Works	4 489	£9 246–12– 6	41/2 ¼
Somerset Works	2 882	£5 340–17– 5	37/1
	7 371	£14 587– 9–11	39/7
Year to 31.3.1898			
Surrey Works	5 874	£11 986–16–10	40/9 ¾
Somerset Works	3 018	£5 702– 1– 4	37/9 ½
	8 892	£17 688–18– 2	39/9 ½
Year to 31.3.1899			
Surrey Works	7 440	£14 536– 6– 0	39/1
Somerset Works	3 206	£5 860–11– 1	36/2 ¾
	10 676	£20 396–17– 1	38/2 ½
Year to 31.3.1900			
Surrey Works	9 892	£18 008–14– 7	36/5
Somerset Works	3 219	£5 793–10– 9	36/ –
	13 111	£23 802– 5– 4	36/3 ¾
Year to 31.3.1901			
Surrey Works	8 238	£16 276– 0– 1	39/6
Somerset Works	3 306	£5 881–11–10	35/7
	11 544	£22 157–11–11	38/5
Year to 31.3.1902			
Surrey Works	10 986	£21 527–19– 8	39/2
Somerset Works	3 116	£5 713– 8– 3	36/8
	14 102	£27 241– 7–11	38/8
Year to 31.3.1903			
Surrey Works	14 117	£28 279–13–11	40/1
Somerset Works	4 047	£7 160–15– 1	35/5
	18 164	£35 440– 9– 0	39/ –

Table continued

Table 7 *Continued*

Sales of earth	*Tons*	*Realising*	*Rate per ton*
Year to 31.3.1904			
Surrey Works	12 331	£26 777–15– 6	43/5
Somerset Works	3 556	£6 354– 8– 1	35/9
	15 887	£33 132– 3– 7	41/8½
Year to 31.3.1905			
Surrey Works	1 1761	£25 608– 5– 5	43/ 7
Somerset Works	3 122	£5 918– 6– 0	37/11
	14 883	£31 526–11– 5	42/ 4
Year to 31.3.1906			
Surrey Works	13 821	£29 660–11– 5	42/11
Somerset Works	3 200	£6 128– 8– 6	38/ 4
	17 021	£35 789– 2–11	42/ 1
Year to 31.3.1907			
Surrey Works	15 648	£33 620– 7– 9	43/ –
Somerset Works	3 240	£6 089–16– 9	37/8
	18 888	£39 710– 4– 6	42/1
Year to 31.3.1908			
Surrey Works	13 115	£28 741– 7–10	43/10
Somerset Works	2 667	£5 182– 0– 4	38/10
	15 782	£33 923– 8– 2	43/ –
Year to 31.3.1909			
Surrey Works	11 936	£25 994– 9–11	43/7
Somerset Works	2 363	£4 560– 3– 9	38/7
	14 299	£30 554–13– 8	42/8
Year to 31.3.1910			
Surrey Works	13 677	£30 533– 9– 7	44/8
Somerset Works	28 40	£5 433– 3– 0	38/3
	16 517	£35 966–12– 7	43/7
Year to 31.3.1911			
Surrey Works	17 622	£40 436– 5– 3	45/11
Somerset Works	2 995	£5 721–17–10	38/ 2½
	20 617	£46 158– 3– 1	44/ 9
Year to 31.3.1912			
Surrey Works	18 324	£42 873– 2– 7	46/9½
Somerset Works	3 710	£6 895– 2– 8	37/2
	22 034	£49 768– 5– 3	45/2
Year to 31.3.1913			
Surrey Works	21 392	£49 138– 1– 8	45/11
Somerset Works	3 209	£6 178– 2– 5	38/ 6
	24 601	£55 316– 4– 1	45/ –
Year to 31.3.1914			
Surrey Works	23 307	£54 591– 9–10	46/10
Somerset Works	2 907	£5 570– 7– 0	38/ 4
	26 214	£60 161–16–10	45/11

Table continued

Table 7 *Continued*

Sales of earth	*Tons*	*Realising*	*Rate per ton*
Year to 31.3.1915			
Surrey Works	24 361	£56 431– 7– 1	46/4
Somerset Works	2 931	£5 533–17– 9	37/9
	27 292	£61 965– 4–10	45/5
Year to 31.3.1916			
Surrey Works	18 826	£45 548– 3– 9	48/5
Somerset Works	4 404	£8 698– 6– 2	39/6
	23 230	£54 246– 9–11	46/8
Year to 31.3.1917			
Surrey Works	19 319	£51 392– 7– 8	53/2
Somerset Works	4 466	£10 637– 0– 3	47/8
	23 785	£62 029– 7–11	52/2
Year to 31.3.1918			
Surrey Works	18 431	£57 020–15– 7	61/10½
Somerset Works	4 545	£12 144–17– 1	53/ 5
	22 976	£69 165–12– 8	60/ 2½
Year to 31.3.1919			
Surrey Works	16 531	£51 797–11– 9	62/8
Somerset Works	3 442	£10 795–16–11	62/9
	19 973	£62 593– 8– 8	62/9
Year to 31.3.1920			
Surrey Works	22 623	£86 054–18– 2	76/1
Somerset Works	3 878	£13 574– 6– 5	70/8
	26 501	£99 629– 4– 7	75/2½

Table 7 kindly provided by Laporte Industries Ltd.

£2 1s 11d a ton, ex works, and the average cost of digging, drying, grinding and sending to market earth raised at Nutfield and Combe Hay was £1 9s 7d a ton. Severe competition was blamed for bringing the price down to this almost uneconomic level, but the general fall in prices after 1873 must have had its effect too. One preparation of earth which used to be sold at £6 10s 0d a ton was then selling at £1 14s 0d a ton. It was the Union's intention to sell lump earth at £2 12s 6d and powdered earth at prices ranging from £3 to £14 a ton, averaging £5 15s 0d a ton. At that time the expected sales were 11,250 tons of lump and 1,250 tons of powdered earth. From these figures it may be inferred that lump earth before the amalgamation cost about £1 7s 6d a ton. Expectations were for an income of £36,718. 15.0, which was not reached until 1906–7, as may be seen in Table 7.

Good prices were not realised at first in the American market. Unexpected competition coming from the fuller's earth of Woburn, where mining had started in 1890, may have kept the prices down. In

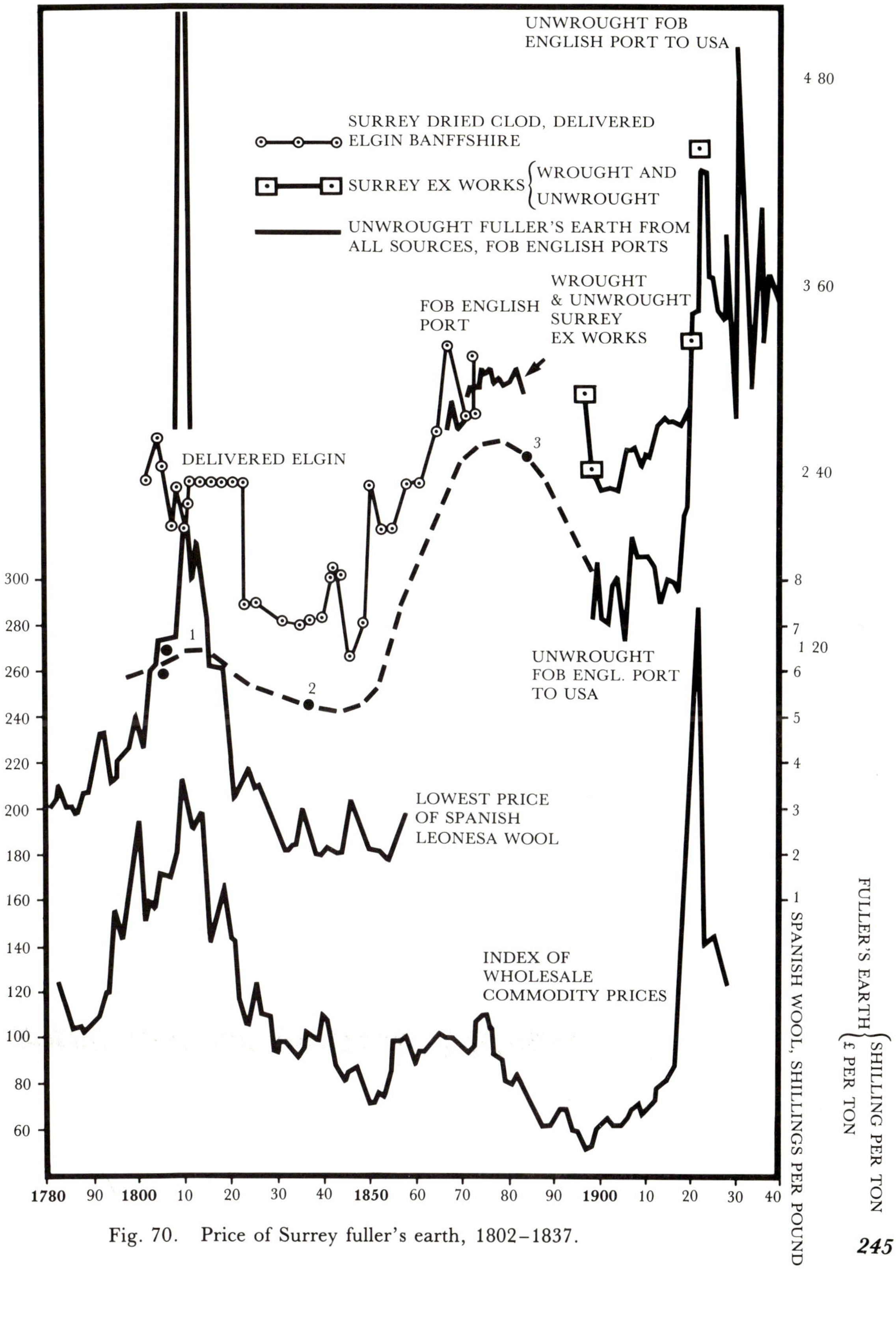

Fig. 70. Price of Surrey fuller's earth, 1802–1837.

33i

any case prices in general were falling from 1890–96, and it was not until 1918 that the 1880 price was reached again.

In Figure 70 the income and average prices cover both dried clod and milled products; they are shown thus □——□. The prices of wrought (milled) earth shipped to America are discussed in chapter 34.

Cameron (1893) writes: *It has been rumoured that should our repositories prove by no means so inexhaustible as was formerly considered, there will be a large rise by-and-by in the price of this commodity.*

33ii

Production figures

NO production figures are known for Kent or Woburn fuller's earth in the early days. The very few figures for Somerset and Surrey possibly reflect the gradual competition from soap and an increase in sales after 1825 when the export ban was lifted.

Official statistics began in 1854 in time to show the dramatic fall in production which occurred during and after the American Civil War. To the figures given by Highley (1972) I have added the average annual production in each decade.

D. E. Highley, Fuller's Earth, Mineral Dossier No. 3, HMSO (1972), p. 17, quoting Ministry of Fuel and Power, Report of the Mineral Development (Westwood) Committee, *Cmd* 7732 (1949) for the years 1854–1947.

Fig. 71 shows that in 1915 The Fullers' Earth Union produced about 88 per cent of the total; and the American market that year absorbed 64 per cent of the total English production. By 1933 very little fuller's earth was shipped across the Atlantic.

Table 8 United Kingdom: Production of Fuller's Earth by decades, 1854–1970

Period	Thousand tonnes	Average tonnes per annum
(1854–1860)	85	12 140
1861–1870	81	8 100
1871–1880	na	na
1881–1890	154	15 400
1891–1900	178	17 800
1901–1910	242	24 200
1911–1920	303	30 300
1921–1930	239	23 900
1931–1940	203	20 300
1941–1950	446	44 600
1951–1960	790	79 000
1961–1970	1,347	134 700
Total: 1854–1970	4,068	

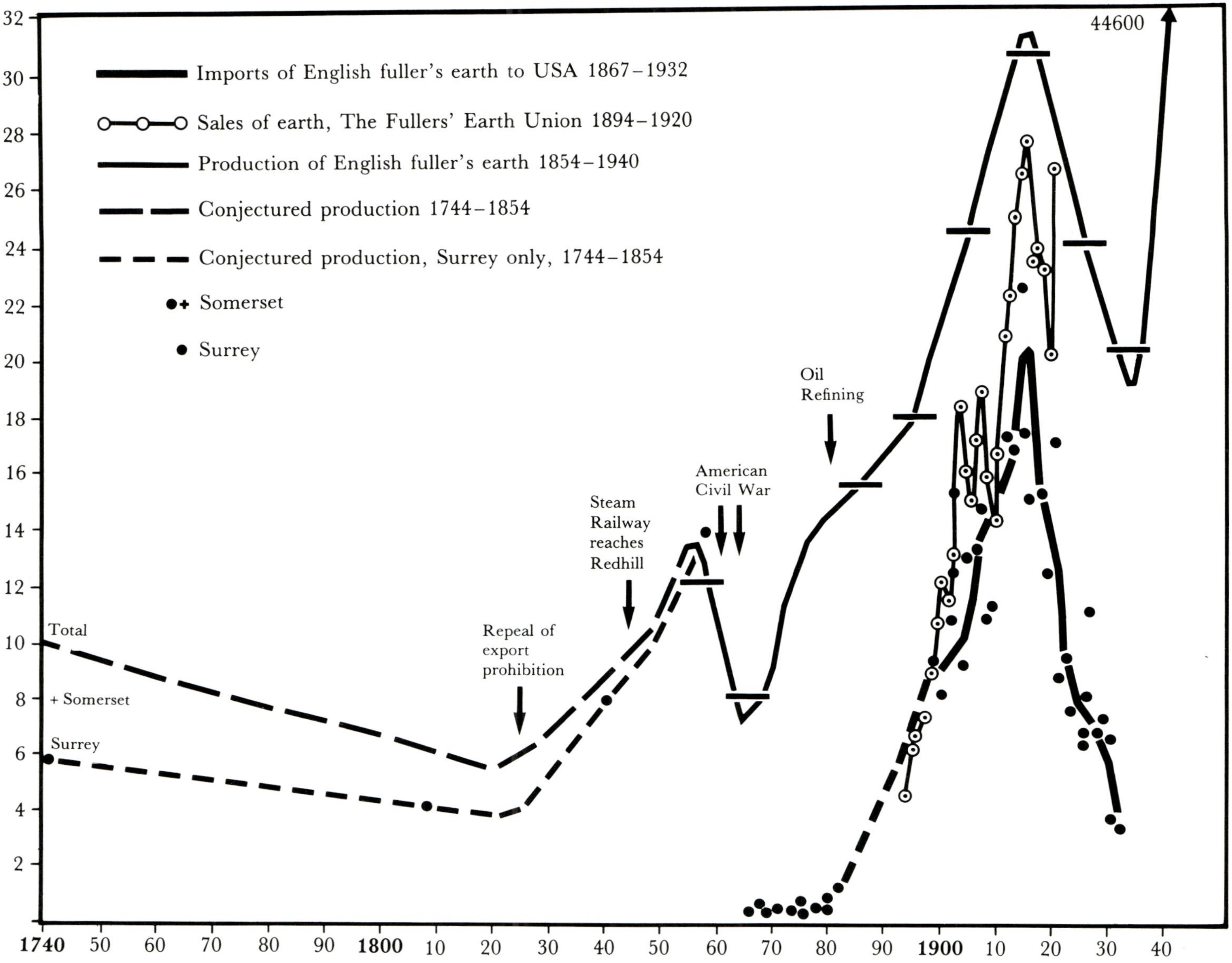

Fig. 71. Production statistics 1744–1940. Various sources.

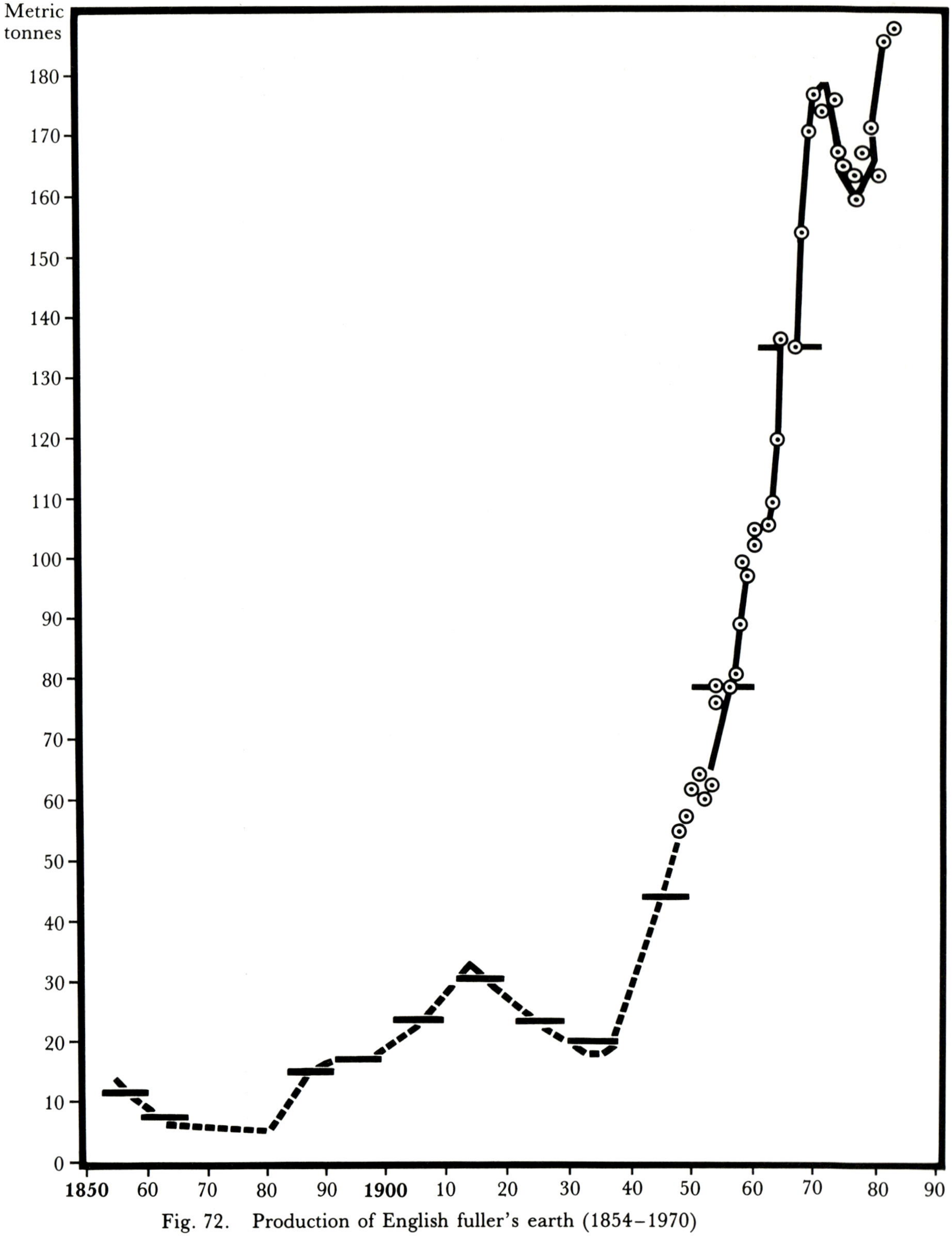

Fig. 72. Production of English fuller's earth (1854–1970)

Table 9 United Kingdom: Production of Fuller's Earth, 1948–1982

Year	Thousand tonnes	Year	Thousand tonnes	Year	Thousand tonnes
1948	55	1960	102	1971	174
1949	58	1961	104	1972	174
1950	62	1962	105	1973	185
1951	64	1963	109	1974	166
1952	60	1964	119	1975	164
1953	63	1965	136	1976	163
1954	76	1966	140	1977	159
1955	79	1967	135	1978	167
1956	79	1968	153	1979	171
1957	81	1969	170	1980	163
1958	89	1970	176	1981	185
1959	97			1982	187

The meteoric rise in production of fuller's earth in England since 1933 is shown in Figure 72 and since 1948 in Table 9, compiled by the British Geological Survey, then the Institute of Geological Sciences, by adding the production figures supplied to them by the producing companies.

D. E. Highley, *op. cit.* (1948–1970), personal communication (1971–1982).

The sharp post-war (WWII) increase in production was largely due to the great expansion in output of activated earths and foundry bonding clays by The FEU Ltd., as well as grades for other developing applications – Berk and then Steetley swelling the total. In 1970 activated earths accounted for about 40,000 tonnes. The fall away between 1974 and 1980 was due to the economic consequences of the rise in oil prices.

34 EXPORTS TO AMERICA

J. Franklin Inst. (1900), *150*, 214–223.

DAVID T. Day gives the imports of English fuller's earth into the United States in long tons for years ending June 30th between 1867 and 1883. Except at the end of this period, most of the earth imported was unwrought or unmanufactured, that is to say lump fuller's earth to be used mainly for fulling or in pharmacy. Moreover the prices are f.o.b. English port.

The London f.o.b. prices are given in the two tables which follow.

Table 10 Imports of Fuller's Earth into the United States.
1867 to 1883, in Long Tons (Wrought & Unwrought).

Year ending 30th June	Long tons	Value $	£/ton f.o.b
1867	280	3 113	2.29_6
1868	211	2 522	2.46_9
1869	324	3 587	2.28_7
1870	239	2 619	2.26_4
1871	290	3 383	2.41_0
1872	274	3 358	2.53_2
1873	251	2 978	2.45_1
1874	277	3 440	2.56_6
1875	300	3 694	2.54_4
1876	246	3 097	2.60_1
1877	400	4 469	2.30_4
1878	335	4 095	2.52_5
1879	361	4 269	2.44_3
1880	578	6 925	2.47_5
1881	268	3 207	2.47_2
1882	908	11 444	2.59_0
1883	1241	14 309	2.36_9

RATES OF EXCHANGE used in converting dollar values into £/ton f.o.b.	
1867–1881	4.84 $/£
1882–1919	4.867
1920–1921	3.93
1922	4.42
1923	4.66
1924	4.35
1925	4.79
1926	4.86
1927–1930	4.867
1931–1933	
1934–1937	4.867

The next table gives the imports into the USA of unwrought and wrought fuller's earth.

Table 11 shows that the imports of English fuller's earth into the United States had been about 267 tons a year from 1867 to 1873; 306 in the next four years; and 385 from 1878 to 1881. The new demand for bleaching earth for animal and vegetable and fat refining was probably about 468 tons in 1882 and 781 tons in 1883.

Table 11 Imports of Unwrought and Wrought Fuller's Earth

All in long tons, U.S. dollars and £/long ton f.o.b. English port.

Year ending 31st Dec.	Unwrought fuller's earth: Long, tons	Unwrought fuller's earth: Value £	Unwrought fuller's earth: £/long ton	Wrought fuller's earth: Long tons	Wrought fuller's earth: Value $	Wrought fuller's earth: £/long ton	Unwrought + wrought long tons	U.S. Production of attapulgite (mainly) Long tons 6161[b] 8814[c]
1897	2 308	14 283	1.27_1	2 138	20 037	1.92_5	4 446[a]	15 279
8	2 038	15 921	1.60_5	6 315	55 123	1.79_3	8 354	13 268
9	3 743	23 194	1.27_3	6 577	46 446	1.45_1	10 320	11 054
1900	2 431	14 750	1.24_7	5 742	50 047	1.79_1	8 173	8 659
1	2 916	17 230	1.21_4	7 850	63 467	1.66_1	10 766	12 600
2	3 785	26 635	1.44_6	9 728	75 945	1.60_4	13 513	10 261
3	3 804	28 339	1.53_1	11 464	92 332	1.65_5	15 268	18 476
4	1 763	9 546	1.11_2	7 363	64 460	1.79_9	9 127	26 321
5	1 522	12 798	1.72_7	11 480	93 199	1.66_8	13 003	22 480
6	2 594	20 129	1.59_5	10 643	88 566	1.71_0	13 237	28 607
7	2 223	16 833	1.55_6	12 425	105 388	1.74_3	14 648	29 331
8	2 110	16 242	1.58_2	8 753	77 171	1.81_2	10 863	26 530
9	1 609	12 492	1.59_5	9 777	88 659	1.86_3	11 386	29 898
1910	1 929	14 399	1.53_4	12 881	118 146	1.88_5	14 810	29 305
1	1 679	10 877	1.33_1	14 592	132 717	1.86_9	16 271	36 337
2	1 759	11 619	1.35_7	15 303	133 718	1.79_5	17 062	29 210
3	1 711	12 344	1.48_3	14 921	133 657	1.84_0	16 632	34 460
4	1 311	9 283	1.45_5	20 990	185 800	1.81_9	22 293	36 590
5	759	5 176	1.40_1	16 599	147 317	1.82_3	17 358	42 670
6	1 011	7 742	1.57_4	13 990	131 922	1.93_7	15 001	60 555
7	1 287	11 718	1.87_1	13 887	164 699	2.43_7	15 173	64 792
8	969	12 636	2.68_0	14 140	213 599	3.10_4	15 109	75 418
9	333	4 301	2.65_3	12 054	185 410	3.16_0	12 387	94 772
1920	1 355	19 793	3.71_6	15 622	202 100	3.29_2	16 978	114 721
1	481	6 172	3.64_2	8 269	113 243	3.48_5	8 700	94 294
2	542	7 413	3.09_5	8 895	128 282	3.26_3	9 437	124 057
3	573	8 252	3.08_9	7 058	105 692	3.12_1	7 631	133 155
4	264	3 385	2.94_4	6 255	89 103	3.27_5	6 520	158 923
5	192	2 619	2.84_8	6 964	108 676	3.25_8	7 158	184 441
6	141	2 290	3.34_0	7 982	121 384	3.12_9	8 123	209 064
7	491	5 679	2.37_6	6 277	103 602	3.39_0	6 768	236 141
8	316	6 710	4.36_2	6 463	125 293	3.98_3	6 779	256 261
9	543	11 583	4.38_4	6 870	140 849	4.21_3	7 413	282 128
1930	335	5 805	3.46_1	6 115	150 715	5.06_5	6 460	299 682
1	121	1 534	2.59_6	3 462	47 430	2.88_0	3 583	257 500
2	86	1 040	2.49_3	3 383	32 927	2.00_0	3 469	225 805
3	15	260	3.51_9					
4	30	406	2.74_8					
5	122	1 876	3.15_1					
6	63	960	3.11_1					
7	40	569	2.91_0					

From *Minerals Year-books*, especially 1932–3.

[a] July to December only

[b] 1895

[c] 1896

Unfortunately import figures were not recorded from then until 1897, as shown in the next table. Here we see that unwrought, that is lump, fuller's earth had risen to 2308 tons, and reached a peak of 3804 tons in 1903. One cannot be sure that milled earth – wrought earth – was not sometimes shipped under the unwrought category so as to avoid the higher import duty, but even so the demand for English fuller's earth for fulling, pharmacy and household purposes fell to a low point of 15 tons in 1930. The decline was accelerated by Dr. Charles L. Parsons (1867–1954), who moved to Washington in 1911 as chief of the division of Mineral Technology in the Bureau of Mines and became chief chemist of the Bureau in 1916. As a government official, he advocated greater use of neglected natural resources. He directed attention to the unnecessary importation of fuller's earth from Europe.

This alleged deceit was the reason for litigation at one time.

On the other hand the wrought or powdered earth for oil refining rose from 2138 tons in 1897 to 20,990 tons in 1914, but fell to 14,140 tons in 1918 and 3,383 tons in 1932.

Fuller's earth, W. W. Adams & C. Galiher, *Minerals Yearbook*, (1932–33) p. 711.

Fig. 73. Charles Lathrop Parsons (1867–1954)

Some comparisons with American attapulgite

AMERICAN output statistics are shown in Figs 74 and 75. There was a peak in price in 1920 and of production in 1930, when a decline set in.

This decline continued until 1940, when new uses were developed. By far the most important were the absorbent uses – cat litter, mopping up oil spills and the like – which saved the industry from collapse. Calcium montmorillonite apparently is not included in this figure.

These figures may be compared with a diagram showing American attapulgite sold or used by producers for specified uses. The world

Fig. 75
Clays, Sam H. Patterson and Haydn H. Murray, *Industrial Minerals and Rocks*.

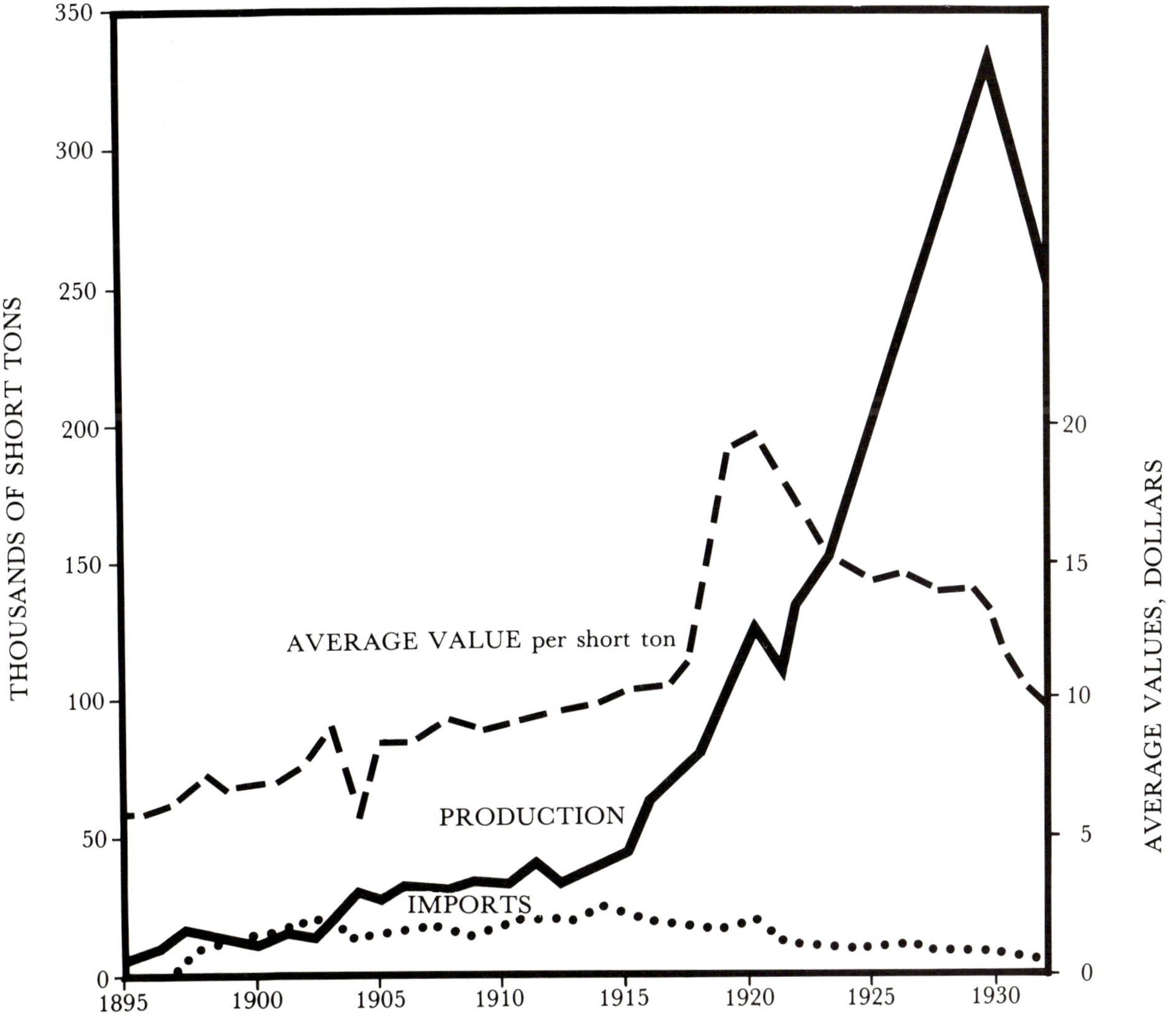

Fig. 74. Trends in production and average value per short ton of mainly attapulgite in the US, 1895–1932, and imports of English fuller's earth.

recession of the 1930s and the decline in the use of granulated attapulgite in the percolation-refining of mineral oil is apparent, but the steady climb in production of English fuller's earth between 1951 to 1959 is not reflected in the U.S.A. where business stagnated till 1962, after which absorbent uses increased phenomenally.

Clays, Sam H. Patterson and Haydn H. Murray, *Industrial Minerals and Rocks*

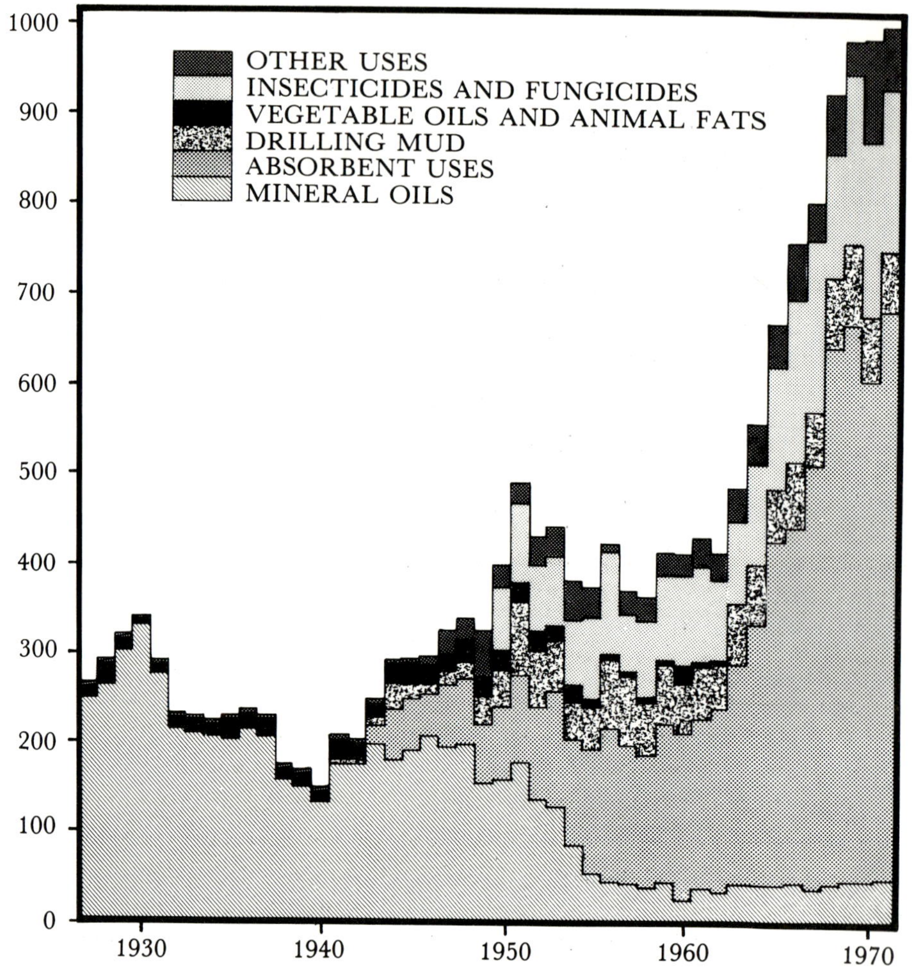

Fig. 75. Production of US attapulgite for specified uses 1927–1971.

A HISTORY OF THE CHEMICAL ANALYSIS OF ENGLISH FULLER'S EARTH

35

Hampshire Fuller's Earth

35i

THE first chemical analysis of fuller's earth was made by Torbern Olaf Bergman (1735–1784), who became professor of chemistry and mineralogy at the University of Uppsala in 1767. He was famous in his day for his improvements in analytical techniques, though he is remembered now more for his founding of the modern doctrine of chemical affinity. His works were collected and printed in six volumes called *Opuscula physica et chemica* between 1779 and 1790, and were translated into French, German and English. Bergman reported the analyses of three *Lithomarges*: Lemnian earth and the fuller's earths from Osmund in Dalecarlia and from Hampshire – on 15th June 1782.

Opuscula physica et chemica, vol. 4, pp. 152-156.

lithomarge, a compact marl.

Of the last he states that it is dug in the district of Hampshire in England, but its export was prohibited. Nevertheless enough of it was available abroad in mineralogical collections. In former days *it was used by fullers but now in various English factories cloth is finished by other means, namely by treading them underfoot with swine's dung and warmed human urine.*

In the chemical analysis of this earth, Bergman found in a hundred parts by weight: silicei farinacei a ferro flaventis 51.8, calcarei 3.3, magnesiae 0.7, argillacei 25, ferri calcinati 3.7, and humidae volatilisque materiae 15½. In the French edition (1792) the analysis is given as terre siliceuse 0.518, chaux aérée 0.038, magnésie aérée 0.007, argile 0.250, chaux de fer 0.037; et eaux et matière volatile 0.155.

Known as the *Manuel du Minéralogiste ou Sciagraphie du Règne Minéral*, Paris (1792), p. 238.

Richard Kirwan (1784) refers to Bergman's work in his textbook. Here the constituents are termed siliceous earth, argillaceous, magnesia, mild calcareous earth, de-phlogisticated iron and water. Kirwan, who did so much to establish the chemical analysis for determining mineral species, can hardly be congratulated for publishing as an analysis of fuller's earth the *average* values for the three "lithomarges" of Bergman, viz. silex 53, argill 18, calcareous 5, magnesia 3, iron 4, water 17, *and marine acid at a medium* that is to say a modest amount of chloride ions.

Elements of Mineralogy, (1784), p. 75.

p. 404.

35i

Bergman and Kirwan were both members of the phlogiston school who believed that many substances contained two components, one combustible or the phlogiston and the other incombustible or the fixed residue. When iron filings were burned they form a *calx*; and phlogiston was held to have passed out of the iron. Hence dephlogisticated iron is what we call ferric oxide and semi-dephlogisticated iron ferrous oxide. It was only when Black and Lavoisier showed that substances *increased* in weight on oxidising that the theory, which had at least stimulated thought and advance, declined and died with the unrepentant upholder Joseph Priestley in 1804.

Outlines of Mineralogy, (1809), *1*, 176.

The discrepancy in the calcium oxide content given in the Latin and French editions of Bergman is not serious, but Dr. Kidd (1809) claims to be quoting Bergman when he gives for Hampshire fuller's earth: silica 51.8, alumina 23.0, magnesia 7.0, carbonate of lime 3.3, oxide of iron 3.7, water and other volatile matter 5.5 – making a total of 94.3 per cent. No reason is given for his altering the figures. The original version is however quoted correctly by Aitken (1807) and Herbert (1840).

Aiken's *Dictionary of Chemistry and Mineralogy* (1807), *1*, 449: Herbert's *Encyclopaedia* (1840), *1*, 570

Brierley, *Hampshire Field Club* (1891), 2, pt.1, 84–88.

J. Brierley published an analysis of fuller's earth recovered from a well-boring at Greyshott Down in 1891: silica 45.56, alumina 14.70, oxide of iron 9.31, lime 5.44, magnesia 6.21, soda trace, moisture and volatile matter 18.80 per cent; in all 100.02. Its magnesia is strangely high and its silica low: it may not have been a smectite.

35ii

Surrey Fuller's Earth

M.H. Klaproth, *Analytical Essays*, (1797); translated from the German in (1801), London, pp. 248–254.

THE first chemical analysis of Surrey fuller's earth was made by Martin Heinrich Klaproth in 1797: silica 53.00, alumina 10.00, oxide of iron 9.75, lime 0.50, magnesia 1.25, sodic chloride 0.10, potash traces; moisture and other volatiles 24.00; "loss" 1.40, in all 100.00 per cent. This analysis was quoted in the *Rural Cyclopaedia* (1847); in Mitchell's *Dictionary of Chemistry; Mineralogy and Geology* (1864); and in the *Encyclopaedia Britannica* (1874).

(1847), *6*, 382.

(1864), 262.

(1874), *9*, 816.

The next analysis (quoted in the *Penny Cyclopaedia* (1838) came from Thomas Thomson's *Outlines of Mineralogy, Geology and Mineral Analysis* (1806). Thomson, who was a Professor at Glasgow University, was the first to express the constituents as atomic proportions:-

(1838), *11*, 13.

London (1806), *1*, 246.

Chemical analysis: Surrey F.E.

	Per cent	Atoms	
Water	24.95	22.17	2.16
Silica	44.00	22	2.14
Alumina	23.06	10.25	1.00
Lime	4.08	1.16	0.11
Magnesia	2.00	0.80	0.07
Protoxide of iron	2.00	0.44	0.04

Although it was not until 1811 that Avogadro published his hypothesis, Thomson did somehow manage to obtain roughly the right proportions for the main oxides, though not for the atoms as stated. He regarded lime and magnesia as impurities. After Avogadro he became one of the leading proponents of the atomic theory.

35ii

Andrew Ure (1867) published an article by the Director of the Geological Survey, Mr. H. W. Bristow, F.R.S., in which analyses of blue and yellow fuller's earth from Reigate were given:

A. Ure, *Dictionary of Arts, Manufactures and Mines*, 6th ed. (1867), *2*, 412.

Chemical analyses, Blue and Yellow F.E.

	Blue	Yellow
SiO_2	42	44
Al_2O_3	18	11
Fe_2O_3	6	10
CaO	4	5
MgO	2	2
Na_2O	5	5
Loss on ignition	23	23
	100	100

The erroneously high soda content was noticed by Brierley. The yellow earth was more particularly described as an upper bed of greyish red earth six feet thick, and the blue earth as a lower bed of bluish-grey earth ten feet in thickness.

op. cit.

Brierley compared these analyses on a moisture-free basis, but no valid deductions could be drawn from such inaccurate results.

Another early analysis, the origin of which is unknown, is quoted by James Malcolm (1805); Nutfield fuller's earth is here said to contain silex 46.5, argillaceous 27.15, lime 8.7, oxyde of iron 1.3, carbonate of magnesia 1.5, water 13.5; adding up to 100.0.

A compendium of husbandry, (1805) *1*, 46.

In the earliest period of chemical analysis, as we have seen, there was a great variety in the way the results were expressed. We have seen the last stages of the phlogiston period and the beginnings of a more modern terminology. The nature of alumina was obscure to many early workers who called it "argillaceous" or Thon in Germany, where it is still often called Tonerde. Dr. Joseph Bonavita Blank (1810) quoted Bergman's analysis as: 51.8 Kiesel, 25 Thon, 0.7 Talk, 3.3 Kalkerde. 3.7 Eisenoxyd, 15.5 Wasser oder flüssige Stoffe. The translation of magnesia by talc is astonishing. Joseph Black first used the word alumina in English (before 1803).

Handbuch der Mineralogie, Würzburg: F. E. Nitribitt, (1810), 215.

Regarding 'alumine', see ch. 52.

Other English Fuller's Earths

35iii

WOBURN, Kent and other English fuller's earths do not appear to have been chemically analysed before the 1880s, when more reliable

35iii

Institute of Geological Sciences *Annual Report* for 1978.

procedures were in use; Kent was analysed in the 1950s, Woburn and Bath in the 1890s, Baulking 1978, and Tillington 1979. Apart from the work carried out within laboratories of The FEU and largely unpublished, the most detailed and accurate analyses of Surrey and Bath were carried out by the Laboratory of the Government chemist in 1978.

35iv

Ionic composition

H. Le Chatelier, *Bull. soc. franç. mineral.*, (1887), *10*, 204–211.

J. D. Dana, *System of Mineralogy*, 6th ed., (1914).

C. E. Marshall, Z. *Krist.*, (1935), *91A*, 433.

R. H. S. Robertson, *Clay Min. Bull.*, (1961), *4*, p. 285.

K. Vogt & H.M. Köster, *Clay Minerals*, (1978), *13*, 25.

S. I. Tsipurski & V. A. Drits, *Clay Min.*, (1984), *19*, 177–193.

LONG after Thomson's attempt at calculating the chemical composition of fuller's earth, Le Chatelier presented the formula, 4 SiO_2. Al_2O_3. H_2O + aq. and this was accepted by J.D. Dana in 1914. C. Edmund Marshall (1935) showed how to calculate cation exchange capacity from the chemical analysis of a montmorillonite or a beidellite; in the former a smaller proportion of the silicon atoms were replaced by aluminium atoms than in the latter. Even so the tetrahedral substitution resulted in the apparent silicon content of $Si_{3.48}$, $Si_{3.59}$ and $Si_{3.83}$ in the examples given. However a sample finer than 2 micrometers and therefore rather pure, gave $Si_{3.98}$.

In 1961 I calculated the ionic composition of ten chemical analyses of English Aptian fuller's earth, having removed from the analyses all impurities thought to be present. The mean figure arrived at was $Si_{3.965}$. However, in recent years I have found that very good chemical analyses of montmorillonites of which the impurities were more accurately determined had $Si_{3.985 \pm 0.015}$, though Bathonian fuller's earth *may* have a higher degree of substitution by aluminium; but even in this clay the underestimation of illite would give the low results obtained ($Si_{3.83}$ – $Si_{3.84}$).

Vogt and Köster found 12 out of 19 ionic compositions of calcium montmorillonites to be within the narrow range of $Si_{3.982 \pm 0.015}$; and I found that a calcium montmorillonite from Kimolos, Greece, had $Si_{3.99}$ in the unit cell. On the other hand in Wyoming bentonite more silicon is apparently substituted by aluminium atoms and in the mineral beidellite tetrahedral exceeds octahedral substitution, though in a fine fraction of Wyoming bentonite $Si_{3.99}$ has been found, the same as in the white bentonite of S'Aliderru, Sardinia. Tsipurski and Drits found 3 bentonites in the USSR had Si_4, two had $Si_{3.98}$ and one $Si_{3.95}$. Perhaps tetrahedral substitution in many industrial bentonites is a myth.

PROCESSES

36

"I was admiring your fuller's earth", said I; "I think that I should be better able to advise you as to your machine if I knew what the exact purpose was for which it was used."
SIR ARTHUR CONAN DOYLE, The Engineer's Thumb, 1891.

THE first mechanical processes to be applied to fuller's earth were dying and milling, followed many years after by the solitary wet purification at Midford. The next major advance in processing was the acid-activation of calcium montmorillonite and, subsequently, the replacement of calcium by sodium ions. These will be discussed under Germany in the 20th century; and the synthesis of a smectite under England.

Chapter 37.

Chapter 38.

The only 'process' to be found in this chapter is the development of quarrying methods.

Development of Quarrying Methods for Fuller's Earth

36i

Nor does the miner neglect the digging of "extraordinary earths", whether they are found in gold mines, silver mines, or other mines; nor do other miners neglect them if they are found in stone quarries, or in their own veins.
GEORGIUS AGRICOLA, De re metallica, (1556), Book 5

Translation by H.C. Hoover and L.H. Hoover, *The Mining Magazine*, London (1912) and Dover Publications, Inc., New York (1950) p. 115.

THE Hoovers in their translation added a note that the "extraordinary earths" were such substances as ochres, tripoli, fullers earth, potters' clay, clay used for medicinal purposes, etc., etc.

The history of extracting fuller's earth in Surrey, may be taken as an example. The Lower Greensand ridge running through Redhill and Blechingley was once heavily wooded: fuller's earth outcrops were near the top of the ridge. It was probably first exposed in the roots of fallen

36i

trees. The first workings would certainly have been along the line of the outcrop and would have been got by simply digging the earth away. As the earth deposit dips more steeply towards the north than the land surface the overburden increases towards the north. The first quarries were made with very little overburden and must have lasted for many centuries. All that could be seen in 1940 along the site of the most ancient workings was a little disturbance of the natural slope of the hill, a feature but barely discernible.

In the late 18th century, the agricultural revival led to a great increase in marling and liming of the land, and innumerable marl pits were dug wherever suitable clay was available. These were often in the form of a narrow shaft which belled out at the bottom, like an inverted mushroom, until the roof seemed to become dangerously insecure. Then a new shaft was sunk and so on. This method was used for getting fuller's earth on Redstone Hill probably in the period 1800 to 1850, though open quarries were also being worked elsewhere at the time.

The quarry nearest the outcrop in the Chartfield area was worked so carefully that all the overburden was replaced as a flat surface and it is difficult to tell that the area has been worked at all. The tidy operator was probably John Hynam.

Fig. 76. Quarrying about 1900 at Redhill

From a photograph owned by The Fullers' Earth Union Ltd.

According to Sowan, adit-mining was undertaken between 1861 to 1898, often in periods when the demand for the earth was low. He records mines at Park, Church Hill, Beechfield, Capenor and Cockley, (an experimental mine was opened in 1962 in the grounds of Robert Denholm House, formerly Capenor. The three 300 foot headings though timbered every three feet, very unstable, were abandoned in 1964 and eventually collapsed). Quarrying at Patteson Court also broke through into underground workings.

Paul W. Sowan, Fuller's Earth Mine discovered at Nutfield: Historical Notes, *Cave Research and Exploration Report*, 12 April 1981. He quotes H. Pearman, Secret Tunnels in Surrey, *Records of the Chelsea Speleological Society*, (1963), *3*, 45.

About 1890 to 1910 the quarries were becoming larger and deeper. The growing overburden began to present a problem in its disposal. In this period it was carried in wheelbarrows by men who steered them over planks held on tall trestles. Fig. 76. These plankways were at a considerable height and the men had to be most skilful to manoeuvre their charges over this greasy and hazardous course. If a fall was inevitable they took care to throw the barrow to one side and to cast themselves over the other. This method of quarrying could be seen in Arnold's silver sand quarry beside Watling Street near Leighton Buzzard as late as 1948.

Old workings were discovered in the 1950s when the Cockley works were being expanded. A large opening running south to north west of the works was perceived as the quarry on that side developed; and there was clear evidence of "Bell Mouth" mining.

The next stage in quarrying technique was the introduction of giant cranes. Figs 77, A,B. The crane in the Copyhold quarry was a familiar object to travellers on the Brighton line. It was one of the largest quarry cranes in England. The jib was 100 feet long and a three-ton load could be hoisted at a distance of 80 feet. The crane was carried on as many as 32 bogey wheels. Every now and then the crane bed of rails had to be moved further north and this represented a major operation for the quarrymaster. One disastrous morning the huge crane ran down the slope from one level to the next, and crashed.

Made by Stothert & Pitt, Ltd., Bath; Charles Baker came from them to Redhill to become quarrymaster.

One disadvantage of this system of quarrying is that the overburden is laid behind the crane in a continuous mountain-ridge which breaks up the natural contours of the countryside and prevents the land from being used for certain purposes afterwards. The Fullers' Earth Union planted conifers in some places, but harvesting was difficult.

Stripping overburden from fuller's earth, Anon., *Pit & Quarry*, (Aug. 1920), 87–88 (with illustrations).

Since 1942 the Company made more use of draglines, diggers and dumpers, and with these machines the overburden could be dumped anywhere at will. Then at the end of the War, The FEU made a deal with Nutfield Parish by which they extracted the earth from recreation grounds, including a sloping playing field, and restored and levelled the area to give a better sports field. Later, a public subscription was raised to build a pavilion and youth centre as a War Memorial and the company was the main contributor.

Charles Baker, having transferred to Copyhold Works, Richard Ramm and his team took over as quarrymasters.

By 1945, the quarries had reached a depth of over 150 feet, whilst higher volume and cheaper extraction were vital. In that year, the Board authorised the large-scale mechanisation of quarrying. The project had been initiated by Col. W. T. Raikes and R. W. Nichols, the Production Director, and they were later helped by Mr. E. G. H. Jeffery, a major in the Royal Engineers during the War and member

46 metres

Mine & Quarry Engineering, (Aug. 1957), 326–333: photos. *Quarry Managers J.*, (1963), *47* (3), 103–106, photos.

of a firm of consulting engineers immediately after it. As a result, bulldozers and draglines and, later, enormous tractor-scrapers were introduced into The FEU operations. The quarry labour force went down to about one third of the original number of 150 men, and, at the same time, output vastly increased. Moreover, the improved methods allowed extensive restoration and landscaping, so that neither scenery nor traditional land uses were spoilt, and the resources and amenities of the neighbourhood preserved.

ILLUSTRATIONS ON FACING PAGE

(Left-hand page, top)
Fig. 77A. Quarry in Folkestone beds and Sandgate Beds, Nutfield. A general view of the Cockley pit of The Fullers' Earth Union Ltd. at Nutfield, showing the removal of the overburden. The main fuller's earth seam is the dark layer at the bottom of the quarry face. The surface of the ground forms the dip-slope of the Lower Greensand.

(Left-hand page, bottom)
Fig. 77B. Fuller's earth in Sandgate Beds, Nutfield. The main bed of fuller's earth in the Cockley pit is about 10 feet thick; it dips north at about 8°.

THE TWENTIETH CENTURY

37–54

FULLER'S earth had long been known and used in many countries in the world, but when its use for refining oils provided a new market, search for the mineral was stimulated. The bleaching power of German montmorillonite, when dried and powdered, was not satisfactory, but it was discovered that treatment with mineral acids increased the bleaching power very markedly. Other countries followed Germany not only in setting up acid-activation plants but also in adopting the conversion of calcium montmorillonite to a swelling clay like Wyoming bentonite by the addition of sodium carbonate.

European countries are arranged in the order in which significant scientific and technological advances were made, starting with Germany. Japan, where early scientific investigations were made comes next, but the USA, which should come third, is deferred to chapter 53, even though acid-activation was started about 1920. The chapters are:-

	Date	ch:		Date	ch:
Germany	1904	37	Hungary	1937	46
Japan	1912	39	Italy	1938	47
England	1925	38	Czechoslovakia	1943	48
France	1932	40	Greece	1946	49
USSR	1933	41	Cyprus	1950	50
Austria	1935	42	Yugoslavia	1957	51
Romania	1935	43	Switzerland	(1945)	52
Spain	1935	44	U.S.A.	1920	53
Poland	1936	45	Other countries	–	54

37 GERMANY

An manchen Stellen im Lande* [*Württemberg*] *wurde die* Walkerde *gewonnen.

MANFRED FRANK, (1949).

Technologische Geologie der Bodenschätze Württembergs, Prof. Dr. M. Frank, Stuttgart: E. Schweizerbart: sche Verlagsbuch-handlung (1949), 436, quoting Fraas (1860), 163.

The place name Walkenried in Brunswick also suggests fulling.

IN FORMER times "Walker's earth" used to be dug at many places in Württemberg, not only in the Keuper marl directly under the Bonebed-sandstone (Balingen), but in the middle Brown Jurassic (Aalen) and in the upper White Jurassic (Heidenheim), all for the fulling of cloth.

im heiligen Kreuz.

In the "Lower Keuper" (Keuperloam) such fuller's earths were worked in the 19th century near Deckenpfronn (from lots [pieces of land] in Müllerweg; at Berg in the Holy Cross; at Egelsee; at Hölderle and other places) and taken to Calw, where they were used for scouring wool. At the turn of the century the last of the loam diggers of Deckenpfronn died. Very little is remembered of this old use of the clay by the local people. The deposits were too small to be worked industrially according to Frank (1949), because with modern communications raw materials can be sent from only a few centres to supply a wide area.

1851 Exhibition (London) Catalogue, *3*, 1132.

The Government Mining Engineers of Nassau in the Zollverein of the German States exhibited fuller's earth at the 1851 Exhibition in London. They explained that in some parts of the country, deposits of this substance had been worked for a long period; and extensive deposits had been recently found in other districts.

Manuel de Minéralogie, Paris (1862), *1*, 204.

A. des Cloiseaux (1862) gives Vaels near Aix-la-Chapelle (Aachen) as one of the chief sources of fuller's earth in Europe, as well as Rosswein in Saxony.

American Petroleum Institute Project 49, Reference Clay Localities – Europe (Preliminary Report No. 4) (1950), Columbia University, p. 84.

Ralph J. Holmes (1950) visited the famous refractory clay area in the Westerwald, a section of the Slate Mountains in Hessen-Nassau. Altered basaltic tuffs in this region were formerly a principal source of fuller's earth but the mining of the "Walkerde" had long been abandoned and attempts to locate some of the old workings and secure samples were unsuccessful. The oldest is at Breitscheid, dating from the late 18th century.

Acid activation

A RECORD of the early days of the acid-activation process in Germany was given by A. Löb of Kitzingen am Main[1] and by Dr. Oscar Kausch in a book on silica gel and the bleaching earths.[2] Other books which tell the story are *Die Bleicherde* by O. Eckart and A. Wurzmüller;[3] F. Krczil's *Untersuchung und Bewertung technischer Adsorptions-stoffe*;[4] and most recently by a commemorative volume on the occasion of "125 Jahre Süd-Chemie AG."[5]

The story really begins with the manufacture of sulphuric acid. Regius Professor Freiherr Justus von Liebig had already made his epoch-making discovery that crop-yields could be enormously increased by the application of artificial fertilisers. A group of entrepreneurs petitioned on 8th July 1857 for a royal concession to manufacture fertilisers in Bavaria, an industry already established in England. Liebig was requested to take the position of managing director in the new company, the Bayerische Aktien-Gesellschaft für chemische und landwirtschaftliche chemische Fabrikate, Heufeld, Upper Bavaria.

Calcium montmorillonite was not discovered until one of the founders of the Pfirschinger Mineralwerke of Kitzingen am Main discovered "white clay" in an excavation for a building in 1904. This "Weisserde" was at first sold for making cheap pigments but about 1908 a yellowish grade was found to have the property of decolorising oils. As much as 1000 tonnes of FRANKONIT earth were sold in 1909.

Meanwhile a Mr. von Delling came across a similar clay in an excavation for a villa at Kronwinkl, and on 12th May 1906 "Erdwerke Kronwinkl Franz Schmidt & Co. GmbH" was founded, August Ostenrieder having a 97 per cent holding. On 17th November 1909 the clay was activated with acid for the first time. In 1910 he was joined by his brother Max and the name of the company was changed to Tonwerk Moosburg A. and M. Ostenrieder (1911). A member of this company, Dr. E. Theobald, succeeded some time in 1909 to activate the clay with hydrochloric acid without any knowledge of Pfirschinger's discovery; and the name TONSIL was adopted well before World War I, when the industry grew since Germany could not import bleaching earths.

An attempt was made in 1919 to patent acid-activation, but as a result of litigation between the two Bavarian firms, the patent was declared void. Hydrochloric was the acid used and it was found that American fuller's earth (attapulgite) did not respond to the treatment. Apparently English fuller's earth was not tried at that time; indeed it was not known by the German operators that an acid-activated bleaching earth could be made from English fuller's earth until 1938,

1) A. Löb, *Seifensieder-Zeitung* (1925), *52*, 1006–1008, 1024–1026
2) *Das Kieselsäuregel und die Bleicherden*, (1927), Berlin: Verlag von Julius Springer
3) [1925], 2nd ed (1929), Braunschweig: Serger & Hampel.
4) Leipzig: Akad. Verlagsges., (1931)
5) *Süd-Chemie Zeitung* (Nov. 1982), *151*, 1–64

See Adsorption of Cationic Dyes.

Von Delling first tried this clay in ceramics and in the cement industry, needless to say unsuccessfully.

This was Pfirschinger's D.R.P 33 9919 of 21. Nov. 1919.

when they heard with great surprise that it had been made for nearly ten years at Nutfield.

Before World War I several thousand tons of English and American bleaching earths were imported into Germany, but German production exceeded imports from 1906 onwards.

The chief German manufacturers of activated earth in 1927 were:-

Trade names of German activated earths

TONSIL	Tonwerke Moosburg, Oberbayern
TERRANA	Siriuswerke A.-G., Deggendorf a.d.D., Niederbayern
CLARIT	Bayrische A.-G. für chemische und landwirtschaftliche chemische Fabrikate, Heufeld, Oberbayern
ALSIL	Bergbaugesellschaft Ravensberg, Baierbrunn im Isartal.
SILHYDROL	Erdwerke München, Otto Lietzenmeyer
ISARIT	Dr. Ivo Deiglmeyer, München
FRANKONIT	Pfirschinger Mineralwerke, Gebr. Wildhagen & Falk, Kitzingen am Main
SILICA R	Bayerische Silikatwerke A.-G., Landau a.d. Isar.
The last had Al_2O_3 24.6 or 35.8%!	Other brands bore the trade names LEUKOSIT, ALBANIT, LUNIT, etc. Rheinische Fullererden, brand name LUNIT S was made by the Andernache Nahrungsmittelfabrik.

cf The Fullers' Earth Union formed in 1890.

See also "History . . . of the bleaching earths", O. Eckart, *Seifensieder Zeitung*, (1926), *53*, 30.

After the inflation of 1923 a price war was waged by the numerous bleaching earth companies. The fierce competition led Tonwerke Moosburg to develop a notable improvement in quality in 1926, when their output was 7700 tonnes a year (the same amount per *month* in 1970); and eventually to a series of much-overdue amalgamations. In 1924 as much as 48,000 tonnes of high-grade bleaching earths were exported from Germany, and in 1927 the production by Siriuswerke, 35,000 tonnes, was reckoned to be a third of the total – about 105,000 tonnes.

About this time the process consisted of making the raw clay into a slurry, adding hydrochloric acid and passing steam into the mixture in large wooden vats. After a period of some hours the acid-treated clay was separated and washed in wooden filter-presses; then dried in a drum or channel drier and milled to about 5000 meshes/cm^2 (about 170 mesh B.S.) Its activity for bleaching oils was two or three times as great as the English or American natural, or unactivated, earths.

At last in 1930 the Sirius-Werke AG Deggendorf merged with Tonwerk Moosburg and Tonsilwerke Schönebeck to form the "Vereinigte Bleicherdefabriken AG" in 1930. In their mining area of Gammelsdorf and Margarethenried the seam, roughly 1 metre thick, was extracted from below an overburden rising to 20 metres. The earth was sorted into qualities and carried to factories at Moosburg, Deggendorf, Schönebeck and Grossenbrode for activation.

The TONSIL range of products in the 1930s was:

Tonsil range of German products

soya, linseed, sulphur olive, rape; and mineral oils, paraffin, train oil, fatty acids and arachis (groundnut) oil.	TONSIL AC
easily bleached rape seed, maize, olive and palm oils, white oils, light mineral oils.	TONSIL ACA
bone fat.	TONSIL K
sunflower, arachis, sesame oils, as well as for mineral oils.	TONSIL SPECIAL (several varieties)
for removing protein and mucilaginous matter from vegetable oils and resinous matter from mineral oils.	TONSIL 6 and TONSIL 13, (absolutely neutral)
also for castor oil and easily bleached sesame and sunflower oil.	TONSIL 13
paraffins, beeswax, difficult hogs fat and tallow; many sulphur-olives and cottons.	TONSIL 60C
light coloured beeswax, cotton and white oils.	TONSIL 35C (rather milder than 60C)
palm kernel, coconut, easy cottons, fresh hog's fat and many sunflower oils.	TONSIL 13C

The TERRANA grades EXTRA, A SUPERIOR, SPECIAL, D (a natural earth), 60, 35 C, 13 C, show how nearly the manufacturing processes had been unified. The Sirius-Werke, however, made GRANOSIL, a highly active granulated earth for vapour-phase treatment of mineral oils or for percolation, and powdered GRANOSIL for use with non-acid treated mineral oil distillates.

Several years before WWII, The FEU in England decided to broaden its horizons and bought into an activated earth business near Tannenberg, in order to consolidate its growing continental trade. The factory, unfortunately, ended behind the Iron Curtain after the War and The FEU interest could not be recovered. The state of the art of the industry remaining in West Germany in 1945 was described by A. D. Rich.

Activated clay bleaching adsorbents, *FIAT Report* No. 810, (13 May 1946).

Apart from the deposits in Bavaria, calcium bentonite has been discovered at Hegau (Lake Constance), in the Eifel, in Vogtland, and in Münsterland (Kromer & Schüller, 1983).

H. Kromer & K.H. Schüller, *Interceram* (1983), (*2*), 35–38.

Bentonite

PROFESSOR Johann Plesch of Berlin was one of the first in Germany to be fascinated by Wyoming bentonite. In 1928 he obtained samples from Ward's Natural Science Establishment, and information from The Hill & Griffith Co of Cincinnati, Ohio, suppliers to foundries, regarding the Wyoming Bentonite Company's Osage Clay Spur Bentonite. In the leaflet then available, bentonite is said to be used in foundry work as an ingredient of core washes, and the Cincinnati firm, who milled the clay wrote that *a very large quantity is used today as a bonding material in foundries, for moulding sand and cores, and as a substitute for the natural aluminium salts to give higher mass production rates than is possible with natural sands*. At this time bentonite was supplied by F. W. Berk & Co Ltd., but was virtually unknown in German industrial circles.

Letter 24.1.1929.

Then, at 1 Fenchurch Avenue, London, E.C.3.

Plesch himself tried for nine years to introduce sugar-coated bentonite compositions in pill form for healing intestinal ailments and the regulation of bowel movements. He abandoned his patents only when he realised that NORMACOL was much more efficient because of its higher swelling power.

2 German Patent Applications, 16 Oct. 1928; 'Tonveredelung' is P58804 VI/80b. Includes sodium carbonate treatment. U.S. Application No. 373, 924 filed 26.6.1929 rejected 18.3.1932.

In the first two years of his enthusiasm he tried to interest Dr. A. Mittasch of the I.G. Farbenindustrie AG, Leverkusen, in bentonite and gave a list of possible uses, and enclosed the US Bureau of Mines Technical Paper No. 438, (1930).

Of the pharmaceutical division, Werk Oppa.

"Bentonite".

On 23 June 1930 Dr. Mittasch replied at length saying that American bentonite was of no interest to the I.G. and for many other purposes its value was unproven. As an emulsifying agent they themselves used a clay from Central Germany, namely Teutschetal clay because of its very high water absorption.

Hamburg 1, Curienstr. 8 1. Dec. 1931.

On the commercial scale Willy Bodensiek, who handled chinaware, had imported 25 tons of bentonite from America in 1931, but had failed to secure a substantial order, but a refinery had used small quantities for drying an oil. He had learned too late that he should have approached foundries. For other purposes, because of the high price, he could make no headway against the German bleaching earths.

Struktur und Quellung von Montmorillonit, *Z. Kristallographie*. (1933), *86*, 340–348.

Angewandte Chemie. (1934), *47*, 539.

British Patent 447 710 (12/6/35), Ger. (16/6/34).

Industrial Minerals, (1969), (27), 33

The crystal structure and swelling of montmorillonite was scientifically studied in 1933 by K. Endell, U. Hofmann and D. Wilm, who in the following year wrote their famous paper on the X-ray and colloid chemical investigations of clay. Endell and Hofmann's process for synthesising bentonite from calcium montmorillonite was patented by S. Erbslöh, 16 June 1934. Erbslöh's kaolin works had been founded at Geisenheim am Rhein in 1892 by C. Hugo Erbslöh, so production of sodium-bentonite was begun there. It was not until 1953 that the firm Erbslöh & Co. set up a large works at Landshut, nearer the raw material. They prospered and eventually designed the plant for the factory called "Bentomak" in Yugoslavia.

Na_2CO_3

At first the calcium-bentonite was agitated with 2 to 2.5 per cent of soda ash in a thin slurry and the product washed and dried. The resulting product, sodium montmorillonite + calcium carbonate, gave a thixotropic gel with 15 to 20 times its weight of water. The expensive drying stage was eliminated when it was discovered in 1936 that the soda ash could be added in a dry, finely divided condition to the clay containing 20 to 30 per cent of moisture and intimately mixed.

British Patent, 458, 240 (1936), addition to 447, 710.

Tonindustrie-Zeitung, (1950), 74, 122–129; with an obituary.

It is fitting that Kurd Endell's last and posthumous work was about bentonite, its properties and uses. Born 1 February 1887 at Stade near Hanover, he became a Professor in 1919 and, at the height of his powers, overcome by the frightful catastrophe which had overwhelmed Germany, took his own life 22 March 1945. He was a great loss to science. Ulrich Hofmann fortunately lived for many years and published many papers on clays and their applications.

The Vereinigte Bleicherden Fabrik merged with Bayerische AG

für chemische und landwirtschaftlich-chemische Fabrikate to form Süd-Chemie AG, München, on 30/4/1941. During World War II, bleaching clay was produced only for the home market. After the war, Süd-Chemie became the largest producers of montmorillonite products in Europe, and helped to establish Tonsil Mexicana SA at Puebla in 1965 and Süd-Chemie Italia SpA at Giba in Sardinia in 1975. Recent history of this company is given in *Industrial Minerals*. Today TONSIL bleaching earths retain that name; GEKO is produced (by adding soda ash) as a moulding and core sand binder; TIXOTON is their highly swelling grade for use in drilling fluids and in civil engineering works; and TIXOGEL is the brand for their gellants for paints, varnishes, lubricants, adhesives, coatings and putties.

Bayerische, founded in 1857, had acquired an acid-activated enterprise in 1925.

Chemical Age, (18/4/1942).

Industrial Minerals (Sept. 1969) (24) 44. Bleaching earths from Bavaria.

In 1969, Süd-Chemie also joined the Société Française des Glycerines S.A. in a joint venture named "Société Française des Bentonites et Dérivés SarL". Since 1971 SFBD has been a wholly owned subsidiary of Süd-Chemie.

Ind. Min. (Aug. 1969).

By 1976 bentonite production in Germany had risen to 628,000 tonnes; Süd-Chemie with its works at Moosburg, and Erbslöh & Co at Landshut, also in Bavaria. An account of "bentonite – a valuable and versatile mineral and raw material" was given by C. E. Hofstadt and R. Fahn of Süd-Chemie at the Second Industrial Minerals International Congress, München, 17 May 1976.

Ind. Min. (Aug. 1978), 25.

Süd-Chemie A.G. 8 München 2,|Lenbachplatz 6.

The present range of twenty grades of activated earth are given in Süd-Chemie's "TONSIL – Highly active bleaching earths". In addition, the Company manufactures the so-called G- and K + E-Hydrogenation Catalysts.

Further information is given in the Company's Annual Reports, published in English. Thanks are due to Dr. Hans Baur for his help with this chapter.

38 ENGLAND

Violet Powder
Fullers Earth
Tooth Powders
Seidlitz Powders
Pink Powders for Children
Head and Stomach and Various other pills
Italian Borax
Pickling Spices
Epsom Salts
Senna Leaves
Chamomile Flowers
Cough Lozenges
Antibilious Pills
Raspberry Beer Powder
Ginger Beer Powder

Chemists' Sundries distributed by ALFRED BIRD & SONS before 1900.

Thanks to R. P. Verrall, Research Service Manager, General Foods Ltd., Banbury, Oxon., OX16 7OU (17 June 1983).

ALFRED Bird established his business in a small chemist's shop in Bell St., Birmingham, in 1837 and sold a variety of medicinal and toilet preparations, now so strange to us. With the 'invention' of Custard Powder and Baking Powder the repacking of these other commodities for retail sale declined and by the early 1920s had ceased.

When Alfred Bird & Sons (1900) was incorporated they carried out their business at Devonshire Works, Floodgate St., Birmingham, and at Vulcan Buildings, Upper Thames Street, London E.C., not far from warehouses where fuller's earth was held in store. They also bought earth from the Midford Fuller's Earth Works and sold to druggists all over the country, a market which the fuller's earth producers would have found costly to reach. However, even when the modern wholesale pharmaceutical industry took over packaging and distribution, sales volumes were never very large, although fuller's earth powder and cream continued to be used by many for the treatment of sore and chapped skin, particularly of babies.

Continued from Chapter 16.

The textile industry's requirements for fuller's earth was declining in spite of the efforts of the agents Wm. Haigh, drysalters, Prickleden Mills, Holmforth, and W.P. England & Co., drysalters, (Established 1790), William Street, Huddersfield.

Mr. J. S. Highfield, of Highfield & Roger Smith, Consulting Electrical Engineers, joined the Board in 1915 and became Chairman in 1919, following his father-in-law Theodore Stretch, a founder-subscriber who

After World War 1 the Chairman of The Fullers' Earth Union Ltd. was worried lest the importation of German acid-activated bleaching earths would result in disaster for his company, relying as it was on the sale of natural fuller's earth for edible oil refining. He did not have far to look for someone to develop an activation process in Surrey. For a number of years the Osmosis Company, Ltd., had been trying to

Fig. 78. Lt. Col. Wilfred Taunton Raikes. From a photograph in the possession of Miss Dorothy E. Penny.

joined the Board in 1893 and was Chairman 1912 until his death in 1919. Co-director in the thirties W. E. Gray was also a founder subscriber.

Jim Carvell had three sons. John died during RAF service in WWII. Eddie worked in the labs and Ernie as an electrician – both retiring in 1983. Jim, Eddie and Ernie were for a period all members of the 25-year service club at the same time.

History of Clay Research in Great Britain, R. H. S. Robertson (1983), Clay Minerals Group, in the press.

The Nutfield Manufacturing Co Ltd was taken over by RTZ c. 1970 and closed in 1984.

Large pressure vessels had been made for Bergius in Germany.

The "Vortex" main mixers were made by The Kestner Evaporator & Engineering Co Ltd., whose founder J. Arthur Reavell, died in 1973 at the age of 100, a great pioneer in chemical engineering. Kestners also made the first drum driers, but the "doctor" knives scored the drums. Pumps were made by Ralph Smart and others.

extend the work pioneered by Dr. Graf Botho Schwerin of Frankfurt am Main. Fairly promising results had been obtained in purifying gelatine and making stable silica sol from sodium silicate solution – both by electrodialysis. Kaolinite concentrated by cataphoresis held out more promise. Mullite crucibles made from this kaolinite mixed with alumina were so tough that they could be dropped on to the floor without breaking. No ceramics firm could be persuaded to manufacture long-life crucibles, understandably, but Thomas Morson & Sons, Ltd., obtained the rights in selling OSMOKAOLIN, a pharmaceutical product, and 'electro-osmosed' ball clay found an application in the leather industry. Lt.-Col. Wilfred Taunton Raikes was appointed to manage the production of these clays at a plant erected in some disused buildings at Park Works of The FEU. The process was operated by his one-time batman, Mr. James Carvell. He joined The Fullers' Earth Union in 1924. Despite some early resistance from British oil refiners to the idea of "chemicalised clay" after their long satisfaction with the English high-quality natural earth, German competition increased. Consequently Mr. Highfield invited his kinsman, the "Colonel", as he was always called, to design and erect an acid-activation factory at Cockley Farm, the area in which had been sited the Cockley Works once owned and operated by C.W. Cawley. In this remarkably tough assignment, which he successfully accomplished in the period 1925–1928, he was guided at times by his wife "Gentle" (her middle name), who had been in the chemical industry during the war.

He did not follow the German process at all closely. To begin with hydrochloric acid was very expensive in England; sulphuric acid on the other hand could be supplied by the Nutfield Manufacturing Co., Ltd., at South Nutfield, where hydrofluoric acid was made. Even so, a saving in the proportion of acid used was advisable. The Colonel decided to carry the reaction out under pressure in a lead-lined autoclave provided with a stirrer. No vessels of this size had ever been made in Britain. The first one made to the Colonel's specifications was a failure for it leaked where the stirring shaft entered. The Directors nearly abandoned the expensive project but decided to carry on. A second autoclave, ordered from William Neale & Co., was a complete success. Widnes Foundry made the big digesters now in use (13.6 m^3 in volume).

After acid-treatment the product was mixed with water in a large tank provided with a turbomixer. The slurry was pumped to one of a number of tanks where the solids settled and the supernatant acid liquor was drawn off through wide funnels or drainers lowered by hand from time to time. After several washes the acidity was tested and if low enough the earth was concentrated in another tank and finally dried on a film (or drum) drier latterly made by Manlove, Alliott & Co. Ltd. The product "237" was named from the number of the experiment in the Colonel's logbook. It was notable for quick filtration of oils

Raikes family tree

Robert RAIKES b. 1690 d. 1757 printer, founded the "Gloucester Journal" = (2) Anne MONK

Robert RAIKES b. 1736 d. 1811 pioneer of Sunday Schools = Anne TRIGGE
2s 6dau.

William RAIKES b. 1737 d. 1808 Valentines, Essex = Martha MATHEW

Thomas RAIKES = Charlotte FINCH

2s 1dau.

A long line of CRIPPS in Cirencester

Robert RAIKES b. 1765 d. 1837 Welton, Yorks = Anne WILLIAMSON

Thomas RAIKES b. 1777 d. 3.7.1845 dandy & diarist = Sophia BAYLY

Henry RAIKES b. 1782 d. 1854 divine archæologist =

Henry William CRIPPS b. 20.3.1815 at Parmoor, Bucks = Julia his cousin

Thomas RAIKES b. 1790 d. 1866 Welton = Elizabeth ARMSTRONG

Henry Thomas RAIKES Judge of the High Court of Calcutta. Diary 1856 & 1858

3dau.

Henry RAIKES b. d. of Llwynegrin, Flint =

Henry William CRIPPS Q.C. b. d. 14.8.1899 Beechwood Marlow =

Robert RAIKES b. 1818 d. 1901 Treberfydd = Frances TAUNTON

Rt. Hon. Henry Cecil RAIKES b. 1834 d. 1891 Barrister, Tory M.P., P.M.G., 1886 = Charlotte Blanche TREVOR ROPER of Plâs Teg
5s 4dau.

Theresa POTTER = Charles Alfred CRIPPS b. 3.10.1852 cr. Baron PARMOOR 1914 d. 1941

Julia CRIPPS = Theo. STRETCH First chairman of The FEU Ltd

Rosa Margaret CRIPPS b. 8.1.1854 d. 29.8.1923 (2) = Robert Taunton RAIKES, J.P., b. 1843 d. 1919 Treberfydd — introduced as original share-holder by Theo. Stretch 1890

Marina Julia STRETCH b. 1874 (2) = John Somerville HIGHFIELD b. 1874 d. 15.8.1946 Chairman, The FEU Ltd

Admiral Sir Robert Henry Taunton RAIKES b. 23.8.1885 d. 24.5.1953 = Ida Guinevere EVANS b. 21.10.87 d. 1984

Lt-Col. Wilfred Taunton RAIKES b. 4.7.1892 d.15.3.1957 Man. Dir., The FEU Ltd = Elizabeth Gentle KENNEDY b. 1893 d. 1979

Rt. Hon. Sir (Richard) Stafford CRIPPS b. 1889 d. 21.4.1952

(Robert) Martin RAIKES, D.S.C. b. 21.11.1918 Man. Dir., The FEU Ltd

Major (Robin) Duncan RAIKES b. 1921 d. 5.7.1984 at Treberfydd

The grade 237 was still on sale in 1983.

decolorised by it. Some customers however wanted a finer powder which gave a quicker bleaching action; this was "249", made by milling "237" in a KEK pin mill.

That an amateur, however skilled a businessman, should have developed such an advanced chemical engineering process is truly remarkable, especially since no essential changes were made in the next half-century.

The first nine units were built in groups of three: after these more driers were added, a new tank farm and new effluent treatment plant.

Crooks died of TB in 1938; F. W. Bird, who later won the MBE for his work on the development of foundry bonding clays during the War, died 22 Jan. 1974.

When the first unit of two digesters and their accompanying washing and drying plant had been installed Freddie W. Bird was engaged at the age of 16 to carry out control tests. By the time the author arrived in October 1933, Bird and his colleague W. A. E. Crooks had standardised many test procedures including the all-important acidity and "oil tests" to show how effective the earth was in decolorising different kinds of oil. The importance of bulk density as a measure of activity was already realised.

R. H. S. Robertson, Studying clays for industry, *I Konferencja Minerały i Surowce Ilaste*, Bolesławiec 1978, Warsaw: Polska Akademia Nauk (1978).

W.H. Bragg, London: G. Bell & Sons Ltd., (1926).

I feel at this point I should honour the memory of those who inspired me in my youth. I attended Sir William Bragg's Christmas Lectures "adapted to a juvenile auditory" (as Faraday had put it) in 1925 at the Royal Institution on "Old Trades and New Knowledge" (Bragg 1926) and am fairly certain that my lifelong interest in using science to create new products and industries began thus when I was 14. In 1930 E.A. Hauser gave me a sample of bentonite after lecturing in London on thixotropy. Already intrigued by montmorillonite I was asked by Mr. Highfield (a staunch member of the R.I.) if I would like to try to find new uses for fuller's earth. The Colonel took me on and my first research, not surprisingly, was on the thixotropic properties of calcium and sodium montmorillonite.

I started badly, because I was unable to translate German legal documents!

In 1933 fuller's earth clod was dried to an average moisture content of about 4 per cent on steel checker plates above a fire which was lit each afternoon and burned out during the night. Even in the coldest weather men would strip to the waist to enter the kilns in the morning to shovel the dried earth to store. These ancient kilns were not all replaced by Rotary Louvre Driers in Surrey until about 1955.

Rotary Louvre driers made by Dunford & Elliott, now Newell Dunford Engineering Ltd.

Research work about 1941–2 showed the 'thermal regime' required to dry earths to the optimal moisture content of ~14%. The Louvre Drier exactly fulfilled our needs. (Huge chunks of metal were flung a great distance once when the fly wheel broke, but no one was hurt).

The earth was ground at Copyhold Works in a giant Krupp mill driven by a somewhat unreliable single-cylinder low-speed gas engine with a $5\frac{1}{2}$ ton flywheel and a rather hypothetical capability of 2 to $2\frac{1}{2}$ tons an hour of 100 mesh powder. This was later replaced by Dragon and Johnson ring-roller mills of more advanced design.

Steam wagons were used until about 1938 or 1939. The principal driver was Jim Earl, whose son Jack was also a driver for many years. Bertie Turk was yard foreman of all despatches, Povey, the joiner – later Carpenters' Foreman – had his workshop in an attractive old building which had probably been one of the outbuildings of the original Copyhold Works before 1890 and which lay between the Colonel's house The Copyhold and the works.

The office and laboratory were in idyllic surroundings; the spotted flycatcher and the blackcap made their nests on the building, and when the telephone was pointed to the open window the nightingale could have been heard in Berkeley Square.

The lab staff grew from two to ten in the years 1933 to 1942. John Carvell came as an assistant and Tony Callard joined in 1936. S. George Henderson from 1941 made a speciality of foundry sand problems. He not only studied the many and varied foundry practices but, with Freddie Bird, developed moulding clays to suit each one. The successful link between supplier and the foundry floor or the "chemist to chemist" approach was the basis of our philosophy of selling industrial minerals without benefit of salesmen.

John Carvell, missing believed killed in air raid over Germany in 1942.

A. G. Callard retired 1979.

During World War II, production of refined edible fats grew strongly although The FEU suffered a blow when some petroleum oil refineries, so vulnerable to attack, were closed until they could be defended. Meanwhile the huge increase in demand for iron and steel castings led to large orders for foundry grades of fuller's earth products to reduce the importation of U.S. bentonite.

Peter V. Danckwerts who joined The FEU from Oxford soon left for the RNVR and was awarded the George Cross (1940) for defusing bombs in the London Docks: he eventually became Shell Professor of Chemical Engineering at Cambridge (1959–77).

B. 14 Oct 1916 D. 25 Oct 1984 (interviewed 13 May 1939).

The first chemical engineer to be taken on was "Jim" Frisken. In those days the plant at Cockley Works was managed by Albert W. Boorman, assisted later by Tony Callard. After him Bill Merry was Production Manager for many years. Still remembered (1984) are the prodigiously strong Jim Leadbetter, who alone could lift a digester lid unaided, and Reeves who specialised in removing scale from film drier drums.

Joseph Frisken was interviewed 30 Jan 1939.

No important alterations were made to Cockley Works until it was realised that the partly-limed effluent could no longer be completely neutralised by pouring it onto the mountains of calcareous sandstone which had once been overburden in the quarry. This nightmarish performance had to cease when acid water began to seep into a tributary of the River Mole. Henry Hathaway and I were given the job of designing an efficient treatment plant in which slaked lime reacted with the acid sulphates to form a colloidal mixture of aluminium and iron hydroxides and gypsum, which, after filter-pressing, formed a butter-like solid we called FULLOGEL. Sir Robert Paterson said it was one of the most daring and imaginative designs his company had ever had to build from. During the war we sold a small amount of FULLOGEL for fireproofing the timbers of historic buildings against incendiary bombs, but most of it was returned to the worked out quarries.

Obituary, *Nature*, (30 Sept. 1944) *154*, 422, by R.H.S.R. He was development chemist at The FEU from Jan. 1940 to July 1943.

of the Paterson Engineering Co. Ltd.

Also known as GYPSOGEL in connection with S. G. Henderson's Brit. Pat. 680, 234 of 12.1.1950 and DBP 914 951, filed 22.5.1950.

Also in 1939 one Dr. Barbara S. Beer wrote from Hungary to say that she could improve our grades of activated earth. At an interview we could hardly ask for details of how she would set about this, but she had the qualifications and we had not tried to optimise the Colonel's

Later Mrs. S. George Henderson.

process. We agreed to employ her, but before she could emigrate she had married and the war had started. Very special permission had to be obtained to get her out of Hungary – now as Mrs Emődi. Assisted by Maureen Whiteside, 'Dr. Barbara' lost no time and became, through the years, a leading authority on the analysis, processing and application of clay minerals and, in particular, montmorillonite. Her immediate task on arrival in this country was to show how English activated fuller's earth could be greatly improved and result in savings of acid and edible oil. These highly successful products, later called FULMONTS, were first seen as a means of countering post-war competition from the Continent.

Sir William Bragg's inspiration to apply science to the management of innovation led in a few years to the foundation of many technical advances in the industry. The FEU laboratories had studied by the early 1940s what was not published by others until decades later. These were vintage years.

Several large oil refineries were built after the war to make the country less dependent on foreign supplies of refined products and to replace installations outmoded and outworn in the war.

Parallel changes were taking place in the foundry industry; although demand for castings was steadily increasing to meet the needs of the automobile industry, new foundry processes not requiring fuller's earth were beginning to gain acceptance, although the total demand for foundry bonding clays continued to rise.

At the same time, competition was becoming more intense in the foundry market since F.W. Berk & Co Ltd's Woburn deposit lay nearer the major foundries than the Surrey and Somerset deposits. Moreover, the near-monopoly of The FEU in supplying earths for refining edible oils and fats began to be eroded by the 1950s when the German producers were becoming active again, challenging The FEU in several lucrative export markets by offering very high quality products. In this they were greatly assisted by the sulphuric acid shortages of the early 1950s, which bore particularly heavily on the activation output of The FEU and painfully restricted the tonnage available, especially for export.

The result was a ferment of post-war activity at Redhill, to control costs and improve quality of the products, including activated earths, by adopting the process worked out years before by Barbara Emődi, now Mrs. B. S. Neumann, while studying strategic expansion.

New quarrying techniques had first to be introduced, as told in Chapter 36. We have seen how C. R. Dames had installed an edge runner mill to disperse the freshly mined wet clod for use at Midford; one of the first post-war innovations was to use such mills, coupled with very fine screens to remove extraneous material, to make a feed for the digesters and thus save the cost of drying, grinding and bagging.

Chapter 22.

With the construction of the big petroleum refineries in the U.K.

after the War, demand for the 'S' grade of activated earth for treating lubricating oil sharply increased and led to open digesting for the first time at Cockley. This development was accompanied by a new batch counter-current washing system, which greatly economised on water usage and consequently lightened the load on the treatment plant for neutralising all works effluent[1]. Also associated with the spray-drier installation were 250-ton bulk silos, into which the products were pneumatically transferred. From these silos were loaded bulk road vehicles, air-slide vehicles initially, which carried earth, mainly to the petroleum refineries. High-speed bagging lines were later incorporated.

After test programmes with new types of filtering and drying equipment, Moore filter presses were installed at Cockley from 1956, the low-moisture cake then being fed to Atritor mills. With the spray-drier, this new combination replaced practically all the old film-driers and further economised in wash-water.

The broad result of these changes was to quadruple the production of activated earth between 1948 and 1960, with no increase in the labour force and through a period when labour costs, as well as fuel and other costs, were rising fast.

In the production of milled natural earth, a drying Atritor mill was installed at Park Works to boost production during seasonal peaks. Louvre driers were progressively installed in the main Copyhold Works, with large silos, and high-speed packing equipment was installed there too; these were followed by an extruder-Atritor plant to make better bentonites for well-drilling and civil engineering uses.

A driving force behind these innovations was R. Martin Raikes, DSC, a nephew of Colonel Raikes. Martin Raikes, after distinguished service in the Royal Navy, during W W II, joined The FEU in 1946 and the Board in 1950, with special responsibility for technical development.

A factor of the highest importance was the relationship then developing between The FEU and Laporte Industries. There had been contacts between the two companies for some years. Senior production engineers had met when exchanging know-how during the War on fuel utilisation techniques under Government fuel economy initiatives. Lt-Col. J. L. Harvey MBE, who managed a company supplying acid to The FEU and who knew Laporte well, joined The FEU Board in 1945. Mr L. P. O'Brien, FRIS, Chairman of Laporte, joined in 1946 and became Non-Executive Chairman two years later. In all these links, it became evident, not only that representatives of the two companies got on very well together, but that a great deal of common ground existed in the technologies used and the industries served. In 1954, The FEU Ltd was merged with Laporte Industries.

It is probably true that this friendly take-over was welcomed by The FEU in another sense. By this time, the company had developed into

Open digesting was normal practice in Germany.

Construction by Taylor Woodrow Ltd; consulting engineers, Cremer & Warner.

[1] A large spray drier, made by Kestner, was brought into operation in 1953, making obsolete many of the film driers. A Ruston gas-turbine and electrical generator were coupled to it several years later, the gases from the turbine feeding the spray drier.

ATRITOR Mills by Alfred Herbert Ltd., Coventry.

The film driers remained on the site for several years before being sold for scrap. Several lingered in a yard near the railway at Croydon in the late '60s, but others were used for drying pre-boiled potatoes for the convenience food market.

Sand accumulates in the mill chamber of ring roller and RAYMOND Mills, turning them into attrition mills. Pin mills are less damaging to the rheological properties. RHSR.

RMR read Natural Sciences at Cambridge just before W W II.

Chem. Age. 11 Sept. 1954 and 9 Oct. 1954.

one of the largest producers in the world of activated earths and other montmorillonite and bentonite products but was still vulnerable to acquisition. The Laporte move closed these possibilities and offered the larger company's financial and technical strengths and international marketing experience.

At The FEU and as a director of Laporte after the take-over, Colonel Raikes presided over all the activities of these burgeoning years. He had planned to retire early on 30th June but, tragically, died in March of that year, after 33 years' service with his company. R. Martin Raikes became Managing Director of the fuller's earth operation.

Despite all these advances, the outlook was clouded with some uncertainty in the early 1960s. The demand for 'S' grade declined abruptly, with the adoption of new refining techniques for lubricating oil which did not use earth. Catalytic processes to improve the quality of light distillates such as kerosene were modified so as to treat lubricating oil. The base stock could then be decolorised and stabilised against oxidation without the need for bleaching earth, or the need to dispose of spent earth, which had always annoyed the refiner. The loss in profit could be made up only by new product development or by winning export markets against intense international competition. The business had to be directed to meet these pressures, to open new markets and to effect further economies.

Chapter 58ii to iv.

The originality of the work coming out of the new laboratories and pilot-units at that time was evinced in a completely new approach to the activation process. Hydrochloric acid was well known in activation but was relatively expensive, demanded more resistant materials of construction and created problems of effluent disposal. Supported by the enthusiasm of Martin Raikes, a research team led by G. K. Jones began, in 1958, to seek a way through these daunting difficulties. The chemical engineering aspects were entrusted to Laurie Preston, and Tony Callard was heavily involved in the pilot-work.

In the event a first installation was commissioned in 1962 which was ahead of its time. After activation with hydrochloric acid, the earth was washed in a fully automated Moore filter. The relatively low volume of filtrate was further concentrated and finally evaporated in a massive Amman reactor, which produced pure HCl gas and solid oxides, mainly of iron. The gas was absorbed and recycled. The chemical and thermal balances were very efficient.

Activating clays with acid: Brit. Patent 939,937–8 (A) 19 Dec. 1958 and 11 Aug. 1959, (B) 29 Feb. 1960. The process also eliminated wet effluent treatment.

Main contractors, Taylor Woodrow & Co 1962.

Chemical Age (10 July 1965) 43.

The process was not, however, designed to cope with the fourfold increase in fuel costs from 1973 and, at that time, was retired in favour of the sulphuric acid process which had, meanwhile, also been considerably improved.

The Works was progressively enlarged through the decade from the early 1960s, one feature being the installation of specially designed big digesters.

On the organisational side, too, there were changes. In 1965, Derek Oliver moved from the Laporte site near Grimsby to become General Works Manager at Redhill, where he remained until retirement in 1979. During this period, The FEU ceased to exist as a separate company but product development and optimisation of costs continued unabated. Large-scale tree planting and landscaping of the quarried areas was also carried out.

In the late 1970s, Laporte came to the decision that a new activation plant was best located on one of its existing sites at Widnes. There, in 1981, a new and completely re-designed plant was commissioned, with a capacity of 30,000 to 35,000 tonnes a year of FULMONT bleaching earths and FULCAT catalytic grades. Cockley Works finally closed down.

As has been said at the beginning of the book these are Registered Names.

Table 12 Earths for edible and mineral oil refining

Laporte Brochure, (1981), p. 14.

Edible oil refining	
Fulmont Première	New ultra high performance bleaching earth specially developed for the most demanding applications, including the bleaching of oils prior to physical refining.
Fulmont AA	Superior activity grade for difficult to bleach oils and fats in particular linseed, palm, rapeseed and fish oils.
Fulmont 700C	Excellent bleaching power for normal and difficult edible and technical oils.
Fulmont BB	High efficiency general purpose bleaching earth for medium quality vegetable oils.
Fulmont 237	General purpose bleaching earth specially recommended for animal fats and high temperature bleaching of palm oil.
Fulmont 300C	Specifically designed product for maximum control of free fatty acid levels and suitable for bleaching oils such as groundnut, sunflower and soya.
Mineral oil refining	
Fulmont XX	Maximum bleaching power for mineral oils of all types, and especially re-refining of used lubricating oils.
Fulmont MQC	For general purpose decolourising of mineral oils and waxes especially treatment of electrical oils.

The success of the Widnes plant, as referred to in the Company Reports, was part of a larger investment in fuller's earth. Already, in 1973, Laporte had acquired a holding in Minas de Gádor SA in Spain and, in 1976, a 40 per cent interest in Bentonit União Nordeste in Brazil. In 1982, a radical £4 million modernisation was completed of the Copyhold plant for natural earths, foundry clays and similar products. By 1985, Laporte had raised its stake in Minas de Gádor to 100 per cent and had acquired a 76.5 per cent interest in Brett Bentonite.

Chapters 44 and 54.

Chapter 25.

38i Laponite, a synthetic hectorite

W. Noll, *Chemie der Erde,* (1936), *10*, 129.

H. Strese & U. Hofmann, *Z. Anorg. Chem.*, (1941), 247, 65–95.

W. T. Granquist and S. S. Pollack, *Clays and Clay Minerals*, Monograph No. 9, Earth Science Series Pergamon Press, (1960), pp. 150–169.

SYNTHESIS of clay minerals was first accomplished by Noll (1936). To make montmorillonite, very high pressures were required, but Strese and Hofmann (1941) made a "two-dimensional gel" by stirring together solutions of magnesium and boiling potassium silicate. Then Granquist and Pollack, in work sponsored by Baroid Division National Lead Co. made a better crystallised product – hectorite – by reacting silica gel, magnesium hydroxide, sodium and/or lithium hydroxide or lithium fluoride.

The Research and Development Department of The Fuller's Earth Union Ltd at Redhill realised that, by modifying the preparative techniques involved in the Granquist and Pollock paper, improvements could be made in the crystallisation of the hectorite. Research work was started under the direction of Dr. B. S. Neumann and G. K. Jones in 1960 and a process for manufacturing a synthetic hectorite was patented in June 1962 under Dr. Neumann's name. This was followed in December 1963 by a patent for the preparation of an organoclay gellant based on a synthetic hectorite and published under the names of G. K. Jones and B. S. Neumann.

Brit. Pat. No. 1 054 111.

Brit. Pat. No. 1 121 501.

A pilot plant for the commercial manufacture of these synthetic hectorites was built at Redhill in 1962 and managed by Tony Callard. In 1963 the product name of LAPONITE was registered by Laporte Industries Ltd. Further developments, also covered by patents, led to the building of a production unit at Laporte's Stallingborough site in 1968. This plant had a capacity of 50 to 100 tons a year.

Brit. Pat. No. 1 213 122.

In the late 1960s following further research by Dr. Neumann at Redhill, a hydrothermal process was found to give improved properties. A plant for making this kind of LAPONITE was constructed at Stallingborough in the early 1970s with an initial capacity of 30 tons a year. Further plant extensions for both types of LAPONITE were undertaken in the late 1970s to meet the increasing demand and in 1984–85, following the sale of this factory site to an American company, the production units were moved to Widnes.

LAPONITE is a high purity synthetic hectorite with controlled gelation characteristics. It swells rapidly when dispersed in water at low solids content (2–4 per cent by weight) to give a thixotropic gel which, because of its high clarity, has a much wider range of uses than the gels made from natural hectorite. LAPONITE is used, for example, in

- non-drip emulsion paints.
- conventional and clear gel toothpastes.
- household products such as oven cleaners, hard surface cleaners, paint strippers.
- shampoos (anti-dandruff).
- fluid drilling of germinating seeds.

Almost the whole of Chapter 38 has been contributed by past and present members of Laporte to whom I am very grateful.

JAPAN

Sansei-Hakudo (Acid Clay), 酸性白土
K. KOBAYASHI, 1st. edition, Maruzen, 1919.

Clays and Clay Minerals of Japan, T. Sudo and S. Shimoda, Editors, 1978, Elsevier's Developments in Sedimentology 26.

IN 1899, a clay having a strong oil-decolorising effect was discovered at Odo, Niigata Prefecture and K. Kobayashi, a Professor of Waseda University, began to test the importance of this clay, the so-called Kambara Earth, to the petroleum industry. He first discovered that the clay gave an acid reaction; and began, with many colleagues, what was to become his life-work. He and his group thus laid the foundation for the acid clay industry, and stimulated a broad interest in studying clays in other areas such as geology, mineralogy, and soil science.

漂布土
Hyōliakudo fuller's earth.

About 1908 bentonite was first worked in Yamagata Prefecture: in 1930 it was shown by R. Shigemune to be very similar to Wyoming bentonite.

蒲原土
Kanbarado Kambara earth.

When Kambara earth had become an established industry in 1912, Kobayashi introduced it to readers of an American journal as being effective for bleaching mineral oils and S. Ueno reported on its bleaching effect on other oils three years later. By the time Kobayashi's second edition of his book came out in 1927, Japan had become the world's most active centre of research on montmorillonite.

J. Ind. Eng. Chem., (1912) *4*, 891 and Seifens Ztg, (1913), *19*, 511

J. Ind. Eng. Chem., (1915) *7*, 596.

The Hojun Yoko Co. Ltd., were established in 1914 to produce bentonite.

Over extensive areas in the north of Japan Green Tuff of Miocene Age has been weathered to calcium montmorillonite; at the surface, however, the clay has been naturally activated and converted partially to the hydrogen-exchanged form. T. Okazawa noted that the acidity varies in different mountains but is usually greater nearer the sea coast. Where the soil is most acid, it is still feebly acidic at 12 metres, but in weak cases the acidity ceases 2 metres down. Further down the clay is neutral and then alkaline. The acidity gradient is developed mostly on mountains facing the Sea of Japan, where the wind is strong in autumn and winter and snow rich in salts falls up to 4 to 5 metres in depth, so that no acerose trees (maples) can grow there. Kabayashi confirmed that alumina was released from acid clay by treatment with neutral salt and approached modern ideas of these exchange reactions.

Bull. Inst. Phys. Chem. Res. Japan, (1923) *2* 189.
J. Inst. Phys. Chem. Res. Tokyo (1928), **7**, 776.

K. Kobayashi, *J. Chem. Ind. Japan*, (1920) *23*, 543; *Mem. Faculty Sci. Eng. Tokyo* (1927) *4*, 1.

H. Isobe & Y. Yendo, *Bull. Inst. Phys. Chem. Res. Tokyo*, (1928) *7*, 805; Engl. Ed. *1*, 72.

A fruitful partnership was formed by K. Kobayashi and K. Yamamoto. These two wrote many papers, but in the period 1924 to 1936, the following also made outstanding contributions:- S. Kawai, H. Inoue, H. Isobe, M. Kimura, S. Ohdake, K. Ono, G. Tomita, S. Uchida, M. Tanaka, N. Kameyama, T. Kuwata, Y. Tanaka,

M. Nakamoto, M. Nakamura, T. Watanabe, R. Nakai, K. Kashima, N. Kameyama, S. Oka, S, Iimori, T. Okuno, T. Ogawa, H. Ishikawa, J. Abe, K. Ishimura, M. Murata, Y. Tanaka, K. Ikeda, G. Kita, K. Suzuki, T. Suzuki, J. Sameshima, N. Morita, and others.

Among the pioneer works of these scientists were many studies in *catalysis* – on camphor, turpentine, α-pinene, citronellal, cyclohexylamine, nitriles, and many other compounds: *pyrolysis* studies were not confined to the cracking of heavy mineral oil and paraffin but were applied to the decomposition of calcium cyanamide, dehydration of anthraquinones, aliphatic primary alcohols and ethers, cresols, aromatics, low temperature carbonisation; *colour reactions* with liver oils and vitamin A, carotene, benzidine[1] (11 years before Hendricks and Alexander, 1940), malachite green leucobase HCl; *adsorption* of oryzanin from yeast, colour from oils and fats, alkaloids (1924), gasoline, benzine and other solvent vapours, indophenol and paranitraniline dyestuffs and the *sorption* of gas and water vapour.

Other studies included water purification, saccharification of starch, effect on portland cement, recovery of oil from spent JAC, and a comprehensive series of measurements of the physical properties of JAC.

If Kambara earth was the first to be studied, Japanese acid clay from Odo and Itoigawa were important sources by 1932, but clay from Tsuruoka, Kaminoyama, Nata, Akita, Kamo, Yamashira and Hiroshima, were compared with fuller's earth from England and Formosa, Floridin attapulgite, and German and American acid-activated earths. The JAC of Koto was of some interest too, apparently, because its water content was studied; Nakamoto also measured the moisture adsorption of Odo acid clay (1933).

K. Kobayashi began X-ray studies of JAC in 1932 and was still active in this field in 1965.

Not surprisingly this great research effort has resulted in a large industry, for the production of bentonite swelling clays, acid clay itself and acid-activated earths. In 1976 the producers were as shown in Table 13 below:

(1) N. Kameyama & S. Oka, *J. Soc. Chem. Ind. Japan*, (1929) *32*, 87 B.

J. Amer. Soc. Agronomy, (1940), *32*, 455–8

K. Yamamoto, X-ray investigations of JAC VII, *J. Soc. Chem. Ind.*, (1932), *8*, 482 B.

M. Nakamoto and G. Sano, *J. Soc. Chem. Ind.* (1929), *50*, 473 B.

M. Nakamoto, *J. Soc. Chem. Ind.*, (1933), *54*, 717.

J.S.C.I. Japan, (1932) *35*, 482 B and *Nature* (London), (9.10.1965), *208*, 179

Ind. Min. (July 1977), 17 *et seq*.

Table 13 Sodium-treated bentonite in Japan

		Prefecture	tpa
Nihon Bentonite Co.	North coast	Fukuoka	96,000
Nihon Bentonite Co.	Iwani area	Shimane	18,000
Nihon Bentonite Co.		Aomori	18,000
Kuroishi Bentonite Industry Co.	Kuroishi City	Aomori	39,600
Kunimine Minerals & Chemicals Co.	Tsukinano area	Yamagata	84,000
Kanto Bentonite Kogyo Co. Ltd.	Kanben mine, Shirosaka	Niigata	19,200
Hojun Kogyo Co. Ltd.	Haraichi mine, An-maka City	Gumma	
Hojun Kogyo Co. Ltd.	Tomioka mine, Tomioka City	Gumma	18,000 (total for the three Gumma mines)
Hojun Kogyo Co. Ltd.	Myogi mine, Myogi-mati	Gumma	
Nippon Activated White Clay Ltd.			
Toyo Clay Ltd.			

All are in the island of Honshu except the first which is in Kyushu. By 1984 the largest producer was Hojun Yoko Co. Ltd.

Acid Activation

MIZUSAWA Industrial Chemicals Ltd. was formed in 1937, as a subsidiary of Takeda Chemical Industries Ltd., then the largest pharmaceutical company in the East, to produce products made from acid clay. Their original works was in Yamagata Prefecture. In 1944 they added the Nakajo works in Niigata Prefecture. They have two acid clay mines, the Oohiro mine in the former Prefecture and the Hagura mine in the latter. In 1976 they made 3600 tonnes of acid clay for use as a filler, an insecticide carrier and as an oil bleaching earth.

M. I. C. Ltd., Kinsan Bldg 5, 4-Chome, Muromachi, Nihonbashi Chuo-Ku, TOKYO-103.

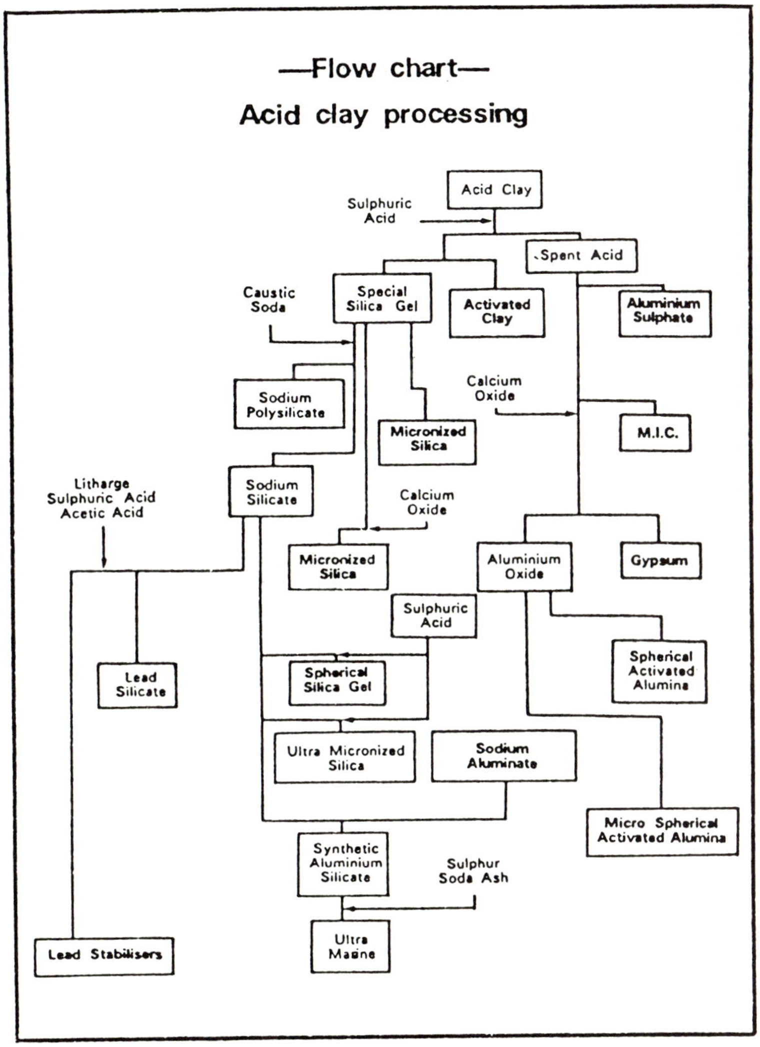

(On left). Fig. 79. Flow-chart – acid-clay processing in Japan.

39

Their main products are activated bleaching earths going by the trade name GALLEON EARTH – about 50,000 tpa. As by-products they make a range of fine silicas (SILTON), about 20,640 tpa, and silicates – sodium, lead, aluminium – and ultramarine, as well as 34,200 tpa of aluminium sulphate.

Ind. Min. (July 1977) 20.

SILTON is used in the manufacture of no-carbon copying paper; SILTON B is a sodium alumino silicate used as a builder in detergents and MIZUKASIL, a flatting agent for paint and ink.

Ind. Min (Mar. 1978), (126), 79.

The total production of bentonite and activated earths in Japan amounted to 398,364 tonnes in 1976; and will be much over 400,000 now.

FRANCE

40

A peser icelles laines le dit sergent le doit esgarder et en oster suing, gars, crotins, esconssures et coteriaulx
1410. St. de la drap. de Chauny, Arch. mun. Chauny.

The meaning of *gars* is unknown; I suggest it was an old word meaning *burs* or *burrs*. However, cf. OScots *gar*, ONorse *gare*, filth, slime, from *c*. 1420.

THE purchaser of wool did not want to buy dirt – he insisted on paying for cleaned wool. Here we have a description of all the most important kinds of dirt which had to be removed: *when weighing the fleece the said sergeant should examine it and remove from it suint, 'burs', droppings, hidden dirt and knots.*

The earliest records of the cloth industry at Rouen concern, not weaving, but dyeing and fulling. A document dating from 1199 rules that if the River Robec bursts its dykes, every fuller and dyer settled on its banks who owns fulling troughs and cauldrons at the waterside must either himself go to the site of the disaster or send a servant to assist the carpenters in mending the breaches. Fulling is also referred to in records of the thirteenth century. Thus, an edict of Louis VII in 1224 gave the inhabitants of Rouen permission to take the earth needed for fulling and dyeing purposes in the forest of Roumare.

H. Wescher, The Medieval Cloth Trade, *CIBA Review*, (1959), *12* (135), 7–9.

In Douai, on the other hand, a regulation was made in 1245 insisting that all those who have fulling and dyeing equipment near the walls of the city . . . should have it removed during the feast of St John. *Tout cil ki unt les foulenes et les teintures sour les forteteces de la vile . . . ke il les dient fait oster dedans le feste saint Jehan.*

Bans aux échevins, QQ f° **14**, Ve, *Arch. Douai*.

'sour' can mean 'on' or 'near by'.

In 1283, a building was made over to the town [of Rouen] by royal order to serve as a warehouse for fuller's earth. The importance of the ancillary trades points to the conclusion that cloth making had already assumed considerable proportions by that time. There appears to be evidence of its existence from the twelfth century throughout the area bounded by Rouen, Louviers, Evreux, and Pont-Audemer.

H. Wescher, *loc. cit.*

Louviers is in an area where clay rich in montmorillonite has been identified, the best quality being at Les Andelys on the Seine. The question remains whether this clay was ever shipped down to Rouen.

P. le Berre & B. Rozès, *et al.*, *Bur. Rech. Géol. & Min.*, (1980), SGN 178MTX.

The cloths of Rouen, both striped and plain, were second in quality only to those of Paris. Louis IX (St. Louis) had presented the cloth weavers with their first cloth hall in 1256. Fulling, still apparently always by foot, was now fairly widespread in France, as may be seen in the following short references:

1262: *vauterins li foullonz* (Metz):
1265: *fouleor* (Rente, ANS 5175, pièce 46).
1274: *Lou folun* (Fontenay, Arch. Côte-d'Or).
1304: *une foule et un batour assis dessoz la dicte maison sux la riviere* at Doubs near what is now the Swiss border.
1372: *Les drappaleirs et les foliours* (Arch. Fribourg).

About this time the fullers, wool combers and cloth shearers were allowed to join the weavers' guild in Rouen, *provided they submitted to the same examinations, plied their trade in the city, and worked locally woven stuffs exclusively.*

"Terre à foulon" is not recorded by Godefroy, and it is probable that few clays worthy of that name had been discovered in France. The cloth was simply "clayed" as many times as may have been deemed necessary, as the following passage describes: *aussi pourront mettre ou faire mettre en un drap qui sera de fort euvre a fouler, sunt ou lye de vins, ou autre chose profitable pour le drap, par en prenant congié a cil a cui icellui drap sera, et se pourra l'en* bouer *par tant de foiz comme bon semblera a faire, et meuront lesdis maistres et ouvriers leur eaux bien et deuement.* (1424)

Fulling mills were recorded in France from 1298, whereas the first water-powered fulling mill in England was set up in 1175; the phrase used being 'Les moulins foulerez'. There was "La foulerie de Saint-Ayoue" (1300), and there were several other fulleries noted in Provence in the "Censier des Cordeliers de Provins" in 1314 and 1315. There were others at Blois (1319); in 1340 we read of *un foulerie appartenant aux diz religieux*; and of the "folerie d'Orleans" (1409). Some others were:
1274: *Puent faire au moulin un bator ou follour* (Theuley, Arch. H.-Saône).
1317: *Moulin foulereiz*, (Cart. du Chap. d'Evreux).
1318: *Deux moulins fouleres* (Gray, Arch. Doubs).
1319: *Molins fouleres* and *molins folleres* (Blois).
1326: *Moulin foulieres* (Arch. JJ 64, f° 248 1e)
1362: *Des moulins foulerez* (Chap. de Chartes).
1390: *Moulin fouleour* (Denombr. du baill. de Constentin).
1390: *Trois moulin's, deux a blez et un foulour de drapz*, (*ibid.*)
1405: *tissot, affetiours, folliours, rameours, favres*; in this phrase fullers are followed by beetlers, unassisted mechanically.

In Nice, the word for fuller was "foulounié" according to Pietro Carles in his little dictionary, but the word "baissiatour" was also used in the County of Nice (La Countea de Nissa), which was part of La Provence until 1388, when it recognised the overlordship of the Duke of Savoy. In 1402 the fullers from Nice finished four sorts of cloth – from Narbonne in Languedoc, Châlons sur Marne in France, Brussels and Florence.

The word "baissiatour" comes from "baxiator", or "baxator" and, according to Ducange (c 1100), this in turn derives from the Latin

Dictionnaire de l'Ancienne Langue Française et de tous ses dialectes du IXe au XIVe siecle, Frédéric Godefroy, (1885) **4**, Paris; F. Vieweg.

1304 Ch., des compt de Dole, Arch. Doubs, cart. 44, paq. 44.

H. Wescher, *loc. cit.*
Terre à dégraisser and terre à laver were more commonly used.

(1424), *Ord.*, XVIII, 71.

sunt, now *suint, comes from suer* to sweat or ooze, from Latin *sudare.* Lye de vins, if vinegarised, might assist in the thickening.

1298, *Ord.*, Dupuis CXXXIV, 45 Richel.

Arch. Fribourg.

Piccolo Vocabulario Nizzardo-Italiano (1866).

Nice sousle Premier Siède de la domination de Savoie, by Caïo de Picilas, Turin (1898).

"baxa" or "baxea'', a sort of shoe, which suggests that fulling was not only done by foot but that the action was promoted by these hard-soled shoes without a heel.

Ducange's *Glossarium ad scriptores mediae et infimae latinitatis* and F. Gaffiot's Latin Dictionary, Hachette (1934).

Baxa.

A fulling mill was "lou paradou" or "paràire" from the verb "parà" (to protect, to full), as in *Item que si lo senhor ha molin d'oly ho paradour a parar draps que los dis homes sien tengus de modre lurs olivas. . . e parar lurs draps al parayre del dich senhor* meaning "Likewise, if the overlord has an oil-mill or a fulling mill to full cloth, the said men must grind their olives and full their cloth at the fulling mill of the said overlord". This text by Cabris (1496) explains that the fulling mill was the property of the lord and the people who lived on his fief were obliged to use his mill and had to pay a fee. If there were none, they could use whichever mill they wanted in the neighbourhood.

Cabris, *Acte d' habitation* de Cabris, in *Feuillets pour notre histoire*, Fonds Cappati, Musée Masséna, Nice, N.D.

The first mention of a fuller in the County of Nice dates back to 1156 in the village of *Drap*, famous in the county for its textile activities, as its name suggests.

For the information about Nice I am indebted to Jacques Martin, himself a Niçois.

About 16 km NE of Nice.

André Compan *et alii*, *Le Comté de Nice*, Segliers (1980), p. 186.

Even after the Hundred Years' War and the Peace of 1453, which had greatly disturbed the cloth industry, guilds were still strong and conservative. In the Statuts des Draps de Bourges (1466) concerning the "mestier de foulerie et drapperie de draps", rules are given for the fullers' and clothworkers' profession. The cloth trade, which had attained its peak in the fourteenth century, revived once more. *Thus, when Louis XI had conquered Artois and was anxious to inject new life into the languishing textile trades of the city of Arras, Rouen sent five cloth weavers, seven fullers, and 19 wool workers.*

H. Wescher, *loc. cit.*

New threats to the old and highly-organised industry arose when the nobility developed a taste for foreign silks and, much worse, when in the last quarter of the fifteenth century, many of the water mills at Darnétal, near Rouen, had been converted to fulling; their operations were more efficient and cheaper than those of Rouen where fulling was still done by trampling the cloth underfoot. *In 1493, the Wool Guild petitioned the authorities either to have the fulling mills pulled down or made accessible to all masters in the trade. Many buyers preferred the lighter cloths of Darnétal to the sturdy Rouen stuffs.* The workshops of Darnétal eventually encroached on the banks of the river, although originally shared by the two towns. Rouen was forced to have its cloths dyed in Paris and Meaux. And when an enterprising clothier named Baudry had five bolts of cloth dyed with genuine Spanish madder at Martainville, in 1530, the guild impounded the goods; but the Court, seeing the brilliance of shade, gave their verdict in favour of Baudry and the privilege of the dyers' guild was finally repealed.

H. Wescher, *op. cit.*

The decline of the wool cloth industry in the second half of this century was accelerated by imports of Italian silks and serges. In Rouen the weaving of silk, gold, and silver cloths had been established since 1531, and linens were becoming a staple article. The numbers of workers employed in the wool trade fell from 40–50,000 in the 16th century to 15,000 in 1661, the wars against the Huguenots having had a disastrous effect upon the industry. By this time Rouen was unable to compete with the smaller places such as Darnétal and Dieppe. In the 18th century, Elboeuf and Louviers developed into new centres of the cloth weaving trade.

For this quotation I thank Dr. G. Novelli, Livorno. And for all translations I thank Dr. Jacques Martin, Arras.

Regarding Bernard Palissy's *Discours Admirable*, Paris: Martin Ieune (1580), see Properties/Clays in Agriculture.

Fuller's earth was sold in the streets of Paris in 1545. A. Truquet in his book *Les Cent et sept cris que l'on crie journellement à Paris* gives the street-cry:

Terre à laver pour dégraisser	*Washing-earth to absorb grease –*
Je la prends dedans les carrières.	*I bring it from the quarries.*
Qui n'en voudra se tire en arrière	*Those who don't want any must move away*
Qui en voudra j'en ai assez.	*For those who want some: I have enough.*

(1600), II, 1vii 393.

Richard Surflet, C. Estienne and J. Liébault's *Maison Rustique*, or the *Countrie Farm*, was translated in 1600. The phrase "Walker's earth or other scouring earth" occurs in the English version.

Ann. Off. Nat. Comb. Liq., (1936), *11*, 863–943.

This clay may well have come from Ivry, a southern suburb of Paris. It is known to have been worked for some centuries for fulling, according to Professor Jacques de Lapparent of Strasbourg. He showed however that it was not calcium montmorillonite but a mixture of 77 per cent halloysite and about 20 per cent attapulgite.

🜃 *Mem. de L'Acad.*, (1718), pl. 8, p. 21: *Endeavour*, (1962), *21*, 94.

E. F. Geoffroy (1718) in his "Table of Affinities" included *Terre absorbante* and assigned to it a symbol [Geoffroy (1672–1731) tried to show the order of affinity of a base for a series of acids ($H_2SO_4 > HNO_3 > HCl$)]

S. De Luca, *Recherche chimique, sur une argile . . . trouvée dans . . . Pompeii*, Section Société Réale de Naples, 1878, année XVII.

La Bentonite, by M. Deribéré & A. Esme, (1951) Paris: Dunod, 45.

As late as 1878 S. De Luca stated categorically that there were "numerose località in Francia", yet we have been unable to find any comprehensive lists of these occurrences. One of these which had been used for fulling for a long time was quarried at Cailleret near Alzon (Gard), according to Deribéré and Esme (1951). Records of old "fuller's earth" quarries remain remarkably elusive in spite of what De Luca wrote.

See 'the naming of the essential clay mineral of fuller's earth', ch. 70ii.

Muséum d'Histoire Naturelle, Paris.

The original material from Montmorillon noted by Mauduyt and analysed by Damour and Salvétat in 1847 had an attractive rose red colour and was used for a while in cosmetics. The deposit was lost under houses and now specimens may be seen only in museums, though The Fullers' Earth Union managed to acquire over 400 grams just in time.

Manuel de minéralogie, (1862), *1*, 204, Paris.

Des Cloiseaux, 1862, mentions only one French locality for fuller's earth – Condé-suz-Vègre near Houdan, Seine et Oise. In the absence of clays rich in montmorillonite the French woollen cloth industry had to use what was available, often inferior to a true fuller's earth. De Lapparent examined many of these earths in his determined attempt to make France independent of imports of bleaching clays. He did not succeed, regretfully coming to the conclusion that France would have

op. cit.

to rely on the high-grade montmorillonites of North Africa. As already mentioned he made a mineralogical analysis of the clay from Ivry: that of Merlançon, near Brignoles, Provence, of Senonian, Upper Cretaceous age, was similar but with more kaolinite. A clay at Lagny, Seine-et-Marne (Ludian, late Eocene) was a mixture of 59 per cent attapulgite, 24 per cent montmorillonite and 13 per cent kaolinite.

He found some clays were of the same mixure but montmorillonite dominating. These were from St. Radegonde near Apt and Font-blanque near Pernes, both in Vaucluse and of Oligocene age; and from Caramy in Provence in Liassic dolomites.

A clay rich in montmorillonite was found by Dufrénoy (1856) at Confolens, Vienne, and at St. Jean-de-Côle near Thiviers, arr. de Nontron, and called by him confolensite; this was confirmed by de Lapparent to be montmorillonite.

A. Dufrénoy, *Traité de Minéralogie*, Paris, (1856), 2nd ed, 3, 583.

E. S. Dana, *A system of mineralogy*, 6th ed., (1892), 690.

Other discoveries of smectitic clays in France include "steargillite", a pistachio-green, waxy clay, which occurs near Virolet on the Rochelle railway and in a tunnel near Poitiers (Vienne). It was first described by Meillet (1862), and shown by J. Orcel and Mlle S. Caillère (1933) using differential thermal analysis to contain 30 per cent kaolinite.

From στέαρ, fat.

In des Cloiseaux, *op. cit.* *C.R. Acad. Sci.* (1933), *197*, 774–777.

A rose-coloured clay derived from albite at Landevan (Morbihan) was named by Comte de Limur, described by A. Lacroix (1895) and shown by Orcel and Caillère to be almost pure manganiferous montmorillonite.

Minéralogie de la France et des Colonies, (1895), *2*, 480–484.

The French and Swiss housewife had used for generations Terre de Sommières for all the household uses of "fuller's earth". This clay from near Salinelles, Gard, consists of rather pure sepiolite. At Mormoiron, Vaucluse, a clay had been used for fulling for a long time. When de Lapparent visited the quarries in 1942 they had recently been taken over from Monsieur A. Constant by Monsieur Rondani, who later sold them to Société Expansia. De Lapparent had analysed a rather pure montmorillonite occurring in the overlying gypsiferous beds in very thin layers; but it was not realised until Expansia's time and the research work of Henri Pisani that there were three main types of clay at Mormoiron, attapulgite containing 25 to 30 per cent of what may be potassium montmorillonite at the top, rather pure montmorillonite at the bottom, and a mixed clay in between. Before this montmorillonite is described, we shall continue with the early searches for bleaching clays.

Itinéraire rapport relatif au voyage accompli par M. de Lapparent en Afrique du Nord pour la C.E.C.A. du 28 Août au 3 Oct. 1942.

The French chemical industry could not wait for the discovery of a domestic clay suitable for acid-activation. In 1932 the Société de Carbonisation et Charbons Actifs de Paris, (CECA), the makers of ACTICARBONE, began to make acid-activated bleaching earths from calcium montmorillonite imported from Marnia in Algeria.

In 1937 some bentonitic or "colloidal" clays were discovered at Cadouin, Dordogne, and worked in 1942 by M. Lèrefait. The clays

Chem. Age, (9.1.1937). J. de Lapparent, *Compt, rend. (1941), 111–114. See also Schweiz. Ton-Industr.*, (1951), *54*, (10), 7, on French bentonite 'recently found in the Dordogne'.

contain variable amounts of halloysite or kaolinite. By 1981 production was about 12,000 to 16,000 tons a year, mostly for use in animal feed pellets.

8.7.1937

STARGIL. Under this name was formed a Limited Company to exploit bleaching earths in general and to purify and whiten fats in particular. The offices were at 5 Rue Paul-Cézanne in Paris and the capital was 3 million francs. The Directors were three manufacturers, David and Francisco de Mendialdua, Betrand Mure and a Company Director, Hubert Ouizille.

IIème Congr. mond. Pétrole, (1937), *2*, 395–398.

H. Roche described a refining process for desulphurising kerosenes, consisting of adding a small amount of dry active clay previously impregnated with copper sulphate and sodium chloride (ratio 1:4) to the bleaching vessel. CECA then produced copper-impregnated earths under the name CUPROSIL.

B.P. 604, 919, of 30.9.1938.

By 1937 the company were able to offer activated earths for export and a larger factory at Honfleur was built in 1939 to make decolorising earths for treating both vegetable and mineral oils. Their trade name was CLARSIL.

Chem. Age. (1.4.1939).

CECA quarried since 1932 a calcium-bentonite near Cassaigne, Algeria, because this clay was suitable for conversion to bentonite, which was called CLARSOL.

Their production from near Bergerac in 1975 was 14,000 tonnes.

Bleaching earths for lubricating and other oils were also to be made in a factory in course of erection in 1939 at Marseilles by the Société Française de Glycérines of Paris. The trade name was ACTILSIL-T for decolorising and purifying mineral and vegetable oils, greases, tallows, waxes, solvents, etc.

Chem. Age (1.4.1939)

The firm also made swelling grades of montmorillonite under the names BENTONIL C and BENTONIL 7C, the latter with a very high "yield" in a drilling mud.

A new school of study had arisen in France under Professor Georges Millot at the University of Strasbourg. In his paper with P. Elouard, J. Lucas and M. Slansky (1960), he discusses a sequence of neoformed, marine clay minerals: montmorillonite, attapulgite, sepiolite. On the continental shelf of West Africa and in land-locked inland waters, oxides released during the intense early Tertiary weathering and transported to these sites of deposition, precipitated as neoformed minerals. First the potash removed alumina and silica as illite; then montmorillonite was formed; further away from the source, as there was less alumina, attapulgite, as the French call palygorskite, was precipitated; and finally sepiolite, which contains little alumina, and saponite and stevensite.

Bull. Gr. Fr. Argiles, (1960), **12**, 77–82.

Géologie des Argiles: altérations, sédimentologie, géochimie, Masson et Cie, Paris. (1964).

Geology of Clays, Chapman & Hall, London etc. (1970).

The evolution of the clay minerals in the Mormoiron basin was described by Millot's students Jean-Marie Triat and Norbert Trauth. They confirmed that the clay which had long been used as a fuller's earth consisted essentially of smectites. As might have been expected neoformed montmorillonite does not behave in quite the same way as

Bull. Soc. Fr. Minéral. Cristallogr., (1972), **95**, 482–494.

hydrated volcanic ash. It requires less acid to be activated; it disperses readily and has promise as a sealant.

Both Mormoiron and St. Radegonde fuller's earths have been sold to "Amour et Vie", an international fraternal association, whose book *L'Argile, terre curative, pour guérir toutes les maladies* describes many ways of using fuller's earth both externally and internally, (chapter 69).

According to B. Rozès the Bureau de Recherches Géologiques et Minières has located in the Paris basin smectitic clays that appear favourable for sodium carbonate activation. Twelve areas of several square kilometers, where the grade of over 85% smectite seems promising, were discovered in Sparnacian beds.

The first meeting of the French *Groupe des Argiles* was held in Paris in March 1947 under the Presidency of Professor Orcel. The secretary was S. Hénin. An important event was the publication of *Minéralogie des Argiles* by S. Caillère, S. Hénin and M. Rautureau (1982).

The Institut Henri Beaufour, le Plessis Robinson, have used both French and Sardinian montmorillonites to make medicines to control and alleviate diarrhoea in infants and in adults. The smectites are purified and chemically treated by Laboratories Beaufour and IPSEN S.A.

Also N. Trauth, Thèse, l'Université Louis Pasteur de Strasbourg, (5.7.1974).

Les Éditions Amour et Vie, (1956), 36 rue de Lancry, Paris 10e.

Chronique de la Recherche Minière, no. 460, (May–June 1981): *Ind. Min.* (Nov. 1981), 59–63.

Editions MASSON in the series I.N.R.A.

41

USSR

Burrough set off in search of two lost ships from Colmogro (? Namsos) N. of Trondhjem. He reached Finmark where he met 'Kerils, Lappians and Russes'.

Friday at a Southsouthwest Sunne, wee departed from this Salthouse. It is to be noted that foure miles to the Northwards of Dogs nose, there growe no trees on the banke by the water side: and the bankes consist of fullers earth. Over the cliffes there grow some trees: so that Dogs nose is the better to be knowen because it is fullers earth, and the like I have not seene in all that Countrey.

STEPHEN BURROUGH, 1557, in Hackluyt's *Voyages*, (1598), 327.

Facsimile edition, Cambridge University Press (1965).

Collected in 1711 ± 16 AD.

Now Feodosija in the Crimea.

THERE is a specimen of *Keffekil* in Woodward's collection, Cambridge University, which is recorded to be *Earth with a Cast of Green; native, and just as taken out of the pit, near Caffa, a principal city of Lesser Tartary; where it is found in great quantity. The Turks and Tartars call it Caffaca; and make use of it in their Baths, it being smooth, unctuous and detergent; and indeed one of the finest Fullers-Earth I ever saw. 'Tis absorbent and adheres to the tongue. Dr. Picanini of Rhaetia brought it from the Place.*

Mineralogi, eller Mineral-Rikets Upp stallning, Copenhagen, (1758), 79.

Mem. Soc. Nat. Moscou, (1811), *1*, 60.

E. S. Dana, *A System of Mineralogy*, 6th ed., (1892), 680, 696. Kil, cf. *Min. Abs. 2*, 131, 465.

The Crimea has long been known as a source of fuller's earth. A. Cronstedt (1758) described Keffekil Tartarorum as a yellowish white lithomarge used as a substitute for soap. Fischer (1811) refers to a pearl-gray to grayish white lithomarge from the Crimea as Keffekilite from a place Kaffa and *kil* the Tartar word for clay. Dana, however, referred the former to sepiolite and the latter to kaolinite and halloysite, mixed together.

In Pliny, *Nat. Hist.*, Bohn's edition, vol. VI, p. 300, note on Cimolian earth.

(1915) No.1, 25pp; 2nd ed., (1916); *Min. Abs. 2*, 131.

Bull. Univ. Tiflis, (1923), *3*, 329, *Min. Abs. 3* 68

Chem. Apparatus, (1930), 240.

Mineral Mag., *27*, 269.

Masloboino Zhirovoe Delo, (1929), No. 8, 10.

According to Bostock and Riley (1857) fuller's earth is found at Alexandrowsk – some 60 miles north east of Moscow. A. E. Fersman wrote on USSR deposits of fuller's earth and allied substances – *Matter on the natural productive forces of Russia*. A. A. Tvaldchrelidze discovered a large deposit of calcium montmorillonite in Georgia in 1916, and prospected it in 1926. His first paper was in 1923 – fuller's earth in Georgia (Imeretia and Guria) and his second in 1930. He named this clay GUMBRIN from the town of Gumbri, published a chemical analysis and showed that the clay may be used for purifying kerosene and vegetable oils, but he unfortunately thought that the clay was like Floridin, the American "fuller's earth" or attapulgite.

A. Markman and M. Kovalenko headed their paper *The infusorial earth from Akhaltzikh and the Floridin from Kutais*, the district in which Gumbrin occurs, and from then on "Floridin" was used indiscriminately. D. S. Belyankin and V. P. Petrov put the record straight in 1945 when they showed that Gumbrin is a bentonitic clay consisting mostly of montmorillonite and a product of alteration of a volcanic ash of Upper Cretaceous age.

The petrology of Georgia, Petrography of USSR, Ser. I, Regional petrography No.11, *Acad. Sci. USSR*, (1945), Table Ib No. 12 (analysis).

Activation experiments were carried out by V. V. Zhadin, E. P. Ulyashchenko and V. I. Astafiev in 1928. The clay was heated with 59–60° Bé sulphuric acid to 120 °C. The dried mass was well washed, dried and pulverised, (1500 meshes per sq. cm.). The product "has the activity of FLORIDIN and TONSIL". Potash alum and aluminium sulphate were recovered from the effluent.

J. Chem. Ind., (Moscow), (1928), *5*, 864.

A. Markman and F. Wyschnapolskaja (1932) did not think highly of the oil decolorising effect of Gumbrin and other USSR clays, but showed that activation is best done by sulphuric and hydrochloric acids.

Oel und Fett Ind., (1932), 45.

G. Norkina and G. Gutchtschin showed that the heat of wetting of earths with oil was proportional to the bleaching power of the earth.

Seifens. Ztg. (1934), *61*, 835.

A method of refining naphthenic acid by treatment successively with sulphuric acid and Gumbrin was proposed by A. Rabinovitsch and T. Osenova.

Maslob. Zhir. Delo, (1934), *10* (12), 41–43.

A number of 'bentonites and floridins' in Azerbaidzhan were investigated by S.A. Kovalesky, but they showed only a small bleaching value, as they were prepared at that time.

Azerbaidzhanskoe Neftyanoe Khozyaĭstvo; Amer. Chem. Abs. (1931).

On the other hand other clays in Georgia proved to be commercially valuable, particularly ASKANITE from Askana, Guria. This clay, mainly montmorillonite, was probably formed by hydrothermal alteration of a biotite tuff of Middle Eocene age by emanations from a syenite intrusion.

D. S. Belyankin & V. P. Petrov, *op. cit.*, (1945) *Miner. Mag.*, *25*–622. *Miner. Abs. 7*–426.

M. L. Blagodarov describes how Askanite was activated with acid recovered from sludge (20% H_2SO_4) in a petroleum oil refinery; the product was highly regarded (1933).

Neft, (1933), *4* (12), 14–15.

KRUIMSIL, mined in the Crimea, was found by N. I. Chernozhukov in a comparison of many Russian and foreign clays to be the most efficient for treating various petroleum products (1933). R.D. Petty wrote that at that time Russia had two types of bleaching clay on offer, natural clays, comparable with fuller's earth, for neutralisation of oils, and activated earths, comparable with PALEX and FILTROL No.1 from the USA (1933). The former had been shown by L. A. Ssuschizki in 1931 to be considerably improved by treatment with 10% HCl. On the other hand the activated earth KRYMSIL had a greater bleaching power than FLORIDIN, TERRANA A, and silica gel on mineral oil products.

Amer. Chem. Abs., (1933), 2289.

Amer. Chem. Abs., (1933), 2289.

Mineral-Rohstoffe, (1931), *6*, 527.

G. B. Kagan drew attention to the wide distribution of clays suitable for removing resinogenic substances from lubricating oil and gave instructions for purifying oils.

Ukrain. Chem. J., (1936), *11*, 80–83.

NALCHIKIN was a newly discovered Soviet decolorising earth, announced Kuadzhe in 1941. It was named after Nal' Chik in the

Masloboino Zhir. Prom., (1941), *16* (516), 21.

41 *Compt. rend. Acad. Sci.* USSR, (1941), *33*, 259–261.

The Bentonites of Uzbekistan [*Bentonity Uzbekistana*], Acad. Sci. Uzbek SSSR, Tashkent, 1963. Ed. K.S. Akhmedov.

Minerals of Shorsu, Tashkent, 1935.

According to *Clays* by S. H. Patterson, and H. H. Murray, USGS (1974), 538.

Internat. Clay Conference, (25–30 June 1972), 113–114.

A. G. Seidov and Kh. A. Alizade, The formation and mineralogy of bentonites in Azerbaidzhan, *Clay Minerals*, *6*, (3), 157–166. (N.E. in Upper Chalk times; in S.E. in Middle to Upper Chalk).

F. D. Ovcharenko *et al.*, (1967), *Investigations of the physicochemical mechanics of clay mineral dispersions* Israel Program Science Translations, Jerusalem, 146 pp.
Concerning the collection Bentonitic clays of the Volga Regions, *Lithology and Mineral Resources*, Consultants Bureau, New York, *7*, (3), 393–394.

Ind. Min. (Jan. 1968), 30.

Caucasus Mountains, NE of Elbrus, in the Kabardinian ASSR. In the same year (1941) S. M. Yusupova described the bentonite clays of Changhyr-Tash.

According to M. Z. Zakirov (1963) bentonitic clays were known in Uzbekistan under the names "gilmoya", "tog yogi" (gornyi zhir-mountain fat), "tosh savun", Imylnyi kamen – soapstone, etc. Gilmoya, it is said, was used by craftsmen for purifying oils and grape juice, degreasing wool, pottery manufacture and in building materials. Clearly the name was given to a variety of clay minerals which cannot be identified by these descriptions. Certainly palygorskite is also known in the region of Samarkand, Bokhara and Tashkent, though bentonites (called Keffekilites) were stated to occur in that country by A. S. Uklonskii in 1935.

These smectites occur mainly in Lower Tertiary rocks, according to *The Bentonites of Uzbekistan*. P. T. Tazhibayeva and M. S. Galiyev discussed the genetic features of South Kazakhstan bentonites at the Madrid Clay Conference (1972): they seem to have been altered from hydromica in an alkaline and deoxidising environment in inland seas in regions of arid climate and are for this reason unique.

Most of the Azerbaidzhan bentonites, of Cretaceous age, were formed by the alteration of volcanic ash which had fallen into an alkaline sea. It is interbedded with tuff and calcareous sedimentary rocks. In the north-east Minor Caucasus the volcanic episodes occurred mostly in the Upper Santonian and Campanian, but in south-east Great Caucasus the activity was mostly in the Turonian and Coniacian.

Softer, more soda-rich bentonites of Middle Eocene age are also of industrial importance. These so-called 'koun' bentonites occur in the Kobystan-Shemakha region. The vernacular name of these clays is gilabi in Azerbaijan and oglanlite and gebelite in Turkmen; and others have been given already.

Several large bentonite deposits of Lower Miocene age occur along the border between the Cherkassy and Kiev Regions of the Ukraine, according to Ovcharenko *et al.*, (1967), who also identified a layer of attapulgite there.

The principal bentonite deposits in the USSR are in the Volga Region, western and central parts of the Ukraine, Trans-caucasus Region, Kazakhstan and Central Asia according to V. I. Lebendinskii and L. P. Kirichenko, (1972), and an article on the bentonite clays of the Volga Regions, [Trans. Kazan Geological Institute, Ministry of Geology of the USSR, (1970), 1175 pp]. Smaller bentonite resources occur in eastern and western Siberia, Crimea and in northern Caucasus. The eastern Siberian deposits were developed in 1968 at Piervomaiskoie and Tal-Youryakhskoie near Magadan, on the sea of Okhotsk.

The principal uses of bentonite in the USSR are in the manufacture of heavy clay products (in what capacity it is not stated), drilling mud, and iron-ore pelletising, and minor quantities used for purifying oils, water and wine. In 1967 Soviet scientists reported that many years of experiment had shown that the addition of bentonite for pelletising and bonding animal feed increases average weight in cattle and stimulates egg production in poultry.

Ind. Min, (Nov. 1967), 29.

At that time (1967) only about 1.5 million tons were produced, though about 3.4 million tons were consumed, the difference being made up by imports from Italy, Czechoslovakia and Hungary. Since then the industry has greatly expanded, to well over 4 million tons a year and the country is virtually self-supporting.

Lebendinskii and Kirichenko, *op. cit.* (1972).

"Bentonite",

Acad. Sci. USSR, (1980), Kindly presented by Academician F. V. Chukhrov.

42 AUSTRIA

Manuel de Minéralogie, A. des Cloiseaux, Paris, (1962), *1*, 204.

FULLER'S earth was well known in Austria in the last century. In 1862 des Cloiseaux recorded four notable localities

Fohnsdorf	Styria	47.13 N	14.40 E	
Eibiswald	Styria	46.42 N	15.16 E	
Reifenstein	Carinthia	47.1 N	14.44 E	[Reichenfels]
Friedberg	Styria	47.30 N	16.4 E	

B.P. 478, 911 (8/2/1937), Austria (22/2/36).

About 1935 montmorillonitic clays were acid-activated at the Skoda-Wetzler Works at Moosbierbaum, where sulphuric acid was manufactured by Oesterr. Dynamit Nobel, whose patent was licensed to Hungary and Czechoslovakia. The activation process consisted of treating the clay with sulphuric acid and the formation of calcium sulphate was prevented by the additon of hydrochloric acid or common salt to the sulphuric acid.

J. Soc. Chem. Ind., (20/11/36), *55*, 938.

The following year negotiations were expected between the Austrian Board of Trade and the Austrian chemical industry for the institution of protective tariffs on phosphoric acid and bleaching clay imports.

Chemical Age, (17/7/1943), 64.

After the Anschluss the greatest chemical works of Austria, the Pulverfabrik Skoda-Wetzler A. G. (Moosbierbaum) and the Wagemann-Seybel A. G. (Liesing) became the property of the I.G. and were much enlarged, producing heavy chemicals including inorganic acids and salts, gunpowder, explosives, fertilisers, fuller's earth, etc.

Miner. petrogr. Mitt., (1953), Ser. 3, 3, 167–180.

From 1953 there was a short spell of activity when Austrian montmorillonites were re-examined. Erich Neuwirth of the Institut für Mineralogie und technische Geologie at the Technische Hochschule in Graz wrote an interesting paper on the petrography of some bentonite occurrences in Styria, Carinthia and Ponza (Italy). Bentonites were collected from Piehling bei Stanz, Sö. Schönegg bei Stanz, Passail and Ratten in Styria and Mühldorf in Lavanthal in Carynthia. He also wrote in similar vein on bentonite from Deuchendorf.

Berg- und Hüttenmann Monatsh. montan. Hochschule Leoben. (1954), *99* (12) 235.

Oesterr. Ziegl Z., (1954), *4*, 61.

About the same time A. Hauser devoted the greater part of his *Investigation of Clays*, Part XI, to thixotropy, giving briefly the results of his own experiments on the thixotropy of Styrian clays, among 16 types of montmorillonite on which rheological measurements had been made.

Procédé de construction de diaphragmes imperméables à grande profondeur au moyen de bous thixotropiques.

It was Professor C. Veder of Graz University who developed from 1950 his method of making continuous concrete diaphragms formed in a trench. They are particularly useful in forming retaining walls for deep excavation in built-up areas. He presented his ideas to the Third International Conference on Soil Mechanics and Engineering in 1953. Further information on this application is given in Chapter 63.

ROMANIA

FULLER'S earth or smectite has been known for over a century in NW Transylvania. Des Cloiseaux (1862) mentions its occurrence at Oláhlapos, Offenbánya, Mühlbach, and Nagy Bár. The first is now called Lapusul Romin: Mühlbach is Sebeş; but Nagy Bánya is Baia Mare.

Manuel de Minéralogie, I, A. des Cloiseaux, Paris (1862). p. 204.

In 1933 Romania was importing most of its decolorising earth from Germany, so efforts were made to find a suitable raw material for acid-activation. A clay at Tomesti was thought to be the first occurrence of bentonite found in Europe but Dittler and Kirnbauer reported to the Committee on Sedimentation in Washington that they did not consider it to be a true bentonite, although it was also examined by Hofmann, Endell and Wilm in their pioneer paper on the X-ray diffraction of montmorillonite – they showed it contained too much cristobalite.

E. Dittler & F. Kirnbauer. *Z. f. prakt. Geologie*, (1933), **41**, 121.
U, Hofmann, K. Endell & D. Wilm, *Z. angew. Chem.*, (1934), **47**, 539.

It was an altered liparite.

About this time attention was paid once more to the clays of Transylvania. In 1934 an advertisement appeared in the *Journal of the Society of Chemical Industry* enquiring for a plant capable of activating 100–200 tons of earth a year. Progress was rapid since in October 1935, Mr. Ovidiu Bolgiu offered Romanian bentonite to The Fullers' Earth Union Ltd, and sent a sample of acid-activated earth prepared from it. The sample was only about 80 per cent as active as it might have been, but was thought to be a good guide to the possibilities. The bentonite gave a very good bleach on cottonseed oil. Activation could be done equally well with sulphuric or hydrochloric acid.

E472

E473

In 1936, Erdheim published the results of comparative tests on a number of activated earths, and SONDAFIN, as the Romanian acid-activated earth was called, proved to be at least as good as the best German earths in decolorising a mineral oil mixture. He stated that SONDAFIN was already much used in Romania and Poland, but its sale declined in Poland when Polish bentonite came to be activated there commercially.

E. Erdheim, *Petroleum Z.* (1936), **32** (11), 5–8.

Shell-Mex had tested SONDAFIN and considered it superior to all German earths (1936).

Learned on W.T. Raikes' visit to Shellhaven 9 July 1936.

Also in 1936 Erdheim measured the bleaching power of the Romanian activated earth VEGETALIN on colza and linseed oils and found it better than the German earths TERRANA, FRANKONIT and SILITONIT.

The Phoenix Chemical Company of Baia Mare, which had started supplying SONDAFIN to the Romanian mineral oil industry in 1935, were thinking about developing exports in September 1936. By

Chemical Age (5/9/36), 35, 212.

December, Hungary was using some Romanian earth but was taking steps to manufacture activated earth from their own raw material.

Dr. Erdheim later joined the Phoenix Chemical Co. at Baia Mare and continued to write numerous, interesting and important papers on activated earths. He showed that the siliceous residue from the extraction of alumina from clays by hydrochloric acid was not suitable as an adsorbent, except perhaps as a preliminary bleach. With A. Berczeller he showed that a certain amount of water added to animal and vegetable fats before earth treatment improves the bleach, especially of natural earths. The effect was not found on mineral oils. Erdheim also showed that active carbons but not natural or activated earths bleached crude glycerine.

Petroleum Z. (1936), **32** (20), 1–3.

Petroleum Z. (1937), **33** (4), 6.

Seife, Kosmetik, Öle, Fette, Wachse, (1937), (8), 2.

Erdheim published many papers on the mode of action of bleaching earths, their effect in different media, methods of testing earths and so on; and was author of the firm's excellent brochure.

The Romanian firm were the first to produce a granulated activated earth, GRANULIN. In August 1938 Dr. Arthur Weiser and Dr. Nikolaus Weiser visited The Fullers' Earth Union and proposed that 10 to 50 per cent of English natural fuller's earth could be blended with their "Spitzqualität" SUPRA-VEGETALIN so as to equal TONSIL AC, ACC, etc from Germany; that their raw material could be activated in England; that their products could be sold in England and that they could activate English fuller's earth "if it is capable of being activated". They had no idea that The FEU supplied about 12,000 tons a year of products which were found to be just a good as the Romanian!

In 1958 the State Company for Foreign Trade, CHIMIMPORT, Bucharest, advertised in Britain "Bleaching Earth VEGETALIN, used as decolourizing agent for the filtration of vegetable oils, in the soap and wax industry".

Chemical Age. (2/8/1958), 203. PO BOX 525.

Production in 1975 was 62,800 tons of bentonite; sales being handled by *Impexmin*.

SPAIN

44

Let the six fulling hammers be transformed into so many giants.
PETER A. MOTTEUX' Cervantes' History of Don Quixote,
tr. 1700–03 (1722, 1733)

FULLERS clay, used in the Antequera cloth and baize factories was shown at the 1851 Exhibition in London by the Inspector of the Mines of Málaga. The provenance of this clay is unknown, but it might have come from the Sierra de Níjar or from the region of Cabo de Gata, both near Almería.

1851 Exhibition (London) Catalogue, **3**, 1329.

J. L. Martín-Vivaldi *et al.*, *Clays & Clay Minerals*, (1956) **4**, 181–183.

O. Burghardt is believed to have gone to Spain from Germany in 1929 soon after completing the process research which led to his patent for the automatic neutralisation of activated earth. On his advice Minas de Gádor SA set up an acid activation plant, which started production in 1935. Formerly they had made calcium and magnesium carbonates and magnesia. By 1939 they were able to supply the country's needs of decolorising earths. Sales justified an extension to their activation plant in 1946 and in the following year they started making bentonite by their own patented process and this was in full production in 1949. Additional plant was installed in 1955 and modernised in 1965. The preparation of granular absorbants, which were sepiolitic, was begun in 1969. Further enlargement and modernisation took place in 1971. At that time their two raw materials came from La Serrata and Los Trancos, the latter containing mostly exchangeable sodium ions. The high cation exchange of these clays, 97.6 and 97.8 meq/100g respectively, suggests a high montmorillonite content. Their deposit at Majada de las Vacas is held in reserve.

DRP 472 958.

Brochure, *Minas de Gádor, SA.*, founded 14.5.1925.

In 1973 Laporte Industries Ltd acquired a 40 per cent holding in Minas de Gádor SA.

Increased in 1983.

The firm of TOLSA S.A. was founded in 1957 by a group of industrialists, some of them still on the Board of Directors in 1982. They engaged the services of Dr. V. Savelli, son of Dr. Francesco Savelli of Italy, as General Manager. He developed the sulphuric acid-activation of certain of TOLSA's smectites but the process did not prove commercially profitable. On the other hand their calcium- and sodium-montmorillonites, known as B-grades, are still used in foundries and in civil engineering. TOLSA also make saponitic products and since they began to make sepiolite absorbents in 1968, they have become the world's largest producers of this clay – pro-

44

Sepiolite – a versatile raw material, R. H. S. Robertson, *Chem. & Ind.* (1957), 1492–5.

'SEPIOFLORINA'

J. Cláraso, personal communication 7.12.1955.

ION, Revista Española de Quím. Ap. (Jan. 1947); 15–19: *Chem. Age*, (24.5.1947) 674.

Trans. 4th Internat. Congr. Soil Science, *1*, 92–96.

Ind. Min. (1969), (24), 14.

M. Gilbert, *Soil Sci.*, (1970), **109** (1), 23–25.

Ind. Min. (Mar. 1978, (126) 82.

gress has been astonishing since 30 July 1953 when R. E. Grim and I visited Vallecas and saw great potential there.

The firm of BYSSA – Bentonitas y Sepiolitas SA – had also worked deposits of montmorillonite containing a proportion of sepiolite for some years before 1950, between Toledo and Madrid, and supplied both calcium- and sodium-clays for foundries and many other purposes. They ceased production in 1975.

A bentonite from Villafranca de Panadés, Catalonia, was at one time used in pharmacy.

A Southern Moroccan bentonite was discovered in 1943 and three years later was available "in considerable quantity and high quality". Quarried and processed at Tidinit, it was called COLOIDITE. Its use in soaps was suggested by Carbonell. The firm of Anclá SA of Málaga were the selling agents. Various stages of transformation from a leucocratic dacite to montmorillonite were studied by E. Gutiérrez Rios and F. González García in 1950.

In Northern Morocco, however, output of bentonite was 7,500 tonnes in 1967 and nearly 10,000 tonnes in 1969. It was mostly exported – to Nigeria, Netherlands, France, the United Kingdom, and Portugal.

The production at that time of the bleaching clay from near Camp Berteaux (Melg-El-Ouidane), some 96 km from Melilla was 25–30,000 tonnes a year. The cation exchange capacity of this clay is higher than that of Wyoming bentonite.

Spanish production rose to 80,000 tons or so in 1975.

In 1985 Laporte acquired the whole of Minas de Gádor and agreed with Tolsa SA to handle smectites while Tolsa would concentrate on hormites – sepiolite and attapulgite.

POLAND

45

SINCE the 14th century fulling mills are more and more frequently mentioned in the literature. There were guilds and the techniques used became progressively more complicated as they did in other European Countries – wool was carded or combed, cloth was fulled, sheared and tentered on frames. In the second half of the 16th century the wool industry was well developed in Silesia, 'Grande Pologne' and Pomerania and there were 22 sheep per 100 inhabitants. It is not improbable that fuller's earth was used among other detergents in this industry.

Zarys historii włókiennictwa na ziemiach polskich do kónca XVIII wieko Ossolinium Wrocław-Warszawa-Kraków (1966), p. 631.

As has been mentioned in chapter 17ii, J. Fr. Henkel (1679–1749) wrote in 1747 on medicinal earths which included those from Lemnos, Lemnitz and Striegnitz.

Lemnitz is now Łomnica, Lower Silesia. (Striegnitz is in the German Democratic Republic near Dresden.)

Although smectitic fuller's earths may have been used in Poland before this century, modern interest in the mineral dates from 1934, when Z. Sujkowski found that the bentonite from the vicinity of Krzemieniec, South East Poland, was similar to French and American montmorillonites. B. Piaskowski showed the following year that this clay from Volhynia, when treated with a mineral acid, was effective for decolorising paraffins.

Arch. min. Soc. Sci. Varsovie, (1934), **10**, 98–116.

Przemysł Chem., (1938), *22*, 450–453

Erdheim (1936) compared the bleaching properties of Polish (SILITON) and Romanian (VEGETALIN) acid-activated clays with the German earths TERRANA, FRANKONIT and SILITONIT, using the Lunge-Roth photometer for determining the colour of the purified oils. The Polish product was not as effective as VEGETALIN.

Petroleum Z. (1936). *32* (40);

Motorenbetr., *9* (10), 9–12.

According to Prof. A. L.-Kuźniarowa bentonite is mined at Milowice from Carboniferous strata. Her thermograms give a hint of illite and calcite; other workers, (a) A. Bolewski, Z. Michałek and S. Z. Stopa and (b) Z. Michałek, have confirmed the presence of mixed-layer illite-montmorillonite in montmorillonitic clays of this age. A Miocene bentonite from Chmielnik had also been examined (c).

Letter 5/12/1969.

(a) *Prz. geol.* (1963), (7), 333–335,
(b) Zeszyłu Naukowe AGH, Rozprawy (1965), 39.
(c) A. L.- Kuźniarowa. *Termogramy Minerałow* Iłastych, Warsaw 1967.

MINEX, POB 1002, 79 Krak. Przedmieście St., Warsaw 1, wrote that their POLBENIT is supplied to the domestic foundry industry and production was increasing. The composition was given as SiO_2 57.65, Al_2O_3 17.55, Fe_2O_3 3.50, MgO 2.60, CaO 4.30, Na_2O 1.45, K_2O 1.42, P_2O_5 0.08, loss on ignition 2.16, moisture 7.34%. As sold it had 2–5 per cent moisture content which may be lower than the optimum. Calcite was lower than 4 per cent, but the illite content could have been as much as 24 per cent, judging from the potash content.

The bentonites and montmorillonitic clays of Poland were reviewed

45

Polska Academia Nauk, Warsaw (1975), pp. 108, 55 figs, 19 tables, 7 plates.

Ind. Min. (Mar. 1978), (126), 82.

in 1975 by Zenon Kłapyta; production that year was about 50,000 tons, the sales being handled by *Minex*. The purest montmorillonites were of Cretaceous, Eocene, Oligocene and Miocene age.

HUNGARY

46

THE construction of the plant for processing fuller's earth at Nagytétény, belonging to the Terrachemia Chemical Company, was in full swing at the end of 1936. This firm was formed in July jointly by the Hungaria Synthetic Manure Company and the Hungarian subsidiary of Dynamit-Nobel. Dr. Ing. Harry Berczeli wrote that the clay is grey green and soapy, is of great extent and derives from a biotitic andesite tuff in Sarmatian limestone.

Activation in Hungary, *J. Soc. Chem. Ind.*, (26/12/36), **55**, 1046.

A Magyar Mérnökes Epítész-Egylet Közlönye (official organ of the Hungarian Engineers' & Architects' Society), Budapest, (19/12/1937), **71**, 353.

At first, to replace activated earth being imported from Germany and Romania, manufacture was done in Nobel's factory at Bratislava, formerly Pozsony (Czechoslovakia), whence it was sent back to Hungary and to other countries. The Nagytétény plant however started up in 1937, with a capacity of twice the Hungarian demand.

In 1940 Dr. Barbara S. Beer, now Neumann, travelled to England, to The Fullers' Earth Union Ltd. Her notable work there has been described already.

Montmorillonite processing was more intensively studied in Hungary than elsewhere at that time. Indeed in 1952, nine Hungarian scientists and engineers received awards totalling 100,000 forints for their researches into Hungarian bentonite resources, then believed to be second in importance to that of the extensive bauxite deposits. Hungarian bentonite had been used in foundry work for the first time in 1941. In 1949 a bentonite research committee was set up to find out how best to exploit these products during the five-year economic development plan, beginning in 1950.

Chemical Age. (11/10/1952), 500.

As a result of this work, the use of bentonite had spread to 42 different industries, and there was a surplus for export. In addition to the bentonite processing plant at Mád, a large new factory was due to come into production quite soon.

Many scientific papers were written by members of this team. K. Nagy described the bentonites from Istenmezeje, Bánd, Komlóska and Mád. The montmorillonite content was determined by d.t.a. from endotherms at 700°C and at 600°C or 550°C, microscopy and XRD. They followed Foster's method of working out crystallographic formulae:

Tonindustrie-Zeitung, (1957) 20.

Istermezeje: $(Si_{7.16}\ Al_{0.84})(Al_{2.56}\ Mg_{0.97}\ Fe_{0.33})\ O_{20}\ (OH)_4$. 1.12
Bánd: $(Si_{7.62}\ Al_{0.38})(Al_{3.22}\ Mg_{0.54}\ Fe_{0.20})\ O_{20}\ (OH)_4$, 0.64
and both were classed as montmorillonite-4 (after Sedletskii).

Árkosi and Barna measured the particle size distribution of 5 calcium-montmorillonites in the natural state. All were similar, with mean lamellar diameter 100–300 μm. Their thickness, with the excep-

Acta chim. hung., (1954), **4**, 169–180.

tion of Istermezeje is less than 10 μm. The clays of Mád-Koldu and Komlóska, which are close together, are very similar.

Budapest X, Bihari útca 6.

A second spurt of activity occurred when the Central Institute of Building Material Research took out patents for various uses of bentonite and for processing methods, e.g.:

Brit. Pat. applic. 19726/60 of 3/6/1960; DRP 1106670; BP 907043; FrP 1258891; Nor P 101268.

(1) Lightweight concrete building blocks, according to Károly L. Szepesi and János I. Majer, mixing, for example, 50 kg sodium-bentonite, 400 kg fly ash, 100 kg Portland cement, 200 kg ground lime, 800 l. water and then autoclaving.

B.P. 881517 of 6/2/1958.

(2) Improvements in fertilisers, in which Szepesi proposed a filling of montmorillonite or glauconite, to introduce cation exchange.

DRP 1062683
FrP 1192942

(3) Bentonite synthesis, whereby the calcium ions are replaced by sodium ions when the aqueous paste is warmed and sheared or kneaded. When disclosed to F.W. Berk & Co. about 1963 the process was considered novel, but no patent application was made in Britain.

(4) Bitumen emulsions, in which bentonite replaced casein, a process known in Britain at the time.

(5) Acid activation of montmorillonite. The calcium-clay is expanded (sic) before activation with hydrochloric acid.

Dr. Libor later claimed to convert cheap illite-montmorillonites containing less than 30 per cent montmorillonite to swelling clays by simple addition of reagents. An English firm had taken an interest in the process, but the product was deficient in surface area and required heavy dosage rates in use.

Thermogramy Minerałow Iłastych, Warsaw, (1967)

A. L-Kuźniarowa published thermograms of bentonite from Istenmezeje – main endotherm at 675°C for the calcium and 645°C for the sodium-clay – and from Komlóska – main endotherm at 700°C.

Figs. 53, 54 and 57.

Industrial Minerals, (1969) (26), 38.

Hungarian bentonites are exported by MINERALIMPEX, the Hungarian Trading Company for oils and mining products, Budapest 62, PO BOX 130, Telex 961 962.

Silikáty, (1973), *17 (1)*, 39–53.

I. Novák (1973) made a study of acid-attack on four different montmorillonites. Changes in porosity, specific surface and residual octahedral atoms were followed on increasing the time of the attack. The basal spacing rose to 30 Å and eventually even wider, but the specific surface rose to a maximum value, and then fell as the octahedral atoms were all leached away, confirming work at Redhill a third of a century earlier.

By 1975 Hungarian output had risen to 87,900 tonnes.

ITALY

The earliest mediaeval record of fuller's earth.

ONE of the reasons for the growth of Prato as a wool town was its connection with Lucca, known to be a centre of cloth dyeing in the 7th century; another is the presence on Monteferrato, close by, of a dark slimy earth suitable for fulling. This fuller's earth was recorded as early as 1038 AD. and the first fulling mill in 1108 AD. The first fulling mill in Florence was built in 1164 AD. The statutes of the fullers of Prato were agreed in 1285 AD, the Breve dei Gualchieri (= walkers) being still preserved in the Archivio di Stato, Florence.

The Merchant of Prato, Francesco di Marco Datini, by the Marchesa (Iris) Origo, London: Jonathan Cape (1957), Peregrine Books (1963), p. 55.

See also Book 280 of the *Archivio Datini*.

Date of fulling mill given by Alfred Leix, *CIBA Review*, (Sept. 1937), *1*, 10.

The finest cloth requires the longest treatment.

THE Processing of Wool in the Florentine Cloth Industry.

The textile industry of Florence owed the leading position which it held for centuries not to particularly favourable sources of raw material or to any special technical process, but rather to shrewd financing and to a division of labour which in many respects has not been surpassed by the most modern methods. Whereas in other medieval towns, in Strassburg for instance, the wool-combers not only cleaned and carded, but frequently dyed and spun the wool, in Florence each of these activities was assigned to a special group of workmen, distinguished according to their various functions, which also decided the social status of those who performed them. A record of the year 1427 shows that in the processing of wool from the raw material to the finished cloth between twenty and thirty different categories of workmen were employed.

One of the principal sources of our knowledge of the technique of cloth-making in Florence is a 15th century manuscript contained in the "Codex Riccardianus 2580" in the library of the city. This "Trattato dell'Arte della Lana" was published with a detailed commentary by Alfred Doren in the first volume of his "Studien aus der Florentiner Wirtschaftsgeschichte" (Studies in the Economic History of Florence). The following sketch is based in the main on Doren's interpretation of the Trattato.

Walter Reininger, Basle, The Florentine Textile Industry of the Middle Ages, *CIBA Review*, (Nov. 1939), (27), 960–963; largely from Alfred Doren, *Die Florentiner Wolltuch-industrie vom 14. bis zum 16. Jahrhundert.* Stuttgart 1901 and *Das Florentiner Zunftwesen vom 14. bis zum 16. Jahrhundert*, Berlin 1908, the latter about guild affairs.

After the raw wool imported from abroad had been weighed and the duty on it had been paid, it was superficially cleaned and divided by the "sceglitori" into three groups according to its quality. The three groups were lana fina, mezzana, and grossa (fine, medium, and coarse wool). It was then washed in lye and water in order to remove the grease, and after drying in the sun was sent to the workshop of the clothmaker. There it was again cleaned, being spread out on boards, where particles of dirt were removed by hand. It was then treated with small shears and other instruments until completely freed from foreign substances and ready for weaving. Wool which was to be spun and woven in its natural colour, was spread out on frames, beaten, sprayed with water from time to time, and finally soaked in oil. In the next stage of the process it was rolled up and combed out again by the counter-movements of two large combs manipulated by the wool-carder (pettinatore). The wool which was combed out in long strands was called "lo stame" (the carded wool), the remainder which fell to the ground in small tufts was called "palmella" (flock wool). The carded wool was wound on to blocks of wood by the "appennechino", and sent to the spinners, whereas the flock wool had to be subjected to a varied treatment before spinning. It was first placed on a frame made of twigs, cleansed of particles of dirt etc., loosened, moistened with water, and oiled. After this the "scardassieri", the most numerous group of artisans employed in the cloth industry, worked the wool with two implements fitted with fine wire hooks until it was completely separated and carded. After that it was spread out on boards, made up in bundles of the same weight by the "lanini" (the wool carriers), and sent to the spinners. The warp-yarn was spun dextrally, whereas the weft-yarn was given a leftward turn, a spinning wheel being used for the latter only. After spinning, the yarn was taken to the workshop of the "lanaiuolo", where it was sorted according to quality, the different grades being distinguished by marks (crown, lily, rose, etc.)

Wool intended for dyeing was beaten with sticks and rods after being cleaned and loosened; it was sprayed with water, then spread out to dry, and finally packed in sacks to be sent to the dyehouse. After dyeing, the wool was combed and spun in a manner similar to that described above.

The spun threads were spooled and the warp prepared by the "orditore" by means of a warping frame similar to that used in handweaving until this day. After this, the warp was checked. The quantity of yarn used by every weaver was recorded by officials of the guild, as for fiscal reasons the amount of cloth which the textile manufacturer of Florence was allowed to produce yearly was carefully regulated by the masters of the guild. Only after the yarn had been checked was the weaver allowed to begin dressing the warp. The dressing used was usually gum or the refuse from corn-grinding. The form of the loom probably corresponded to that seen on the relief by Andrea Pisano (1290–1348) on the campanile of the cathedral of Florence.

cf. *CIBA Review*, No. 16, p.554.

The finished fabric was returned to the cloth maker in rolls, where it was roughened with small knives and pincers, and where superfluous tufts were removed. The process began with washing the cloth and cleansing it of grease in the so-called "purghi" or "purgatoi". According to the Trattato it was first washed for two hours in hot water, then with soap or putrid urine and lye-water "until it foamed". Finally, the cloth was rinsed in cold water and then wrung out. There followed half an hour's treatment with fuller's earth (terra di purgo).

This process served to remove any remaining grease, and took place in pits filled with boiling water. After that, the fabrics were washed in running water, usually in the river Arno, and hung up to dry against a wall with the felted side facing inwards. Then they were taken to the fullers for further felting, and subsequently cropped by the "cimatore". When the cloth had been stretched in buildings known as "tiratoi", the first stage of the process of dressing was concluded. The whole procedure, which, by reason of the inclusion of fulling, resembles the manufacture of worsted cloth, was then repeated once more. During the repetition the front surface of the cloth was roughened with carding thistles and then cropped once again by the cimatore. Before the goods were offered for sale, they were first examined by the rimendatore, whose duty it was to repair damaged portions; then they were treated once more by the purgatore and the cimatore. The cloths intended for sale in Florence were folded, whereas those designed for export were first placed in a screw-press and then tested again. Before the fabrics were handed over to the purchaser, they were measured, damped, and pressed once more.

In his analysis of the Trattato Doren emphasizes the unusual mixture of the serge and worsted technique which characterized Florentine cloth-making. Whereas at a later date certain fine, long-fibred wools were used exclusively for the manufacture of serges with smooth unfelted surfaces, and other wools with shorter fibre yielded fabrics the surface of which was rough and felt-like as a result of the process of fulling, the technique described above produced both smooth and rough yarns, the former being used for the warp and the latter for the weft. This ensured the fullest possible use of the raw material, as the by-products were also turned to account. Characteristic of the entire method of manufacture prevailing in Florence was the careful treatment, suited exactly to the material, which was observed in every stage of the process. The factors which decided the popularity of Florentine cloths in international trade were above all the brilliance and the durability of their colours.

Reininger has after (terra di purgo), "which usually consisted of silicate of magnesium". Caley and Richards in *Theophrastus on Stones* (1956) make the same mistake, for they say that Cimolian fuller's earth is a "particular form of sepiolite, a hydrated magnesium silicate." (p. 209), see however R. H. S. Robertson, *Classical Review* (New Series) (1948), *8*, 222–223.

Tuscan fabrics had been made for centuries. Silk and wool textiles were made in Lucca as early as 846 AD; but even earlier than that the convent of Or San Michele in Florence had organised textile crafts in what must have been one of the earliest industrial foundations.

Between 1050 and 1200 dyers, fullers, shearers and tailors are often mentioned in Florentine records, but no weavers, wool combers or spinners, because woollen cloth was at first imported and finished in the town. However by 1223 AD Florentine merchants had gone to England to buy wool; because of the enormous freight costs, they always bought the best quality – at first from monastic houses. The wool was shipped via the Flemish trade centres, such as Bruges. These purchases were made possible by the enterprising banking families who set up branches in London. Indeed Guccio, the first of the Medici family, belonged to the wool guild "Arte della Lana" in 1299.

W. Reininger, *op.cit.* 967

The most important record of Florentine trade in the Middle Ages is the "Practica della Mercatura" by Francesco Balducci Pegolotti, who between 1315 and 1336 travelled over the entire trading area of that period. He reported in detail on the kind and quantity of the principal articles of trade, on prices, methods of packing and payment, and on the transport routes and expenses. His notes were continued in 1442 by Giovanni di Antonio da Uzzano, a young merchant of Pisa. Pisa was finally subjected by Florence in 1406, but as the harbour was silting up, the excellent port of Livorno was acquired from the Genoese. The importation of fuller's earth passing through Livorno may not be only a twentieth century happening. No reference to terra di purgo or the like was found in either Pegolotti or Uzzano, not even among the many articles of commerce available in London.

Giovanni Francesco Pagnini dal Ventura, *Della Decima e di varie altre gravezze imposte dal Comune di Firenze..Della Moneta et della Mercatura de' Fiorentini fino al Secolo XVI.* Contains the records of Pegolotti and Uzzano. Four vols. Lisbon and Lucca, (1765–1766). British Museum Library 17.i.1–3.

At the end of the 15th century some Florentines went to England and disclosed their manufacturing secrets. There were other even more weighty reasons for the decline of one of the world's most famous cloth industries.

Modern Italy

IN 1930, O. Burghardt aroused interest in montmorillonite by his paper in an Italian journal on 'Raising the activity of natural fuller's earth' by acid activation. Only five years later Dr. Francesco Savelli, who came from Follonica in Tuscany, rediscovered fuller's earth in the island of Ponza. The name Follonica, it should be explained, has nothing to do with the woollen industry but refers to the stamping mills where iron ore from Elba was crushed so that the non-ferrous impurities could be washed out as slimes. Both words came from the same Italian word *follare*, since the pounding action of the processes is similar.

Industria chimica, (1930), **5**, 876.

Una nuova risorsa mineraria italiana: La Bentonite scoperta nelle isole Pontine nell'Anno XIV del Littorio, Rome, 1935.

Francesco Savelli's company S.A.M.I.P. organised comprehensive studies on bentonite in Rome and brought in foreign experts. The team compared the swelling, plastic, adsorptive and thixotropic properties with those of Wyoming bentonite; and Italian scientists studied applications of bentonite in foundries, soil sealing, pharmacy, and other uses in the chemical industry.

S. A. Mineraria Isole Pontine.

Atti X. Congr. Internaz. Chim., (1938), *2*, 785-795

In Savelli's time the mixture of natural clay and a solution of sodium carbonate was hand-kneaded and then sun-dried (1935–1936), a process reminiscent of the process which gave its name to ball clay.

cf. articles in *Raw materials of Italy and of the Empire* a magazine published by Baldo Savelli, from 1935 to 1942 (Francesco's son).

Mining was centred on "Cala dell' Acqua" and became in time an important industry. Milling was done in the port of Gaeta on the mainland. At this plant, esecially during the war, large quantities of white bentonite, with and without soda treatment, were used, thanks to several government decrees, in the manufacture of paper and soap in partial replacement of kaolin and fats, respectively. The viscosity of the Italian bentonite in suspension in cyclohexanol was not surprisingly found by H. Erbring and H. Lehmann to be less than that of bentonites from the USA and Germany, which were close together.

Kolloïd-Z., (1943), *103*, 126–129.

Savelli's book *La Bentonite* (1943) is full of interest and is much quoted by M. Deribéré and A. Esme in *La Bentonite, les argiles colloïdales et leurs emplois*, (1951). Savelli died in 1944 and all the installations at Gaeta were destroyed during the war, and those of Ponza were damaged. However, his sons Vittorio and Brando continued their father's activity after the war, restoring production in a former cement factory in S. Marinella on the mainland, nearly 50 km north of Rome.

Ed. Hoepli: Milano,

Paris: Dunod, pp. 35–40.

S. G. Henderson, who had moved from The Fullers' Earth Union to join Colin Stewart, Ltd., introduced Ponza bentonite to British foundries in 1948 and greatly helped S.A.M.I.P. in a most difficult time; but by 1953 Ponza bentonite was used in Italy in large quantities for well drilling. In 1957–8 a production plant was installed in Ponza and, not long after, the S. Marinella plant was shut down, allowing the production to grow in Ponza to over 100,000 tonnes a year in 1977, when the plant was closed since it was felt by the authorities that the mining operations were spoiling an island of great scenic beauty. The deposits however were nearly exhausted. Deribéré and Esme reproduced Savelli's cross-section showing the geology. (Fig. 80).

In UK, sold as QUEST bentonite.

For E. Neuwirth's paper on Ponza earth see chapter 43 (Austria).

The glassy rhyolitic tuff has been converted mainly to calcium montmorillonite and practically amorphous silica. The trade name for the

Deribéré & Esme, *op. cit.*, pp. 35–40.

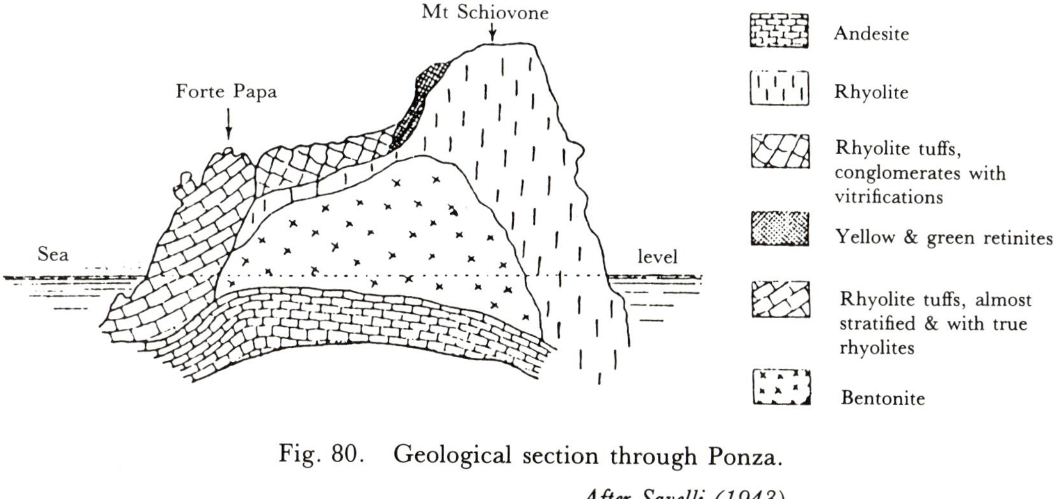

Fig. 80. Geological section through Ponza.

After Savelli (1943)

milled and air classified product was ALBALITE, the whiteness implied by the name being due to the low content of ferrous iron and to the very finely divided silica. The letters G, GS and others were given to the soda-treated gel-forming grades.

See also Ponza bentonite, *Ind. Min.*, (1973), (64), 35.

Meanwhile two other productive areas had been developed on the Italian mainland: one, at first by LITACROM S.A.I. and then by Dr. Settimio Cinicola in Puglia; and the other by a former agent of S.A.M.I.P. Mr. Bicinetti, in Veneto, as Alba SpA.

Industria mineraria, (1943), *21* (4).

Etude röntgenographique de diverses argiles à montmorillonite; *Industria della Ceramica* (1946), *1* (3), 1.

op. cit.

C. Andreatta made a petrographic study of the grey to white clay with a conchoidal fracture underlying a sandy limestone in Puglia, and its mineralogy was established by XRD by M. Rolla in 1946. Quarries have been opened up at Ciccirinella, Pietrino, Marcellino, Macello and Cellenza. Milling was carried out at San Severo in Puglia, some of the clay being acid-activated at Porto Marghera (Veneto); and later the works were at Termoli, Castelnuovo della Daunia (Foggia) and San Severo. Analyses published by Andreatta show that these clays are less siliceous than those of Ponza. The finished products are designated 7C, ULTRAGEL, TISSOLIT 88 and 88B, A.R. and A.R.B.

Also DEPURAL 7c and ACQUAFLOC 7c.

For recent developments of these companies, see *Ind. Min.* (Oct. 1982), (181), 29.

Alba SpA developed two types of smectite near Vicenza (Veneto); the first was obtained from near Arsiero and the second, more nontronitic, in the Monti Berici. Another firm, the VALDOL Company of Maglio di Sopra, also near Vicenza, had started to produce montmorillonitic clays, used also for acid activation. F. Hofmann gave rather promising foundry test results on one of these smectites from near Vicenza in 1956. Six chemical and thermal analyses of clays from this area were included in L. Poppi and M. F. Brigatti's research on Italian smectites. These authors showed that double dehydroxylation reactions in thermal analysis are due to a mixture of nontronite and montmorillonite or to a mixture of montmorillonites of Cheto and Wyoming type. A. Alietti and M. F. Brigatti, using the same samples showed that when the ratio Na,K/Ca, Mg is above 0.28 the d_{100} x-ray spacing is 9.5–10 Å, the clays behaving as if the exchangeable cations were all monovalent. When the ratio is below 0.28 the d_{100} spacing is 11–12 Å, the clays behaving as if the exchangeable cations were divalent.

Giesserei, Techn. Wiss. Beihefte 16, (1956), 857–864.
Miner. Petrogr. Acta, (1976–77), **21**, 43–52.

See Ch. 30, Dunning fuller's earth.

Miner. Petrogr. Acta, (1979), *23*, 145–150.

The Stà. Mineraria Estrattiva Srl operate (1982) a 25,000 tpa non-swelling bentonite operation at Pietracuta de S. Leo, Pesaro, principally for use in foundries.

Ind. Min. (1982) (181) 29.

Acid activation

IN 1938 rumour had reached Dr. Edward Erdheim that an acid activation plant was being set up in Italy, and two years later the activation and decolorising power of some Italian siliceous montmorillonites were reported on by V. Gaglioti and A. Baroni. Indeed, in 1938, G. R. Levi

Petroleum Z., (1938), **34**, (no 15).

and F. Castelli had used the decolorising earth called VAS in chromatographic experiments with opium alkaloids: this earth, presumably a natural smectite, was made by the Società del Caffaro.

Chim. e l'Ind., (1940), *22*, 261–267.

Gazetta, (1938), *68*, 459–470.

Caffaro SpA, Società per l'Industria Chimica e Elettrochimica, eventually became a subsidiary of Ugine-Kuhlmann and Sofdic of France, and in 1970 they completed a new plant for activating clay with hydrochloric acid. They obtain most of their raw material from the north of Sardinia although, in 1982, some bentonite came from an underground mine at Mondaino, Pesaro. In 1982 the production approached 24,000 tpa of bleaching earths, which are sold under the name PROLIT.

Ind. Min. (1970), (32), 44.

Ind. Min. (1982), (181), 29.

The return to Sardinia

INDUSTRIA Chimica Carlo Laviosa SpA was established in 1933 by Comm. Carlo Laviosa to supply foundries with coal dust, graphites and other materials formerly imported. Their small plant was destroyed during the war. They started again in 1945 adding bentonite to their range, at first from other people's quarries but in 1946 they brought in crude from their own property in North Sardinia. Carlo Laviosa pioneered in Italy binders for the CO_2 process, pre-coated resins and desulphurising briquettes. They extended their bentonite markets to agriculture and oenology, rubber and ebonite, water treatment, oil well drilling and civil engineering. In 1956 SAMIP had given them exclusive sales rights for selling Ponza bentonite to foundries, and greatly extended the markets.

From the 'Laviosa Team'.

Good accounts of this company's progress are given in *Industrial Minerals*, (April 1970) (31) 47; *id.*, (October 1982), (181), 29. Much more information is given by G Novelli, Fattori che influenzano la tecnologia di produzione della bentoniti per vari impieghi industriali and by A. Rimaldi & I. Caciagli, I controlli della bentonite nei suoi principali impieghi, *Internat. Congress on Bentonites* (1978), 250–275 and 456–478.

Carlo Laviosa looked more deeply into the Sardinian bentonite in 1959 and shipped it regularly from 1960. Rapid progress was made when they bought a new factory site, twelve times bigger, in 1965. Dr. Giovanni Novelli of SAMIP joined the Company in 1966 when the new plant was well on the way to completion (1967). The way was open in 1966 to make foreign sales, in France, Spain and elsewhere.

The calcium montmorillonite, discovered at Pedra de Fogu, near Uri, between Alghero and Sassari, in Northwest Sardinia, was described by A. Pietracaprina, G. Novelli and A. Rinaldi, (1969), but other deposits such as at Puntenuova were acquired. Mineraria Chimica Sarda SpA began operating a plant in the port of Oristano in 1972. Its capacity is 150,000 tonnes a year. Ships can be loaded at a rate of more than 300 tonnes per hour at its own quay of 10 m depth.

Mem. Soc. Geol. Ital., (1969), **8**, 205–218.

'MI-CHI-SA'.

SVIMISA situated at Villaspeciosa (near Cagliari) was bought by Laviosa from the FIAT group. Some bentonite is ground here though the main products are ceramic clays, chamotte, feldspars and barytes. The Laviosa group own a shipping company (1975) and control their own bulk traffic in M/S "BENTOSHIP".

Laviosa, especially through MI-CHI-SA. also have deposits at Ponte Nuova, Tamadili and Monte Laconi.

The largest reserves of Sardinian fuller's earth occur in the central area called Sarcidano; the area contains the villages of Láconi, Aritzo (nearly), Seui, Nurri and Orroli, Serri, Ísili and Nurallao. Another source in this area is near Villanova Tulo.

Since 1959 Baroid International SpA has obtained its material from Nurallao and also from Giba in the southwest. Their sodium-exchange plant at Sant' Antioco has a capacity of 40,000 tpa; they supply drilling mud especially to North Africa and the Persian Gulf. The clay at Nurallao varies in thickness from 1.5 to 15 metres and has an overburden of only 2 to 4 metres.

Ind. Min. (1972), (53), 35.

In 1971 Süd Chemie Italia SpA announced that they were ordering a sodium-exchange plant from Breda Progetti e Construzioni for erection at Piscinas, Giba. Dr. G. C. de Angelis from SAMIP joined this company, which was headed by Dr. C. E. Hofstadt.

Ind. Min. (1975), (94), 34 and 37.

Ind. Min. (1982), (181), 29 Edamsarda cooperates with Laviosa.

In the Trexenta district at the village of Mandas Edemsarda SpA quarries bentonite for processing to drilling mud quality at its plant in Cagliari. CECA Italiana SpA also have mines in Sardinia.

See Ch. 69i.

A white bentonite was worked for a while by AXIA S.A. in the aptly named Cala Sapone for use by a French pharmaceutical company who liked its whiteness and its low swelling power.

Tixonite, a book of formulations, was published by Laviosa.

Laviosa on the other hand sell in the general market a white product called TIXONITE, a purified white montmorillonite in the form of flakes instead of as a powder. It is used in numerous cosmetic and pharmaceutical formulations.

Bentonite crude is also shipped direct from Oristano and up the Rhône to their factory at Port le Valence, where about 20,000 tonnes a year of bentonite products can be made.

pp 33–37

An excellent bibliography of Italian research on montmorillonite for the period 1934–1978 was presented to the International Congress on Bentonite in 1978.

Ind. Min. (Mar. 1978), (126), 81.

By 1975 Italian production amounted to 280,040 tons of bentonite and 70,488 tons of fuller's earth.

CZECHOSLOVAKIA

48

AMONG the principal localities for smectite or terre à foulon des Cloiseaux gives

Manuel de Minéralogie, (1862), **1**, 204; Paris (Letovice)

Lettowitz in Moravia	49.33N 16.34E
Karthaus in Moravia	49.12N 16.37E
Zwikowetz in Bohemia	50.47N 14.38E

(Nová Ves, Brno)

(Zwickau, Cvikou).

No early accounts of using these clays for fulling cloth have turned up, but according to Loebner he used in 1843 *in Moravia a method he learnt in Belgium of cleansing spinning waste with clay. He packed this waste into a vat and then placed a thick aqueous dispersion of Moravian fuller's earth over it. After the earthed waste had stood for 24 hours workmen entered the vat and trod the waste and earth with their bare feet, after which the waste was rinsed with running water till the effluent was clear. The cleansed waste was soft and open, and spun much better than soda-cleaned waste, in particular giving much less fibre breakage during carding. Soda, however did the job much more quickly and so ousted the use of earth.*

C. H. Loebner, *Studien und Forschungen über Wolle und andere Gespinnts-fasern*, Grünberg i. Schl., no date, [*c.* 1890], quoted by C. O. Clark, Ancient and Modern in Milling and Scouring, *J. Soc. Dyers & Colourists*, (1950), *66*, 187–190.

In 1930 E. Belani described a naturally active bleaching earth, CARLONIT, in glowing terms. He referred to the enormous number of new earths cropping up all over the world in 1925-26 – very confusing to oil refiners: Florian had also dealt with this point and chemists could not test them all, though claims of 20 to 30 per cent higher activity were enticing; but by 1930 only the old and firmly established firms had survived.

Petroleum Z., (1930), **28**, 11.

Petroleum Z., (1926), **21**, 780.

Nevertheless he urged that all refiners should try the newly discovered earth from the volcanic thermal region of Carlsbad. This earth, *high in silica and alumina in the analysis, was more active than any other natural earth known; it had good filtration properties and did not cause colour reversion; nor does it cling to the filter cloth. Containing lime and magnesia it is good for neutralising oils and is therefore superior in this respect to acid-activated earths. For lubricating oils, 3–4 per cent is mixed in the cold with very strong agitation till all inorganic acid has gone; the decolorising temperature is 70–90°C for heavy or 50–70°C for light oils.*

Now Karlovy Vary.

CARLONIT was apparently non-hygroscopic, and, surprisingly, improved in quality on long storage. Oil retention was ~ 30 per cent less than European and American natural earths and 50% less than activated earths, an astonishing claim. Belani recommended Bandau and Köber's process for recovering the oil from the spent earth.

Belani returned to his theme in 1932 and R. Fussteig found that

CARLONIT was suitable for fat refining, "the low alkali content being an advantage in bleaching oils and fats." It had another useful property, it neither adsorbed nor oxidised vitamins A and D in oils. F. Wittka maintained that where the relative humidity is high, as in the tropics, natural earths – and especially the Czechoslovakian earth – with a little active carbon is suitable for bleaching vegetable oils. In spite of these strong claims no more was heard of CARLONIT.

Seifensieder. Z., (1932), **59**, 654.

Olien, Vetten, oliezaden, (1934), **18**, 485.

Seifensieder-Z., (1937), **64**, 393.

Indeed from 1936 Czechoslovakian needs were largely met by the Hungarian raw material activated at the Dynamit Nobel's works at Bratislava; the process they used had been patented by their Austrian branch.

D.R.P. 682, 641, Oest. Dynamit Nobel A. G., Wien.

The deposit of calcium montmorillonite at Braňany, near Bílina, northwest Bohemia, was the first to be worked in modern times – from 1943; but it was not until 1968 that a plant was planned for making products for the foundry, chemical and other industries, the building trade and agriculture. About 100,000 tonnes of bentonite were to be chemically treated a year under the management of the North Bohemian Ceramic works at Most – some 30 per cent of this being reserved for exports. The factory itself was to be at Zelenice. Exports were controlled by Chemapol Ltd.

Chemical Age, (24/8/1968) 28.

Before this M. Gregor pioneered the study of clays in Czechoslovakia. In 1948, he wrote on the power of some Slovakian clays to decolorise vegetable oils. He thought that a mineralogical analysis was more reliable than the usual tests for decolorising power. He and B. Lesko reported on new occurrences of bentonite in Czechoslovakia; the Ca-Mg-montmorillonite content varying from ~24 to 70 per cent, the main impurities being quartz, cristobalite and illite. He described the bentonite found near Fintice and Kusmice-Lastovce in Eastern Slovakia.

Industrial Minerals, (1968), (13), 23 *Chem. Age* (26/8/50) vii *Zpr. čsl. Keram. Sklář. Společ.*, (1948), **24**, 101.

Silikattechn., (1963), **14**, (3), 88–89.

In 1957 Professor Jiří Konta described, in his book on the clay minerals of Czechoslovakia, the calcium-montmorillonite of Braňany, of only moderate quality, of Ruznice (cristobalitic) of Jarčová (illitic) and of Vonšov (containing kaolinite and illite), but the search was intensified for better material. The bentonites of Bohemia are argillised Miocene basanites, olivine basalts and tephrites, pyroclastics and redeposited tuffs or volcanic ashes; while those of Slovakia are argillised Miocene rhyolite and andesitic tuffs.

Jilové Minerály. Československa, Jiři Konta, Nakladatelství Československé Akademie Věd, (1956).

Industrial Minerals, (1981), (162), 32.

In 1969 the main commercial bentonites were BENTONIT SPECIÁL, a Ca-Mg-bentonite from Černy Vrch, in northern Bohemia used for bonding foundry sands, and BENTONIT SABENIL B, a soda-treated bentonite, with cation exchange capacity over 50 meq./100g; its main uses were in foundry moulding sands, pelletisation of iron ore and civil engineering, from the same source. More recently the Na-treated clay has been used in drilling muds. Some of the clay is acid-activated and used for decolorising mineral and vegetable oils: for this purpose only the bentonites of the Jelšový Potok in Slovakia are suitable.

Industrial Minerals, op. cit.

Organo-derivatives of bentonite were studied by C. Kubík, F. Mitterhauszer, I. Novák and J. Krajči, who treated bentonites from various localities with aqueous sodium carbonate; the fine particles in suspension were treated with 2 per cent solutions of Ajatin, octadecylamine hydrochloride or Katexol 298. The products were tested for swelling with $PhNO_2$ and for thickening mineral oils; M. Gregor, P. Mocíková, F. Esterka and C. Kubík prepared an oil emulsion stabiliser from the middle one of these organic reagents.

Chemický Prům., (1967), **17**, 577–580.

Chemický Prům., (1968), **18**, 393–398.

J. Uhřik and J. Mezulánik on the other hand discussed some of the difficulties of using these bentonites in the foundry, with and without the use of additives such as waterglass.

Slévárenství, (1967), **15**, 345–346.

In 1970 I. Novák attempted an objective evaluation of the bleaching capacity of clays on vegetable oils. The light-absorption of the bleached oil gave a linear plot against the concentration of clay used in the oil. The slope and intercept quantitatively characterised the bleaching capacity of the clay.

Chemický Prům., (1970) **20**(4), 155–159.

I. Novák (1973) made a study of acid-attack on four different montmorillonites. Changes in porosity, specific surface and residual octahedral atoms were followed on increasing the time of the attack. The basal spacing rose to 30 Å and eventually even wider, but the specific surface rose to a maximum value, and then fell as the octahedral atoms were all leached away.

Silikáty, (1973), **17(1)**, 39–53.

J. Konta has kindly supplied a list of deposits of bentonite known in Bohemia: Braňany (65–70% montmorillonite with 35–30% kaolinite, though his specific surface measurements would suggest a proportion of 50:50), Černy Vrch near Most (the largest deposit in the country), as well as Střimice, Stránce, Blšany, Velké Třebušice, Kadaň, Hroznětín north of Carlsbad.

Imbibometrie, J. Konta, *Acta Universitatis Carolinae*, Prague (1961).

In Slovakia four bentonite deposits are open: Lastovce, Fintice, Jelšový Potok and Kuzmice (where the bentonite contains a considerable proportion of cristobalite).

Sodium montmorillonite is made in Braňany and Lastovce, and calcium montmorillonite in Fintice and Jelšový Potok. Acid-activation is carried out in Bratislava.

49 GREECE

Argentiera, the Island of Soap

RIPLEY in his pictorial series 'Believe it or not'
Scottish Sunday Express, 25.4.1948

Relation d'un voyage de Levant, . . . Paris, (1717), 4.

M. H. Klaproth, *Analytical Essays towards promoting the chemical knowledge of Mineral Substances*, tr. from the German, London: T. Cadell and W. Davies (1801) **1**, 248.

Franz Ambros Reuss, *Lehrbuch der Mineralogie*, nach des Herrn O. B. R. Karsten, *Mineralogischen Tabellen*, Leipzig: F. G. Jacobaer, (1802), *2*.

There were early 19th century analyses of Cimolian earth by Sonnini, Olivier, Friedler, K. Ehrenberg, Vauquelin and Christomanos.

TOURNEFORT (1717) was the first modern writer to visit Cimolo or Argentiera; he thought the clay there was calcareous though unusually soapy. But it was one John Hawkins, Esq., who, on his voyage to the Grecian islands, collected the genuine earth and "allowed knowledge of this stuff to be revived," according to Martin Heinrich Klaproth, (1801) who received a specimen. He records that it was massive – in vast strata, its fracture earthy, uneven, and texture more or less slaty. In colour; it is light greyish-white, inclining to pearl white, but superficially reddish on weathering. It is opaque, does not stain considerably. Scraped with a knife it has a smooth, greasy lustre. It adheres pretty firmly to the tongue. Though so soft, it is with difficulty broken, and not easily ground. Small milk white grains of quartz were found here and there disseminated in its substance.

Put in water, it breaks immediately into small thin laminae, of a curved slaty form. A thickish cream of cimolite, left to dry on a dish, detaches itself from the surface in hard, ribbon-like and somewhat flexible bands, which are even harder to grind.

Klaproth oiled some silk with almond oil spots, put a paste of cimolite on and allowed it to dry. Then by a slight washing the cimolite, together with all the greasy spots, were removed, without leaving the slightest trace. He thus confirmed its ancient use and noted *that it appears much to surpass the English earth. To revive the importation of cimolite. . . would therefore be desirable* for our clothing industry as well as for common use; *especially for precious stuffs of delicate colours that will not well bear the agency of acids and alkaline soaps.*

Minerals and their uses. London, (1849), 196.

Here called *cimolite.*

E. M. Leycester, *J. Roy. Geogr. Soc.*, (1852), *22*, 224 quoting an account of a French expedition.

The Reverend J. R. Jackson, (1849) in explaining minerals in a series of letters to a lady, follows Klaproth's description rather inaccurately. One would like to know what the lady thought of these stuffily didactic letters. About this time Cimolian earth, according to Leycester, was "still used for cleaning linen".

The recent history of Greek bentonite began when Richard A. Sonder published his geological map of Melos, Polivos and the southern part of Kimolos in 1923 and made a systematic appraisal of the islands' mineral resources in 1925 when he recognised the importance of the bentonite occurrences; but no one else did at the time. Whereas Kimolos had once supplied Mediterranean ports with supplies of detergent, all memory of this famous fuller's earth had died. How it was rediscovered and became once more a raw material of international importance can be told thanks to a letter from Dr. B. E. Vassiliou of Athens ("Vassos")

G. K. Bourlos, who knew Vassos' father, and was a close friend of Professor Stavros Hors, Dean of the Faculty of Chemical Engineering in Athens University of Applied Science, dug out from store (1945) some samples of kaolin and "plastic clays" from Melos, used, the labels said, for various purposes in the small ceramics and refractories industries in Athens. At the end of that year Vassos went to Sheffield University to work for a PhD degree under Professor James White, who during the war had examined by x-ray diffraction many clays, including calcium montmorillonite, for the Moulding Materials Sub-Committee under Professor W. J. Rees.

At the end of 1946 Vassos brought a ton of samples of Bourlos' clays to Sheffield University where the Professor himself helped to carry the heavy sacks up to the Refractories Department. To their surprise they found that some of the samples were "bentonite", that is, calcium montmorillonite. When Vassos wrote to Bourlos and to Professor Hors of his discovery, neither of them had ever heard of bentonite or of its use in foundries.

Vassos examined and published a series of reports on these Milos bentonites under his own name, Jim White's and Joan Goodison's: she was an assistant to the Professor. He thus earned a small one year scholarship from BISRA. Publication brought visitors to the University.

L. H. Dyke Acland, my successor at The Fullers' Earth Union, visited Milos where he is still remembered for his immense capacity for food and drink. He ordered 50 tons of earth from the Anthoussa and Garyfalakena deposits, which by the way still contain close to 2 million tons of good bentonite. But S. George Henderson, whose story is told elsewhere, was cleverer. He ordered 500 tons of Anthoussa bentonite for Colin Stewart, Ltd, at the terrific price for 1950–51 of £5 sterling a ton of raw Bentonite as Mined. This price not only secured an exclusive contract but allowed Bourlos to improve the plant.

The first consignment contained calcite and caused a lot of trouble, but George undaunted and with good practical sense revisited Milos and showed the natives how to test for calcite by dropping acid onto the clay placed in little tubes, and looking for the tell-tale fizzing. From that day on the delivered bentonite was of good quality and eventually reached as much as 5000 tons a year – one customer, Qualcast Ltd., alone using nearly 2000 tons from 1951 onwards.

Bourlos was a lovable dreamer but he was often in financial difficulties and eventually formed a fifty-fifty partnership with Silver &

Zur Geologie und Petrographie der Inselgruppe von Milos, *Zeits. f. Vulkanologie*, *8* (4), 181–237, Berlin.

"The OMO or DAZ of antiquity", RHSR.

Letter from Silver & Baryte Ores Mining Co., dated 13.9.1982.

British Iron & Steel Research Association. See also R & D Report 16/51 MM of British Steel Founders' Association, April 1952, by J. Goodison, P. Williams, S. Judge & J. White, who reported on two bentonites from Melos, one (GAR) from Silver & Baryte, and one from Polivos.

Baryte Ores Mining Company of Athens, and his contract of exclusivity was passed on to the new partner, whose Managing Director George Eliopoulos had himself tried unsuccessfully to sell ground bentonite to Britain. Bourlos' activities had attracted the attention of CECA HELLAS, who acquired some rights and deposits, in Kimolos mainly, and George Eliopoulos began to feel that Greek bentonite had a future. He invited Vassos to be his consultant for 1960–1964, but he sold only a few hundred tons to the Middle East apart from the continuing sales to Colin Stewart.

In 1967 Vassos returned to Greece and joined S.B.O.M. and Bauxite Parnasse Mining Company. At that time Eliopoulos decided to enter the Canadian market and to approach the Iron Ore Company of Canada who were preparing their first pelletising plant. Vassos made two trips to Canada with George Henderson, who tried unsuccessfully to enter this market on behalf of BERKS on a fifty-fifty basis with George Eliopoulos; but the latter decided he had enough money to do it himself. The result was that within 5 or 6 years sales had reached in 1973 200,000 tons a year, a hundred times more than in 1967, when no one had dreamed of such a meteoric rise in sales. SBOM are still the only European firm that exports large quantities to Canada. Eventually sales nearly reached 300,000 tons a year.

Silver & Baryte have extensive shipping facilities in Vouthia Bay, where three ships can be loaded simultaneously and one of them can be up to 12,000 tonnes capacity. ZENITH is the trade name of the drilling mud grade, DIABOND is a foundry moulding clay, and PELBENT is used in iron ore pelletising. The quarry-to-ship costs are exceptionally low, so Milos bentonite can compete with Wyoming bentonite in Canada.

Ind. Min. (Sept. 1969), 13, and (Jan. 1973), 17.

Mykobar Mining Co., S.A. a subsidiary of Dresser Industries, Inc., also have their own quarries and plant on Melos, and a few small operators ship the crude clay in lump form. Mykobar ship dried, activated and granular bentonite in bulk.

R. H. S. Robertson, *Mineralogia Polonica*, (1980), *11, (1)*, 4–6.

A. Alietti & Maria F. Brigatti, *Miner Petrogr. Acta*, (1982), *26*, 3–47 find much more tetrahedral aluminium.

Cimolian earth is rich in magnesia and is classed as of the Cheto type and contains little free silica; its clay mineral has the ionic composition: $(Si_{7.98}\ Al_{0.02})(Al_{3.00}\ Fe^{3+}_{0.16}\ Mg_{0.89}\ Ti_{0.02})\ O_{20}\ (OH)_4.\ Na_{0.42}\ K_{0.03}.$ n H_2O. The analysed mineral has been sodium-exchanged.

CYPRUS

50

BETWEEN 1950 and 1952 Colin Stewart Ltd imported from Berdy Mining Co. trial shipments of crude Cyprus fuller's earth. Then a Mr. Manglesis installed milling plant, but failed financially. Harrison, Fletcher Co. tried to sell CYBOND to the UK foundries in 1953. F. W. Berk & Co. Ltd imported a good grade in 1958, but prospecting, encouraged by S. G. Henderson and W. S. Jolley, showed great variation and few good samples. Virtually the whole production ceased in 1960. The earth made till then contained quartz, chalcedony and diatoms; was often acid from pyrite; and chocolate brown in colour. The addition of sodium carbonate did not make a swelling clay unless sodium chloride were added first.

Cation exchange capacity 81.5 meq./100 g. (CECs 35–48, usually).

Up to 0.5% H_2SO_4.

Export figures speak for themselves:

Exports of clays from Cyprus

1953 49	1954 182	1955 451	1956 958	1957 785	1958 894	1959 1009	1960 219
1961 NIL	1962 NIL	1963 536	1964 1528	1965 3190	1966 2658	1967 4539	1968 9069
1969 9036	1970 12894	1971 12366	1972 10748	1973 8745	1974 4446	long tons	

The Hellenic Mining Co. started processing Troulli bentonite in 1963 and the Berdy Mining Co. supplied, from 1970, earth to Sil-Ka Ltd., in Israel, for conversion to swelling grades. An impressive feasibility study by Robertson Research Co. explored the market for pelletising iron ore (1969). In 1978 Dem. Mitides produced a 100 mesh powder and Peletico Ltd was the largest producer on the island: they soda-treated at Larnaca a clay from Limassol – BENTOCYP in 'granular' form for grinding in the importing country. Two smaller companies mine at Agia Marina and Poliatis, and in the Paphos district. Berdy's firm had become Bentex Minerals Co. Ltd. It is fair to say that whilst the Cyprus deposits are large, their location and on-costs make it hard to compete in Europe, America and the Persian Gulf.

Ind. Min. (Dec. 1970), 37, 39.

Report No. 362, (Nov. 1969).

Ind. Min. (March 1978), 81.

51 YUGOSLAVIA

Map showing, near the centre, the town of Celje in Slovenia. Fuller's earth from near here, formerly called Cilly, was named *smectite* by Brethaupt in 1841. See text.

AFTER Škofja Loka, the town of Kamnik, Northwest of Ljubljana, became the second centre of the Slovenian textile industry. In 1428 this town ran an important fulling mill for the production of woollen "loden" fabrics (bukovina). Here, and in the mountain areas further to the East, craftsmen specialised in the manufacture of cloth from local wool and made the popular "mezlan" fabric from a 50/50 blend of wool and linen.

The History of the Slovene Textile Industry, W. F. Schweizer, *CIBA Review*, (1966), (1), 2.

The greenish-grey fuller's earth for this industry came from Cilly (now Celje), between Maribor and Ljubljana, in Slovenia. This was the original "smectite" of A. Breithaupt, (August 1841). A. des Cloiseaux includes it among the most important deposits of fuller's earth known to him (1862). P. F. Kerr (1932) showed that the clay mineral of Cilly fuller's earth and of an American bentonite was identical – montmorillonite; and the chemical analyses of Jordan in 1849 and Kerr in 1932 are close. De Lapparent tried for a while to maintain the prior name of smectite.

Vollständiges Handbuch der Mineralogie, (1841), *2*, 344, Dresden & Leipzig.

Manuel de Minéralogie, Paris (1862), *1*, 204.

P. F. Kerr, *Amer. Mineral.*, (1932), **17**, 192–198.

Another fuller's earth mentioned by des Cloiseaux was at Zengg (now Senj) in Croatia.

In modern times M. Krajcinovic and J. Zaljesov (1938) made some tests to ascertain the value of certain Yugoslavian clays for refining vegetable and mineral oils. In 1953 R. Pfleger made a similar survey to see if any of the clays could be used for injection purposes in civil engineering; and processing methods were reviewed by Gründer in 1954.

Arc. Hemiju, (1938) *12*, 2–8.

Kem. Industr., Zagreb, (1953), *2*, 212–213.

Tonind.-Ztg., (1954), **78**, 99.

Then with the arrival of x-ray equipment, Yugoslav clays could be analysed mineralogically. S. Karamata and F. Lippmann began with the composition of the 'bentonite' of Ginovci in the region of Kriva Palanka. The clay contained a trace of illite. A sample of nearby Rankovci montmorillonite contained more illite.

Under the direction of Professor V. Stubičan, a number of Yugoslav bentonites were analysed chemically, the presence of free silica as quartz or cristobalite determined by X-ray diffraction, and the cation exchange capacity determined. Suspensions containing cristobalite have decreased structural viscosity and thixotropy. The development of big platy crystals of montmorillonite is made favourable by the conditions which promote the formation of quartz. Stubičan's study of synthesized montmorillonites is outstanding.

Bentonites. II. V. Stubičan, N. Lisenko & M. Wrischer, *Croat. Chem. Acta*, (1956), *28*, 239–248 [Institute for Silicate chemistry, Zagreb].

Clays & Clay Min., (1960), *7*, 295–302.

The bentonites of the Slaviste valley – those of Ginovci and Rankovci – were formed from a sanidine-dacite tuff, which accounts for the presence of free silica in these clays. Some were formed *in situ* (grey) and others were sedimented (green or yellowish-brown).

Dacite = quartz-andesite.

A plant for making soda-treated montmorillonite had been established by January 1957 at Kutina, 96 km south east of Zagreb in Croatia. The firm is referred to as "INA"; it was formerly "METAN." The next bentonite processing plant to be built was that of "BENTONIT" at Petrovac na Moru; their deposit at Belo Polje in

INA – Factory of Chemical Products uses clay from the Poljanska Luka mine, and from the Vȑdnik Brown Coal Mine. W. Gründer (Berlin) & Blagoje Velikić, *Tonind. Ztg.* (1958), **82**, 66–72.

Jugoslawiens Bergbau von Nichterzen, *Montan-Rdsch.*, (1969), *17* (9), 201–205.

Yugoslav's bentonite industry, *Ind. Min.* (1969) (27), 35.

Capacity 30,000 tpa. The clay comes from Ginovci, Ljubinci and Rankovci, of Neogene age, overlying dacite-andesite tuff.

J. Rihteršič, *Geologija* (Ljubljana), (1958), *4* (13), 193–196. *Rudarsko-Metalurški Zbornik*, (1968), (3), 277–286.

Rudarsko-Metalurški Zbornik (1968) (3), 277–286, (4) 405–421; (1969) (1) 25–35; (1971) (2–3) 313–320; *Vestnik Skd* (1971), *18*, 27–32.

Montenegro maybe the largest in the Republic. They have another large reserve at Zaloška Gorica, Slovenia.

According to Tvrtko Odić the reserves of bentonite in Yugoslavia are great, though production was still small in 1969. The best known deposits are those of Bar, Bjelo Polje, Ilino Brdo and Poemin, with reserves exceeding 40 million tonnes; then Ginovci and Rancovci near Kumanova; the Karacevo area and in the Tertiary basin of Zaplanje; Zavidinci, Dubrava and Donji Prisjan.

In 1969 a company called KOSMET MINERAL worked a quarry at Karačevo, near Kosovska Kamenica in Serbia at a rate of 10–15,000 tonnes a year for sale to Yugoslav processors, but INDUSTROCHEM at Pula in Istria ceased to purchase it when the freight rate on the 800 km journey became prohibitive. INA had become the largest producer of bentonite at 40–50,000 tonnes a year, for use in foundaries and well drilling. They also produced acid-activated bleaching earths for their own petroleum oil refining.

The most modern plant at that time was built by BENTOMAK at Kriva Palanka, near Kumanova in Serbia, using know-how and equipment supplied by Erbslöh & Co., of Geisenheim am Rhein, who established a laboratory there and trained the staff. BENTOMAK supply grades of Ca- and Na-montmorillonite to suit all requirements, including iron ore pelleting and civil engineering.

BENTONIT at Belo Polje in Montenegro specialised in fuller's earth for use as carriers for insecticides and herbicides. In the 1960s they exported large quantities of calcium montmorillonite granules to the USSR, but in 1967 the Russians installed their own processing facilities using their own clays.

J. Rihteršič was the first to describe the geology and formation of the bentonite occurring around Celje (1958); and E. Lukacs described the prospecting work being undertaken there in 1968 and reviewed the Yugoslav bentonite industry at that time. His work was supported by Prof. Dr. Boyan Držaj and his colleagues who wrote several papers on Yugoslav bentonites. They showed that heulandite-clinoptilolite contributed to the effectiveness of acid activation: but diminished the effectiveness of alkali activation unless basic magnesium carbonate were added.

Not long before 1969 MONTANA reopened the historic workings near Celje in Slovenia. They had even exported some of their production, which in 1982 included natural zeolites containing a proportion of montmorillonite.

BENTOMAK decided in 1976 to build its next plant in Belgrade using raw material from Kriva Palanka. This plant would double their production of soda-treated clay to 40,000 tonnes a year, half of which they expected to export, through their sales agents Erbslöh & Co. Their works are about 180 km from the port of Thessaloniki, Greece, and about 20 km NE of Skopje University where some scientific investigations have been made; after a serious earthquake Professor Jiří Konta of Prague was invited to be visiting Professor for several seasons. Markets were reached more directly via Belgrade. Bentomak expected to produce about 72,000 tonnes in 1982, 14% above the previous year's output – mostly for export.

Ind. Min. (Aug. 1982), 65.

Large reserves (70 million tons) were discovered by Prof. Duhovnik of the University of Ljubljana at Gušica in Kosolvo Polje (7 km. SW of Vitina): montmorillonite is the main component.

My thanks are due to Boyan Držaj for help with this chapter.

52 SWITZERLAND

The ground on which we walked was composed of soft earth as white as snow, and on examining it closely, I found, to my great satisfaction, that it was a mineral earth called **alumine**, *used by fullers in cleaning wool, and therefore known as fuller's earth. I filled my pocket-handkerchief with as much as it could hold to take home to my wife, which made the boys inquire the use of it. 'It will be of great use to your mother in washing our clothes', I said, 'for fuller's earth is used by fullers to cleanse wool before it is woven, and is also one of the ingredients in making soap'.*

JOHANN DAVID WYSS, Swiss Family Robinson, 1813.

First edition 1813, additions by Baroness Montolieu 1824, translated by Mrs. H. B. Paull, London; Fred K. Warne & Co., 1891.

THE didactic Mr. Robinson may be excused for using the word *alumine* to mean earth or clay because this was not uncommon at the time of the translation, for example William Hamilton (1791) writes of *some earths, principally alumine*, in an article in the *Phil. Trans. The absorbent earths were distinguished into calcareous, magnesia, or alumine or clay;* (1798).

W. Hamilton, Bertollet's Elements of the art of dyeing, tr. 1791 (1792). *Phil. Trans. Roy. Soc.*, (1798), *88*, 16.

A white clay at Caslano, south Ticino, now known to be illite, was mined in the middle of the 19th century and used, mostly in northern Italy, for scouring wool and washing textiles, as well as for ceramic purposes. From 1841 to 1847, 223 tons were mined. It was worked again by Georg Fischer AG of Schaffhausen, from 1945 to March 1946 for use as a foundry bonding clay, but abandoned when imports of good bentonite were resumed.

A number of discoveries of calcium montmorillonite have been made in the Upper Freshwater Molasse and described in a series of papers. The first discovery of montmorillonite at this higher level was by J. Favre, who found it near Le Locle in 1911. In 1947 Hofmann began to search for contemporary ash falls and traced the volcanic source back to somewhere in the east of Switzerland, not to Tannenberg, St Gallen, as he first thought. The deposits increased from 20 cm at Le Locle to 2.5 m at Bischofszell, which must have been not far from the source. Thicknesses at intermediate places were rather variable, but the mean particle diameter of the clay treated with 5 per cent sodium carbonate and dispersed in water gives a rather convincing relationship with distance.

J. Favre, Description géologique des environs du Locle et de La Chaux-de-Fonds, *Diss. Univ. Geneva*, Lausanne,
F. Hofmann, *Eclogae Geol. Helv.* (1956), **49** (1), 113–133.

id., (1958), *51* (1), 65–71.

id., (1963), *56* (1), 147–164.

All the calcium-montmorillonites had a similar cation exchange

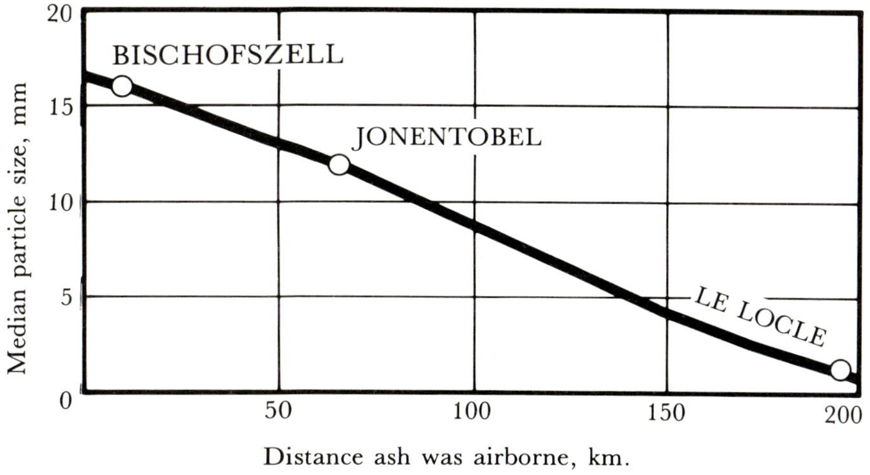

Fig. 81 Median size of ash particles in relation to distance from volcano.
F. Hofmann.

capacity, 61.3 ± 1.4 meq./100 g; and the moisture adsorption varied only from 15 to 15.5 per cent.

Among many contributions Hofmann confirmed that foundry sand compression strength was roughly proportional to moisture adsorption, and by implication also to the smectite content.

The deposit at Bischofszell, discovered in 1947, was mined from October 1949 to March 1950, 70 tonnes being tested in the foundry. This clay and some others of different composition could be used if imports should cease.

In 1980 a 20 to 30 cm thick lens of fuller's earth was discovered in the Bois-Genoud sand quarry near Lausanne, north of Crissier. Unlike all the other former ash falls this one is in the *Lower* Freshwater Molasse. Since this occurence is only 4.2 km north of the Gallo-Roman *vicus*[a] of Lousonna, one might be forgiven for thinking that it was used in a fullery in ancient times; but, even if this thin seam had at one time "outcropped" on the heavily wooded slopes of the valley of La Mèbre, it is most unlikely that it would have been discovered before modern quarrying began.

Yet we are left with the intriguing thought that there must have been a *fullonica* at Lousonna, since the now conduited river Flon, its name hinting at its former use, runs within 200 m. of the warehouses at the port. Ground underlying the roundabout (giratoire de Maladière) cannot now be excavated but the area south and southwest of it might still reveal a fullery. The lakeside town, now Vidy, was founded at the end of the first century AD. It became a busy staging point on the com-

R. Iberg et al., *Oberrh. geol. Ver.*, *(1965)*, *47*, 79–90.

F. Hofmann, U.P. Büchi, R. Iberg & Tj. Peters, *Beiträge zur Geologie der Schweiz, Geotechn. Serie*, no. 54.

Giesserei, Techn. Wiss. Beth., (1956), *16*, 857–864.

F. Hofmann & Tj. Peters, *Beiträge zur Geologie der Schweiz, Geotechn. Serie*, no. 47, 1969.

Niveau de bentonite dans la Molasse grise de Lausanne,
T. Mumenthaler, T. Peters & M. Weidmann, *Eclogae geol. Helv.*, (1981), *74* (3), 639–650.

(a) cf. ϜΟΙΚΟΣ, a house. The old initial digamma = appr. Eng. *w*, so this was pronounced *woikos*

The idea was put to me by Professor Robert Meldau, 3/8/1973.

See Lousonna; *La promenade archéologique de Vidy*, Gilbert Kaenel, Association Pro Lousonna, (1977), p. 4.

munication routes vital to the Empire. Many manufacturing processes were carried out there, in several potteries, bronze and lead works and other workshops; and the commercial company which organised transport on Lac Léman – the *nautae lacu lemano* – had their offices and meeting place in two rooms in the basilica.

Musée romain de Vidy, and Gilbert Kaenel, *op. cit.*

Huge quantities of goods, brought here in amphorae, consisted of wine, olive oil, pulses, fish sauces and other foodstuffs. It is most likely that any fuller's earth brought to Lousonna would have come from Kimolos through the port of Massilia or from the colonia Julia Apta, capital city of the Aptenses and since about 95 AD the seat of Bishops. In either case most of the journey would have been up the Rhône to Lac Léman. Goods were carried by another transport company who owned the boats and managed the land portage where it was needed. The commercial life of the *vicus* came to an end in 260 AD when it was ravaged by the Alamans.

Lausanne dès les temps anciens Rodolphe Blanchet, (1863), p. 12.

A further hint that there really was a fullery at Vidy is given in the deed presenting the woods of Sauvebelin to the town of Lausanne by Leurs Excellences de Berne, dated 26 August 1552, where it is stated "Le bois de Sauvabelin est siz jouxte l'eau, soit le Flon, appelée LAUS et certaines vaux devers orient." It seems that at that period the river had preserved its ancient name of Laus.

"Situated next to the water, namely the Flon, called Laus, and certain small valleys towards the east."

In the 18th century the Flon was diverted in aqueducted lades to serve many mills.

"Lade" = a water course.

THE UNITED STATES OF AMERICA

53i

From fulling to refining animal and vegetable oils.

THE first fulling mill in America was set up at Rowley, Massachusetts, in 1643. It was numbered among the fourteen "Early Colonial Developments, 1650–1950" by Wm. Haynes and L.W. Bass. There can be little doubt that the early settlers would have discovered and used clays which served as fuller's earth in the preparation of woollen cloth; and that their descendants would have continued to use clays in this way until soap became cheap enough to replace them. Patterson and Murray admitted that they had done little to search the historical records for earths used in fulling, but they record that soldiers stationed near Perth Amboy, N.J., used Woodbridge fire clay for cleansing buckskins during the Revolutionary war (Cook & Smock, 1878). Fuller's earth associated with an iron-ore bed near Kent, Connecticut had been mined in the early 1800s; Silliman (1820) compared it with Hampshire fuller's earth, his description of which is given in the chapter on that county. Silliman had called attention to the advantages the American woollen industry might derive from the use of fuller's earth. *The subject is of some practical importance to this country on account of its woollen manufactures, which although checked for the present, must eventually rise and prevail. While they are of small extent it may be better to use soap, but in very large establishments fuller's earth from its cheapness – provided it can be abundantly obtained – is very desirable.*

From American Chemical Chronology, by Haynes & Bass in *Our Chemical Heritage*, Amer. Chemical Industries Tercentenary, 1635–1935, p. 31.

Sam H Patterson & Haydn H. Murray, Clays, U.S. Geol. Survey, (1974), 519–85

G. H. Cook & J. E. Smock, Report on the Clay Deposits of Woodbridge etc, (1878) New Jersey Geol. Survey, Trenton, 381 pp.

B. Silliman, sr., *Amer. J. Sci.*, (1820), *2* (2), 201–235.

The embargo on the export of English fuller's earth was lifted in 1825; consequently until the 1860s it more than held its own against soap along the Eastern seaboard to which it was so easily shipped. When imports ceased during the Civil War, soap was used instead in most woollen factories.

Patterson and Murray also record that fuller's earth near Falls City, Texas, had been used to bleach sugar during the Civil War (Lang *et al.*, 1940). *Also, American Indians were reported to have used bentonite in cleaning blankets, and they probably dug it for other cleansing purposes before the Columbian period.* Highly swelling sodium bentonite tends to form a thin layer on fibres and is therefore more difficult to wash out when the cloth is rinsed. The Indians may have used weathered, superficial bentonite for fulling and the swelling sort for other purposes.

W. B. Lang *et al.*, Clay investigations in the Southern States, 1934–35, (1940), *U.S. Geol. Surv. Bull.* 901, 346 pp.

This is why sodium carbonate, though often used with soap, is never used with fuller's earth in fulling. See p. 31.

An American instruction manual of 1909 seems to reflect a revival of interest in a practice of using fuller's earth for a light scour as an adjunct to fulling in soap:

John F. Timmerman, Chicago: American School of Correspondence, 1909

> **Fuller's Earth.** *On all fine goods, whether they are speck dyed or not, it is of great benefit to use fuller's earth after washing and also after speck dyeing. This has a tendency to purify the pieces of any remaining trace of soap; and it imparts a soft feeling which it is impossible to obtain in any other way. . . . On face goods the bath of fuller's earth should not be omitted, both at the first washing and after speck dyeing. If such goods are intended for piece dyeing give them a bath of earth before sending them to the dyehouse and after they are gigged, and the results will be astonishing.*

to gig is to raise the nap of cloth by a rotating drum of teasels or wire-cards.

The instructions, specifically, are:-

> *To 50 gallons, 2 pailfuls of the earth is ample, and it will be found that the water will not hold more in solution without continual stirring. There will be a sediment with even this amount of earth, but not enough to do harm. When earth is used let it be put on the goods after they are thoroughly rinsed and drained, so that all the surplus water is out of them, then shut all the gates and give each piece two pailfuls of solution, as stated above. This should be well stirred before using, so as to get all the earth well mixed; then apply and let the pieces run for fifteen minutes and rinse ten minutes, after which the goods may be taken out and passed to the next process.*

suspension

The author adds:

> *As the use of fuller's earth is not universal it will be in some instances hard to convince parties of the benefit derived from its use, but invariably a continued use will show this to be a fact.*

C. L. Parsons, *Bureau of Mines Bull.* 71 (1913), p. 6

Yet by 1913 the use of fuller's earth for fulling had almost ceased, owing to the cheapness of good soap – and in spite of Timmerman's enthusiasm.

Jefferson Middleton, Fuller's Earth in 1923, *U.S. Geol. Surv. Min. Resources of the U.S.*, II 99–107 [1924].

Fuller's earth (calcium montmorillonite) in the modern period – since 1880 – was first discovered in the United States in 1891 near Alexander, Arkansas, by John Olsen. This earth was used for a time by the Southern Cotton Oil Co., at Little Rock, Ark., but its use was finally abandoned, according to J. C. Branner. From 1904 to 1907 Arkansas was the largest producer of fuller's earth in the sense used in this book. [Attapulgite had been discovered in 1893 near Quincy, Fla.] Most unfortunate it is that the term "fuller's earth" was extended after 1880 to mean any earth which decolorised oils. Since no distinction could be made between montmorillonite and palygorskite before about 1928, it is difficult to disentangle the two types of clay in the early literature. If Parsons had persisted with his dye adsorption experiments in 1913, he would have distinguished between the two groups.

J. C. Branner, *Am. Inst. Min. Eng. Trans.*, (1913), *43*, 520–522.

(1913); see Ch. 34.

The deposit at Alexander was a Tertiary clay marl, but a more important one was discovered in 1897, again by John Olsen. Arkansas was not added to the list of commercial producers until 1901. Mr. Olsen's mine and mill at Klondyke were still in operation in 1915, but

H. K. Shearer, Report on Bauxite and Fullers Earth, (1917), *Geol. Surv. of Georgia, Bull, 31.*

several other mines in the vicinity, which had been operated for a time, were closed down. These deposits occur in an area of about 3 sq. miles which lies between Hot Springs and Benton. They differ from all others in the US in being derived from an igneous rock in place. The basic dykes, originally of ouachitite and biotite monchiquite, and syenite, are weathered to a depth of 60 to 200 feet. The dykes, never more than $4\frac{1}{2}$ feet wide, were worked in 30-foot stopes until hard rock was reached. The clay was dried in cylindrical driers, ground, and bolted or air-separated to pass 80 to 120 mesh. The bolting screens tended to be blinded when mica was present. *On account of the expensive methods necessary in mining, the Arkansas earth, even of equally good quality, can not well compete with the thick bedded deposits of Georgia and Florida* thought Shearer in 1916, but he overlooked that a true fuller's earth is usually better for bleaching animal and vegetable oils than the clay which was later called attapulgite.

Hugh D. Miser, *U.S. Geol. Surv. Bull.* 530, (1913), 207–220.

H. K. Shearer, *loc. cit.*

Other mining companies there were the Fuller's Earth Co. General, of Wilmington, Delaware, and Fred. Rossner of Little Rock, Arkansas.

The plant was acquired by The Fullers' Earth Union Ltd of England and was run for a while by O. F. Keevil. The main problem was that the deposit was nearly vertical and of variable width and it pinched out at depth. Production ceased in 1922.

In 1911. Mentioned in Miser, *op. cit.*

Fuller's earth was first worked in Texas in 1906 or 1907. This low swelling calcium montmorillonite occurs in states bordering on the Gulf of Mexico and acquired the name of "Southern bentonite", "Wyoming" or "Western" bentonite being restricted to the sodium bentonite found in Wyoming and neighbouring states. In 1913 the Fuller's Earth Co., of Houston, Texas, operated mines at Summerville, Texas. By 1923 Texas was the largest producer of calcium montmorillonite as the state had been for a number of years; and production was increasing. In 1932 fuller's earth was produced in Bexar, Gonzales and Walker counties. Rich (1964) of Bennett-Clark Co. Inc., Nacogdoches, Texas, outlined some basic factors in the bleaching of fatty oils with this type of earth.

Jefferson Middleton, USGS (1924), *Min. Resources*, II (13).

See also J. C. Blake. Note on cotton-seed oil bleached with Texas fuller's earth, *Western Chemist & Metallurgist*, (1908), *4* (6), 174–5.

A. D. Rich, *J. Amer. Oil Chemists' Soc.*, (1964), *41*, 315–321.

Texas fuller's earth or Southern bentonite has been used extensively as a foundry moulding sand, either alone or mixed with the swelling western bentonite; it is also converted to oil well drilling grade by soda-treatment and the addition of polymers, and acid-activated at Beaumont, Tex.

The percolation process with attapulgite granules had one great disadvantage: the first runnings were very light in colour and were of low density; the last runnings were dark and dense – a sort of chromatographic separation. Making consistent grades of oil was achieved only by blending. When acid-activated earths were introduced the different fractions from the catalytic cracker were given refining treatments which suited them, and consistent products made.

In the original system of powdered earth treatment the oil was first treated with sulphuric acid; the acid tar was settled out and bleaching earth was used to decolorise, dry and neutralise the oil.

53i

Dr. C. H. Nicolaus Bensmann, *Bleaching Earths*, Meeting of the Institute for Petroleum Research, Hannover 13 Nov. 1953.

When solvent refining was introduced, a smaller quantity of bleaching earth was needed. It was usually applied before de-waxing. In dry-refining full use is made of the polymerising and selective adsorbing power of the earth, so that no sulphuric acid or solvent treatment is necessary. Finally vapour phase refining required once more a tower filled with granules: here it was found that sepiolite gave a better colour to the oil than attapulgite, and was more robust. As the process was not universally adopted, acid-activated clays continue to be supplied to the petroleum oil refining industry though in smaller quantities.

California State Mining Bureau Bull., (1906), *38*, 273–75.

In 1906 the most important deposit of fuller's earth near the west coast was worked by the California Fuller's Earth Co. 18 miles north of Bakersfield for use in bleaching animal and vegetable oils; and the Western Fuller's Earth Co. once had mines at Vacaville, Cal., (Parsons, 1913). Production in 1932 was from Inyo County, and in the year before from San Bernadino County as well, but no fuller's earth had been produced in the state between 1927 and 1930. The output was from Kern County in 1926.

P. G. Nutting, Adsorbent clays, their properties, production and uses, *U.S. Geol. Surv. Bull.* 928-C, (1943), 128–221.

According to Nutting (1943), since about 1920 the Standard Oil Co. of California had been [acid-] activating the [calcium] bentonite that occurs near Otay, California, for its own use, and the Filtrol Co. of Los Angeles had developed the bentonite deposits near Chambers, Arizona, for the open market. They also had a plant at El Segundo, Calif., which activated the earth from Otay, near San Diego. In 1936 they opened a new plant at Jackson, Missouri, for activating the high-grade bentonite that forms a bed in the Vicksburg group. Their plant at Beaumont, Texas, had been built before 1943; this drew its clay from Riverside, Tex., and elsewhere.

Natural fuller's earths used for refining animal fats and oils occur also at St. Paul, Minnesota, and Creede, Colorado; and other occurrences have been recorded.

Oil & Soap, (1946), *23*, 226.

In the Report of the American Oil Chemists' Society's Uniform Methods Committee, (May 1946), it was said that *the bleaching method for cottonseed oil is modified so as to conform with the one in use for soya-bean oil – 6 per cent of the official earth is added to the cold oil before heating to 120°C, with agitation for not more than 5 minutes. A stock of standardised "official activated clay" is to be provided.* Nevertheless the Society supplied Surrey Fullers' Earth, grade XL 000, as the official standard for many years, though in 1954 there were proposals to change to a U.S. domestic earth.

Journal A.O.C.S., (1954), *31*, 127–131.

The Petroleum Oil Industry: Attapulgite

An Alsatian cigar maker ... called attention to the close resemblance of this clay to the German fuller's earth

Vide infra

IN the previous section we have traced the history of fuller's earth (calcium montmorillonite) in the U.S.A., from fulling to the refining of animal and vegetable oils and fats first with natural earths and finally with acid-activated earths. Now we shall review some important events in the history of petroleum oil and will touch on the very important use of attapulgite in this industry.

The first discovery of oil in America is thought to be a "Notice of a Fountain of Petroleum, called the Oil Spring" in January 1833. However, the vast potentialities of petroleum were not conceived of until James Young examined a seepage of oil in a coal mine at Riddings in Derbyshire, England, in 1847. He and a chemist, E. Meldrum, erected a refinery for producing illuminating oil and a heavy lubricant for the Manchester cotton mills. When the demand grew and the supply of natural oil dwindled Young found a special type of cannel coal near Bathgate in Scotland and on the 17th October 1850 was granted a patent for the production of oil and wax from coal and for the refining of crude paraffin oil irrespective of its source. Within five years twenty-three oil companies were operating in the United States under licence to him.

Amer. J. Sci. & Arts. (1833), *23*.

1811–1883.

From 'candle' because a sliver of it burned so easily.

In the States, petroleum had been discovered at Tarentum, Pennsylvania, in salt borings, but efforts to refine it were unsuccessful. Some kerosene oil was manufactured as early as 1854 by G. H. & G. W. Austin of Newtown-Creek, Long Island, New York, but the first oil company in the world, known as the Pennsylvania Rock Oil Company, was founded in 1854 by George H. Bissell and Jonathan C. Elvereth. Bissell, an industrialist, had been given a sample of *Kier's Petroleum or Rock Oil, celebrated for its wonderful curative powers as a natural remedy.* It was poor Samuel M. Kier who was ruined as a salt producer because oil had seeped into his brine. He tried to sell the oil, but only as a medicine.

T. Margerison, *The New Scientist*, (27 Aug 1959), 289–291.

Bissell, on the other hand, guessed that the oil had a future not only for lighting – the paraffin lamp had already been invented – but for lubricating machinery. What is more he foresaw the possibility of boring for oil. One of his company's first actions was to evaluate the possible uses of rock oil, to which end they sent samples of oil seepage

[See also *The Birth of the Oil Industry* by Giddens; and *The American Petroleum Industry – 1859–99, the Age of Illumination* by Williamson & Daum; and *Trek of the Oil Finders: a History of Exploration for Petroleum*, by Owen.]

T. Margerison *op. cit.*, 289, who instead of bleaching powder wrote 'decolorising earth'.

B. Silliman, Jr., *Report on the Rock Oil, or Petroleum, from Venango Co., Pennsylvania*, New Haven: J. H. Benham's Steam Power Press (1855): reprinted in 1955 by Ethyl Corporation, New York.

T. Margerison p. 291.

Petroleum in California, *Amer. J. Sci. & Arts*, (1865); *id.* (1867).

[a sepiolite was discovered at Quincy in France].

S. R. Funsten & H. L. King, Jr., *Refiner*, (1935), *14*, 108.

J. Inst. Petr. Tech., (1924), *10*, 687–694 and *id.* (1927), *13*, 325–329.

skimmed from a pond on Hibbard Farm, Titusville, to Professor Benjamin Silliman, junior, at Yale University.

Silliman made his report a few months later. It was an exceptionally thorough and business-like document. He had distilled the various fractions of the crude oil and assessed their value as lighting oils, lubricants and waxes. He investigated the effects of acids and chlorinated the oil with bleaching powder. Silliman had also found how to break down or "crack" the heavier components to make lighter ones.

This report led Bissell to order a hole to be drilled by Edwin L. Drake, contractor. Admittedly, gas and oil had been found by G. C. Hunaus at Wietze, Hanover, in 1857, but the world usually recognises the 27th of August 1859, the date when Drake struck oil, as the beginning of the petroleum industry in the U.S.A.

In a short while oil was so plentiful in the United States that there was no incentive to refine it commercially. The crude oil was heated in large iron retorts until the paraffin and lubricating oil somewhat imperfectly separated. *The first runnings – or what we would call petrol or gasolene – were regarded as useless and highly dangerous. They were usually burned or run off into a convenient ditch. The American industry continued in this wasteful way for twenty years or more by which time oil had been discovered in California (1865) and an account had been made of the naphtha and illuminating oil from heavy California tar (1867).*

Attapulgite was discovered accidentally in 1893 at Quincy, Florida, when an attempt was made, unsuccessfully, to burn brick from the clay found on the lands of the Owl Cigar Company. Unfortunately, the term "fuller's earth" was extended to this different type of clay and was redefined as to mean any clay which decolorised or "bleached" an oil. For clarity I refer to the adsorbent earths of Georgia and Florida as attapulgite, though this name was suggested much later.

It was soon discovered that attapulgite was better than calcium montmorillonite at bleaching mineral oils. Contact with powdered attapulgite was soon abandoned in favour of percolation through granules. The percolation process was probably first tried in the Pennsylvania region about 1890. Attapulgite granules came onto the market in 1895, but their sale remained small until the internal combustion engine, especially in cars, provided a growing market for fuel and high-class lubricants.

The practice of refining oils in Gray towers through granules had spread to Europe thanks largely to Hermann Bensmann of Bremen, who registered the name FLORIDIN in Germany on 21 Feb. 1905 and in seventeen other countries thereafter.

The use of attapulgite in the petroleum industry is described by Rauch and Pain (1924) and by Rauch (1927) and by others, and the growth of the American "fuller's earth" industry has been shown in chapter 34.

OTHER COUNTRIES 54

ALGERIA AND MOROCCO 54i

DE LAPPARENT, having found no substantial deposits of calcium montmorillonite in France, came to the conclusion that adequate reserves in Algeria could be drawn upon (1932). A sample was examined by Hofmann, Endell and Wilm in their classic paper on this mineral (1934) and by Favejee (1939). As early as 1932, Messrs. C. C. Wakefield were using an Algerian earth for improving the demulsibility of turbine oils. The French company Carbonisation et Charbons Actifs (CECA) were working a quarry near Bou Rhara when I went with Monsieur Feuche to collect 20 tons of the clay from his property 12.5 km NW by N of Turenne, 19–23 May 1938.

J. de Lapparent, *Comptes rend.*, (1932), *195*, 257; see also *id*, (1937), *204*, 1776.

U. Hofmann, K. Endell & D. Wilm, *Angew. Chem.*, (1934), *47*, 539.

J. C. L. Favejee, *Z. Krist.*, (1939), *100*, 425.

Né Feuchtwanger.

A new deposit was discovered near Oudjda in Eastern Morocco in 1944.

Chem Age, (24/6/1944), 611.

A siliceous "white bentonite" found commercial outlets through the good offices of Berk Ltd. and a much purer grade was used for acid activation in France. In 1950, Omnium des Terres Activées S.A. started up a new factory to produce activated earth at Marnia in Western Algeria. CECA were the majority shareholders and provided most of the management and technical personnel. The Danish Company, Aarhus Oliefabrik, had a small shareholding and The Fuller's Earth Union, Ltd., with 24 (later 28) per cent of the issued share capital seconded H. Leslie Salter to commission and start up the plant. Montmorillonite from local deposits was activated with sulphuric acid in open digesters and the products were sold mainly to vegetable oil industries especially in North Africa and to Middle Eastern refineries for refining lubricating oil. The factory operated successfully but, in the mid-sixties, it was nationalised by the newly independent Algerian government. After some years the Société Nationale de Recherches et d'Exploitations Minières revived the dormant industry and in 1975 22,500 tonnes of bentonite and 60,000 tonnes of fuller's earth were produced and shipped from Mostaganem and Marnia.

BERKWHITE

H. L. Salter was later the Commercial Director of The FEU.

SONAREM

Ind. Min. (Mar. 1978), (126), 77.

ARGENTINA 54ii

THE Gray process of petroleum oil refining was in operation in Argentina in 1935; a possibly zeolitic tuff was tested by heat of wetting as a

D. Glynn-Jones, *J. Inst. Petr. Techn.*, (1935), *21*, 709.

54ii

V. Vigo, *Rev. Cienc. Quím.*, La Plata; (1938), *13*, 151–155.

O. S. Mallea, *Industria y Química*, (1954), *16* (1), 15; A. Feaerici, *id.* 18–19; A. Teran, *id.*, 20.

Ind. Min. (March 1978) (126), 84.

guide to its activity for bleaching these oils. For refining lubricating oil V. Vigo (1938) found that Paraná clay was more efficient than the Comodoro Rivadavia clay then in use.

A good description of the sodium-bentonite CLARIGEL is made by several authors in 1954; and the clay is advertised by Coloidal Argentina Soc. Resp. Ltda. for clarifying sugar cane juice. The clay contains 90% montmorillonite.

By 1975 the production of "fuller's earth" was only 236 tonnes, but of bentonite had risen to 116,000 tonnes. One mine, operated by Geberovich Hnos. in the Neuquen Province, supplies a factory at Zapala where foundry and oil well drilling grades are made (3000 tpa).

54iii

loc. cit. 84.

B.U.N. have manufacturing facilities for bentonite, mica, barytes and other minerals in the states of Paraiba and Bahia.

BRAZIL

BETWEEN 1968 and 1972 Brazil's entire bentonite production came from the Bravo mine in the Boa Vista district of Campina Grande in Paraiba State, operated by Empresa de Mineração Bravo Ltda., mostly for foundry and drilling mud markets. Other companies have been licensed to mine bentonite and several more are engaged in processing. In mid-1976, Laporte Industries Ltd acquired a 40% holding in one of the most important of these, Bentonit União Nordeste S.A. Production at the time was 80,000 tpa but volume and product range have since expanded.

54iv

BURMA

IN 1975 production of bentonite was 914 tons, of which 300 tons were supplied by the Mining Corporation from Kyun Hla for use in state owned industry.

54v

H. S. Spence, An addition to the mineral resources of western Canada, *J. Inst. Petr. Techn.* (1924) *10* (45) MYW & WSD, *Geol. Surv., Canada*, Memoir 163 (1930), pp 20, 37, 139–141.

See also first two references on next page.

CANADA

HUGH S. Spence (1924) was the first to point to the existence of 'bentonite' in Canada. More information was offered by M. Y. Williams and W. S. Dyer (1930). Apparently use was made of Canadian bentonite at Hudson Bay Company posts for washing blankets and woollens. A bed of bentonite was found *c* 1920 between two coal seams at Rosedale, Alberta, but it could not be extracted cleanly in the mine. The tentative beginnings of an industry are traced by Spence (1931). Dresser Minerals eventually opened a quarry near Rosalind, 100 km SE of Edmonton.

Calcium-montmorillonite has been worked by Pembina Mountain Ltd since 1940. It is difficult to work because the clay occurs in six main seams totalling only 750 mm. separated by 250 mm of shale. The Fullers' Earth Union were called in to get the plant on a sound commercial basis. The company's operations were eventually acquired by Filtrol Corporation (1979), who intended to expand the long-established acid-activation plant. This occurrence is in Manitoba.

In 1981 bentonite was also obtained by Avonlea Mineral Industries for soda-treatment at Wilcox, Saskatchewan. Bentonite was raised in a coal mine in British Columbia from 1926 to 1944; and unusually thick deposits occur in North West Territory near Inuvik.

54v

H. S. Spence, and Margaret Light, *Canada Dept. of Mines Bulletin* No. 723 (for 1930) (1931).

Ind. Min. Canadian Supplement, (167), 69.

Albert Boorman went from the Cockley activation plant to Pembina.

CHINA

54vi

THE Pharmacopoeia of the Heavenly Husbandman, the first of about 200 *Pên Tshao* (pharmacopoeias) dating from 100 BC to 100 AD in the Han Dynasty, puts 'fuller's earth' in the Third Rank of therapeutic value. As 46 minerals were described, it is safe to say that 'fuller's earth' had been used in pharmacy long before that time. Water-driven trip-hammers with vertical pounding stocks were certainly in use for hulling rice in the 3rd century AD; and were used for many purposes thereafter. Recent research however suggests that calcium montmorillonite may well not have been used for fulling woollen cloth in most of ancient China.

Wilson (1936) studied the effect of 38 Chinese clays in decolorising vegetable oils. Four were better than a standard fuller's earth in the natural (dried & milled) state, and these all responded to acid-activation.

About that time (Ishibashi, 1935), attention was being paid to a ferriferous montmorillonite from the opencast Fushun coal mine in Manchukuo.

Science and Civilisation in China, Joseph Needham. Cambridge University Press, (1957), *3*, 643. (Better to say calcium montmorillonite).

Colin A. Ronan, personal communication.

E. O. Wilson, *J. Chem. Eng. China*, (1936), *3*, 325–330.

Studies on Fushun green shale, *J. Soc. Chem. Ind. Japan*, (1935), *38*, 477 13.

INDIA

54vii

FULLER'S earth and bentonite have been used for centuries for washing the hair and for fulling cloth. Some of these clays have also been used as a medicine (chapter 69). Although deposits are known in the states of Bihar, Gujarat, Jammu, Kashmir, Rajasthan, Assam, Madya Pradesh and Uttar Pradesh, by far the largest output today is from the Kutch and Bhavnagar districts of Gujarat State. Here the largest occurrence of natural sodium montmorillonite outside the USA is mined, as well as calcium montmorillonite. A full range of products

Ind. Min. (Mar. 1978) (126), 79.

54vii

is made by Ashapura (and Ambica) Minechem Industries, by Gimpex Minerals Pvt. Ltd at their Bhuj mine in Cutch and by five other quite substantial firms.

54viii

Sir George Watt, *The Commercial Products of India*, (1908), 329. *A mineral reconnaissance in Western Iran*, S. G. Henderson and R. H. S. Robertson, Resource Use Ltd, 28 Feb 1958. Cambridge History of Iran, (1968), I, 262. *Ind. Min.* (Mar. 1978), 79.

IRAN

AN earth known as sang-í-basrí, said to be exported to Calcutta, was used in tonic preparations; and a head-washing clay (gil-i-sar shūr) bought in the bazaar at Kirmānshāh said to have come from Sāva, near Qum, could be described as a fuller's earth since its cec was 65 meq/100 g. In 1975 Iran Barite Co. produced 30,000 tons of bentonite from mines at Mehrejan and Zarrin; and half of that amount was extracted from deposits at Grimonj by Magcobar Iran.

54ix

loc. cit., 79.

ISRAEL

NEGEV Ceramic Materials Ltd began working the clay at Ramon Crater about 1974 and was producing about 15,000 tons a year three years later.

54x

loc. cit. 77.

MOZAMBIQUE

IN 1975 the Luzinada mine at Maputo was being worked (1800 tons) by Luzinada Umbeluzi Mina Ltd, following the pioneering efforts of two former companies.

54xi

loc. cit. 84.

MEXICO

MONTMORILLONITE has long been known in Mexico, some exceptionally white and low in iron. Química Sumex SA de CV, a subsidiary of Süd-Chemie AG of Munich, mine calcium bentonite at Tlaxcala (40,000 tpa); Morelos (80,000 tpa); and Durango (15,000 tpa). At Puebla acid-activated clays (TONSIL) and foundry grades are made.

54xii

loc. cit. 77.

NAMIBIA

HOCHMETALS Africa (Pty.) Ltd mined bentonite at Cape Cross Mine.

PERU

54xiii

THE domestic market is largely satisfied by Minera Baribent SA and Cia. Minera Agregados Calcáreos SA.

loc. cit. 84.

SOUTH AFRICA

54xiv

THE Parys deposits SW of Johannesburg are mined by Zimro (Pty) Ltd, part of Anglo-American Corp. of South Africa. Production (1975) 37,549 tonnes. (J. Calf exhibited fuller's earth at the 1851 Exhibition in London).

loc. cit. 77.

1851 Exhibition (London) Catalogue, *2*, 952.

TAIWAN

54xv

THERE is a sizeable operation at the quarry at Mew Lui, the proprietors are the Kowloon Kaolin Co Ltd of Hong Kong.

TURKEY

54xvi

TURKISH fuller's earth is mentioned in the Index of the catalogue of the 1851 Exhibition in London but it was not until the early 1970s that much progress was made. SAMAS-Sanayi Madenleri AS mines natural sodium bentonite at Akdomus, Pertek, and Kaspinar, the plant being at Resadiye. From 1975 Emko Bentonit Ticaret ve Sanayi AS has mined bentonite at Cankiri using plant at Ankara. Recently Dr. Y. Turan, the discoverer of trona in Anatolia, has been developing the white and remarkably pure calcium bentonite, the intended markets being organoclays, acid-activation, ceramics, detergents and others.

loc. cit. **3**, 1388.

Ind. Min. (Mar. 1978), 83.

ZIMBABWE

54xvii

G & W Industrial Minerals (Private) Ltd, an associate of Zimro, mines bentonite in Southern Africa.

AUSTRALIA

54xviii

REPORTS of the occurrence of fuller's earth are to be found in booklets about the resources of several Australian states. The Minerals

loc. cit. 77.

54xviii

Division of Steetley Industries Ltd began to work a foundry grade of bentonite at Wingen, N.S.W., processing being carried out at Sandgate. Small quantities are quarried in Western Australia and by the Dubbo Mining Division in N.S.W.

54xix

NEW ZEALAND

Bull. Imp. Inst. (1928), *26*, 148.
N.Z.J. Sci. Tech., (1919), *2*, 119.

Foundry Tr. J. (1/10/42), *68*, 88.

N.Z.J. Sci. Tech., (1938), *20*, 162B.
(1941), *22*, 275 B.
(1955), *37*, 132 B.

Ind. Min. (Mar. 1978), (126), 79.

Mining J. (13/6/69) 523.

Ind. Min. (1968) (13), 26.

SEARCH for a commercial source of bentonite has been made for many years. Thus a fuller's earth from Whangerei, north Auckland was examined in 1928. Even before that Morgan (1919) described fuller's earths in Glen Massey, Waikate district; Croydon, near Gore; and Hokonui, Southland. Production of bentonite at Hawkes Bay was 205 tons in 1940; this went to foundries and followed a report by L. R. L. Dunn (1938). Other papers on New Zealand fuller's earth reflected the determination to create a new industry. By 1975 Canterbury Bentonite Ltd, a subsidiary of Lime & Marble Ltd., were mining 10,000 tpa or so at Coalgate and making swelling grades nearby. Reserves were estimated at 20 million tons (1969). Production began about 1968.

54xx

PHILIPPINES, BULGARIA, ETC.

Ind. Min. (1978) (126) 79.

SMALL quantities of bentonite are worked in the Philippines, and in other countries, such as Bulgaria where soda-ash treated fuller's earth meets the OCMA specification for oil well drilling; they supply earths for civil engineering use and for foundries.

PROPERTIES AND APPLICATIONS

55–70

APPLICATIONS depend upon the properties of a mineral and the properties, though largely inherent, are enhanced by suitable treatment. These relationships are shown in a mineral use diagram prepared in 1960, Fig. 82. Check-lists of processes, properties and applications are given in a later paper (1973). Chapters 55 to 69 deal with some of the more interesting subjects, but are by no means complete.

Mineral Use Guide, or *Robertson's Spiders' Webs*, London: Cleaver-Hume Press, (1960), p. 13.

R. H. S. Robertson, Industrial uses of clay minerals, *Silicates industriels*. (1973), *33*, 33–43.

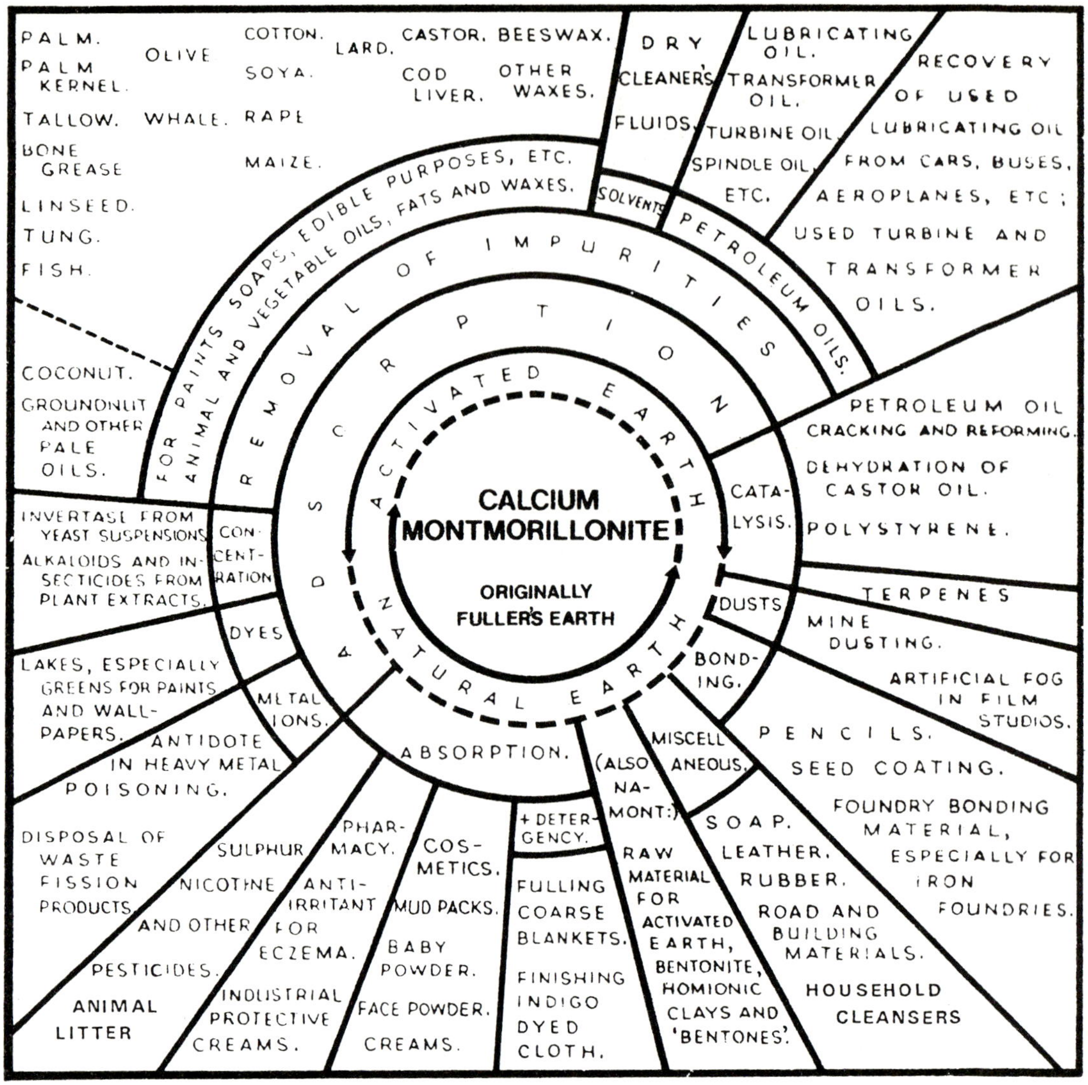

Fig. 82. Properties and uses of fuller's earth.

From R. H. S. Robertson (1960)

SORPTION IN THE GASEOUS PHASE

55

'Twas in the bath, a piece of perfumed clay
Came from my love's hands to mine one day.
"Art thou then musk or ambergris", I said
"That by thy scent my soul is ravished?"
"Not so", it answered "worthless earth was I,
"But long I kept the rose's company,
"Thus near, its perfect fragrance to me came,
"Else I'm but earth, the worthless and the same."
SA'DĪ, Gulistān, Introduction 1.

Gulistān by Sa'dī, translated by Edward B. Eastwick, Trubner, London (1880), Poem quoted in preface.
By this simile, Sa'dī would express his own unworthiness, and the estimation imparted to him by the King's favour.

THUS wrote Sa'dī (Muslih ibn Abd'Allah), c 1184–1291, the Persian poet and moralist, who was born at Shirāz, studied in Baghdad and travelled widely, returning to Shirāz in 1255. His most famous books were the Bustān (1257) and the Gulistān, completed in 1258. (A hair-washing earth, Gil-i-Shirāz, bought in the bazaar at Tehrān, had a cec of only 25 meq/100 g.)

Long before, Pliny the Elder had noted a similar phenomenon. *Baptes, mollis alioqui, adore excellit*, the 'baptes' or 'dipper' has an exceptionally pleasant scent, but is otherwise an ordinary soft stone.

Pliny, *Nat. Hist.*, 37, 55.

Even now a small perfumery industry exists near Kannauj, some 130 km west of Lucknow in India, where baked clay discs are exposed under the open sun during the hot summer months of May and June. Before the wet season they are collected and steam-distilled, the odoriferous compounds being absorbed in sandalwood and sold as 'matti ka attar' or 'earth perfume'.

See note at end of this chapter.

Miss I. J. Bear and R. G. Thomas investigated this 'argillaceous odour' or what they prefer to call 'petrichor': they found that the host material need not be a clay mineral, since "freshly ignited materials rich in iron oxide", exposed in the desert, also produced the odour – the pleasant smell that rises from the soil when the first rain comes.

Nature, (7.3.1964), *201*, 993–995.

Petrichor recovered by these authors in Australia by steam distillation contains cyclic bases like pyridine and quinoline, carboxylic acids, nitro-phenolic compounds, aldehydes, saturated hydrocarbons, both branched and straight-chained. Later I. J. Bear and Z. H. Krantz identified fatty acids[a] and nitrophenols[b] in petrichor, and Bear & Thomas showed that petrichor retards germination and early plant

a. *Australian J. Chem.*, (1965), *18*(6), 915–917.

b. *id.*, (1969) *22*, 2343–2350.

55

c. *Nature* (1965) *207*, 1415–1416.

d. *Geochim. et. Cosmochim. Acta*, (1966), *30*(9), 869–880.

growth[c]. Their final conclusion concerning the genesis of petrichor[d] was that the atmosphere contains, as general contaminants, lipids, terpenes, carotenoids and other volatile decomposition products from animal and vegetable matter. The sorption of these compounds, or their oxygenated derivaties, by rocks and clays is controlled by the properties of the sorbent and the partial water vapour pressure of the atmosphere, low relative humidities favouring maximum uptake. Oxidation and transformation of sorbates take place on the rock or clay surface and are accelerated by warm to hot climatic conditions. The odorous and volatile products of these processes are subsequently displaced from the pores of the rock or clay by moisture when the relative humidity of the air approaches saturation. They suggest that there may be a connection between petrichor and petroleum formation.

Note: Pliny records other pleasant-smelling stones:

N.H. 37, 174

The 'myrrhitis', or 'myrrh stone', has the colour of myrrh and an appearance quite unlike that of a gemstone. It smells like an unguent and, when rubbed, even like spikenard.

N.H. 37, 185

The 'zathenes', according to Democritus, is an amber-coloured stone found in Media, and if it is ground with palm wine and saffron softens like wax and has a most agreeable smell.

N.H. 37, 188

'Narcissitis' is marked with veins coloured like narcissus, and has also its scent.

ABSORPTION OF LIQUIDS 56

THE extensive surface area, the ability of water to penetrate between the sheets of the montmorillonite and the porosity of fuller's earth give it powerful absorbent properties. Some of the applications which make use of these properties have been decribed in the chapters on pharmacy and household uses.

Since 1951 fuller's earth granules have been made at Bath, Redhill and Woburn Sands as an industrial floor cleaning material. Spilled oil is absorbed and danger removed.

By far the largest use of absorbent fuller's earth granules is as an animal litter, introduced in 1959. The most desirable properties in an animal litter are (a) easy recognition of the granules which are saturated or fouled by the animal, (b) the absorptive capacity and the rate of absorption, (c) odour prevention, or destruction, and (d) resistance to 'tracking' by cats – that is granules sticking to the paws by static electricity or by moist adhesiveness.

Two types of material were successful as animal litter granules. Sepiolite which is nearly white and has a high absorptive capacity; and calcium montmorillonite, which swells when wetted. The wetted granules are easily recognised and can be removed daily, a more economical procedure. Both types have their devotees and cat litter provides a large market. Some materials offered as cat litter, such as calcined diatomite, have a high porosity, but do not remove the urine to sites where bacteria cannot follow. After a short while they stink.

57i

ADSORPTION IN GENERAL

W. J. Dubois, *Amer. Chem. Soc. J.* (1907), *29*, 75; U.S. *Dept. Agr. Bur. Chemistry Bull.* (1907), *54*, 23.

B. Segel, *J. Soc. Chem. Ind.*,(1934), *54*, 317 T; 367 T.

P. Lindner, R. Keren & L. Ben-Gera, *J. Sci. Fd Agric.*, (1981), *32*, 1177–1182.

W. T. McGeorge, USP, 1 458 427.

K. Shumburdis, *Kolloïd-Z.*, (1931), *55*, 150.

1) I. M. Eltantawy & P. W. Arnold, *J. Soil Sci.*, (1974), (1), 99.

2) D. L. Carter *et al.*, *Soil Sci.* (1965), *100*, 356.

3) P. S. Keeling, E. C. Kirby & R. H. S. Robertson, *J. Brit. Ceram. Soc.*, (1980), *79*(2), 36–40.

B. K. G. Theng . . . London: Adam Hilger (1974).

EXAMPLES of adsorption from solution are so numerous that one cannot do justice to the history of the subject; moreover the distinction between English and American fuller's earth is seldom made. There was a fuller's earth test for caramel in vinegar as early as 1907; and other processes for purifying beverages were patented, especially in the period 1922–1938. A fuller's earth gel was proposed by B. Segel (1934) for removing protein haze from glucose solutions and molasses fermentation liquors (1934). Lindner and others (1981) showed that bentonite effectively precipitated proteins from potato juice. No doubt the product could be used in animal feed pellets. McGeorge (c. 1922) had found that a small proportion of fuller's earth removed the bitter tasting tannin and polyphenolic compounds from fruit juices, especially from the pineapple, orange, lemon, lime or grape fruit.

The adsorption of isoamyl alcohol was studied by Skumburdis (1931) and thereafter the interlamellar adsorption of glycerine, ethylene glycol[1] and ethylene glycol monoethyl ether[2], the last two being used for measuring the specific surface of clays, though Keeling, Kirby and I have shown[3] that the adsorption of moisture vapour can be just as accurate.

For modern work on adsorption of organic compounds one may consult B. K. G. Theng's monumental work *The Chemistry of Clay-Organic Reactions* (1974).

57ii

The adsorption of cationic dyes

Adsorption on to Ionogenic Surfaces, P. H. Plesch & R. H. S. Robertson, *Nature*, (26. June. 1949) *161*, 1020.

WE have seen how in 1904 one of the founders of the Pfirschinger Mineralwerke in Kitzingen am Rhein discovered in an excavation a white clay which was first used as a basis for making pigments and known commercially as Pfirschinger *Weisserde*. Cationic organic dyes such as methylene blue, rhodamine 3B, auramine, methyl violet B, and malachite green are all strongly adsorbed by cation exchange onto montmorillonite. The 'lake pigment' prepared with these dyes, being unstable in strong light, was restricted in use to cheap wallpaper or ephemeral things like posters.

Methylene blue adsorption was used as a rough diagnostic test of clays at The Fullers' Earth Union before 1933, but it was not until 1948, (Plesch and Robertson) that it was confirmed that two different mechanisms were at work. They found that Michaelis and Ehrenreich had been the first to suggest that the adsorption of large ions such as organic dyes on to ionogenic surfaces takes place by two distinct mechanisms operating simultaneously: one of an electrochemical nature and the other a 'normal' physical adsorption. Later, Freundlich and Posser state explicitly that the adsorption of large positive ions onto clays could be explained by the combined effects of an irreversible ion-exchange adsorption and a reversible physical adsorption, and this idea was further elaborated by Michaelis and Rona. It is interesting for the historian of science that this simple idea was afterwards forgotten but was 'rediscovered' on several occasions prior to 1948. The idea of simultaneous adsorption by two different mechanisms was not mentioned by Swan and Urquhart in their comprehensive review on adsorption, or in text-books. Even Freundlich himself in his "Kapillarchemie" (1930) seems to have lost sight of the implications and usefulness of the idea, but it was formulated in different ways by Stadnikoff[1], Fogle and Olin[2], Hendricks[3] and Grim, Allaway and Cuthbert[4].

Biochem. Z., (1908), *10*, 283.

Kolloid Chem. Beihefte, (1914), *6*, 297.

Biochem. Z., (1919), *97*, 57; (1920) *102*, 268.

J. Phys. Chem. (1927), *31*, 251.

1) *Kolloid Z.*, (1922), *31*, 9.

2) *Ind. Eng. Chem.*, (1933), *25*, 1069.

3) *J. Phys. Chem.* (1941), *45*, 65.

4) *J. Amer. Ceramic Soc.*, (1947), *30*, 137.

Many attempts have been made to measure the specific surface of finely granular matrials by the adsorption of dyes. The first was by Znamenski (1936) who found that the mean specific surface of soils is given by $S = 10.082\ a$, where a is Freundlich's absorption coefficient for methylene blue. It is however a clumsy way of measuring specific surface area compared with one of the glycols or indeed with water. Far more attractive is the use of a dye to measure cation exchange capacity.

Determination by the adsorption method of the mean specific surface of finely granular bodies A. V. Znamenski, *J. Appl. Chem. Russia*, (1936), *9*, 208–216.

Robertson and Ward (1951) compared cation exchange capacity by methylene blue and by the conventional ammonium method and recommended the former to pharmacists as a diagnostic test. In 1957 Fairbairn and Robertson found that cation exchange capacity determined by methylene blue can be regarded as a useful soil mechanical test in its own right. When a large number of liquid limit (LL) tests have to be done to determine variation in clays of the same general type, as on an engineering site, economy of operation could be achieved by carrying out many cec tests and few LL tests. It is only when comparing clays of widely different types that correlation is poor. Physical adsorption was eliminated by reading from a graph the adsorption when the filtrate had a concentration of 0.0075% of the dye.

The Assay of Pharmaceutical Clays, R. H. S. Robertson & R. M. Ward, *J. Pharm. Pharmacol.*, (1951), *3*, 27–35.

Liquid Limit and Dye Adsorption, P. E. Fairbairn & R. H. S. Robertson, *Clay Min. Bull.* (1957), *3*, 129–136.

The method was accurate enough but it required several determinations with different weights of clay to arrive at an end-point. Ulrich Hofmann solved the problem of accurately measuring the cec by a single determination of methylene blue adsorption. Clay is dried at 110°C to constant weight; 0.2 g of the clay is added to 50 ml of 0.2 per cent MB solution and the mixture is shaken for at least 24 h. The solid is removed in 50 ml centrifuge glasses, the speed being 4500 rpm. The concentration of the clear liquor is measured

Giesserei, (1967), *54*, 98–101.

MB = methylene blue

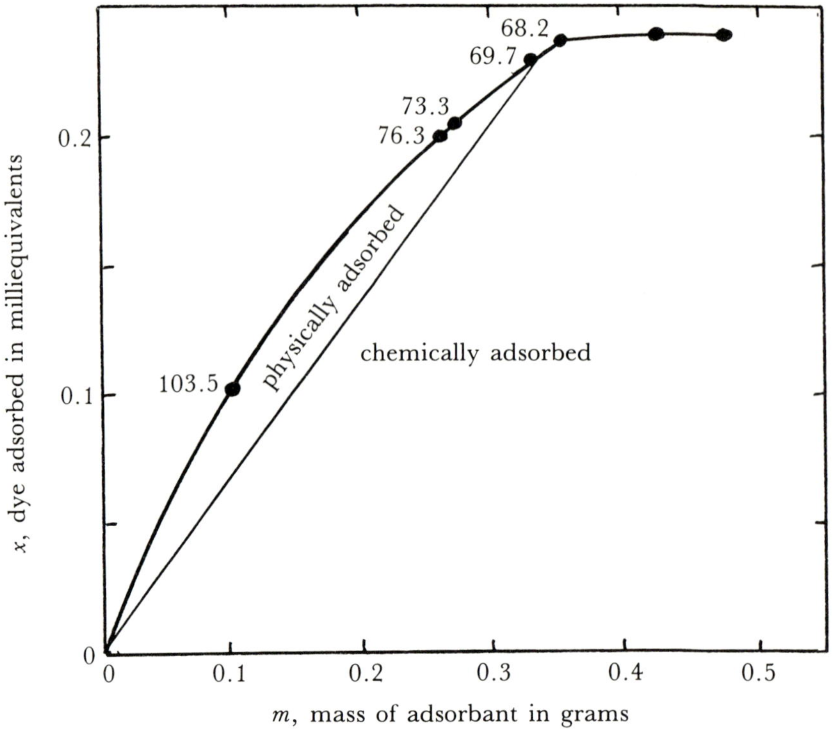

Robertson & Ward, *op. cit.*

Fig. 83. (right). Adsorption of methylene blue on calcium montmorillonite (Surrey Yellow-XLCT) starting with 50 ml of a 0.357% solution of dye.

After P. H. Plesch

photometrically. MB adsorption now, being no less accurate than the standard ammonium acetate method, had at last an unbeatable value as a process control test.

Nevertheless industrial laboratories wanted a much quicker method of measuring cec. F. O. Jones, Jr., made an aqueous clay slurry and titrated it directly with methylene blue dye to an end-point, which was recognised by spot-testing the titrated slurry on filter paper for a sign of free dye. The method was adopted in oil well drilling fluid control.

New Fast, Accurate Test Measures Bentonite in Drilling Mud, *Oil Gas Journal*, (1965), *17* (6) 707–716.

The same procedure was adopted by Baroid Division, National Lead Company, in the control of the foundry floor sands and of ceramic clays. Their Equipment Department made for these industries their Kit 425 consisting of standardised chemicals, equipment and a test procedure for wet and dry samples.

Determination of CEC by MB Adsorption, M. J. Nevins & D. J. Weintritt, *Amer. Ceram. Soc. Bull. (1967) 46* (6), 587–592.

The pioneer work of The Fullers' Earth Union in this field is described under 'Foundry Bonding Clay'. Their laboratory also measured the cec of the fuller's earth in the main seam at very small vertical intervals.

Roy K. Taylor at the University of Durham used the method to show variation in the Mansfield Marine Band cyclothem at Tinsley, near Sheffield, and the British Geological Survey estimated the montmorillonite content of borehole cores in their search for fuller's earth

M. Sedimentary Petrology, (1967), *37*(4), 1221–1230.

at Baulking and Fernham, in 1970, using a method of titrating with methylene blue solution.

The absorption of alkaloids has been reviewed in Chapter 17iii.

57ii

The occurrence of economic calcium montmorillonite (fuller's earth) deposits in the Lower Greensand in the District S.S.E. of Faringdon, Berkshire by E. G. Poole *et al. Inst. Geol. Soc.*, Report 71/12.

58i

THE DECOLORISATION OF OILS, FATS AND WAXES

To lighten the colour of the oils a small amount of fuller's earth was sometimes added. The particles of earth absorbed certain compounds responsible for the colour. This was known as "bleaching" but not a word of it was to be breathed to the public. It was believed that people would not like the idea of having bits of earth in their cod liver oil.

CARINA MANN, A History of British Cod Liver Oils, Martin Books; Cambridge and Seven Seas Health Care Ltd., Marfleet, Hull, (June 1984), p. 19.

[In 1935]

R. J. Forbes, The culinary arts, in *A History of Technology*, edited by C. Singer, E. J. Holmyard & A. R. Hall, Oxford: Clarendon Press, (1954), *1*, 281.

Geschichte des Zuckers, (1890), 135, 295 citing Ibn Beithar, II, 473.

THE practice of clarifying liquids by adding a clay and allowing it to settle must be of great antiquity. In the making of beer in Ancient Egypt, the fermented liquid was filtered into jars. According to Forbes, in some cases fuller's earth seems to have been used to clear the beer – perhaps by Nile mud.

Earths were also used for clarifying wines in classical times according to von Lippman. The Moors probably worked the *tierra del vino* of Lebrija, a mixture of palygorskite and illite. In later years this "Spanish earth" was imported into all the major wine-drinking countries. It was obtained from small pits operated by individuals until 1891, when the first mining concession was granted. The earth was used almost solely for wine fining and the amount dug was therefore small.

Imports of "Spanish Earth" for wine

Year	Tons	Year	Tons
1916	254	1921	363
1917	470	1922	192
1918	146	1923	319
1919	227	1924	300
1920	339	1925	135

The chief importers into England were Boake, Roberts and Company of London.

Many different wine-fining agents have been used. They fall into two groups, flocculants and adsorbants. In the former were isinglass, a jelly extracted from the air-bladders of the sturgeon, gelatine, white of egg and alginates. In the adsorbent group were iron oxide, carbon, light-coloured calcium montmorillonite and hectorite. The use of iron oxide is anomalous, but the other three, alone or in combination, have proved their worth.

I use adsorbant as a noun and adsorbent as an adjective, as in flocculant and flocculent.

Carbon was first suggested by Lowitz for purifying distilled spirits in 1785 and is still in use. As for oil refining, the Royal Society of Arts in London had latterly been much concerned about the need for "edulceration" of whale oil, in reference to which the Minutes of the period 1761–4 carry a liberal sprinkling of adjectives such as "vicious", "stinking", "foetid", "putrid" and "disgustful". At that time only one method of refining the oil, so as to make it suitable for such purposes as soap-making and lighting, was put forward, and that was ineffective.

Chemistry & Industry, (20.3.1954), 320.

Colin Mackenzie, *Five Thousand Receipts*. . .(5th ed.) London: G. B. Whittaker, (1825), p. 297.

Some oils were much easier to clarify than others. Thus in the 1825 edition of *Five Thousand Receipts* it is stated that the mucilage contained in rape(seed) oil may be got rid of in great measure by adding half an ounce of oil of vitriol to two pints of the oil. This process is what would now be called de-sliming rather than decolorisation. Fuller's earth is sometimes used for this purpose. Rape oil could also be made the equal of spermaceti oil for use in illumination by violently agitating the oil with spring water, or, even better, with seawater, and allowing the emulsion to separate overnight – *the oil swimming to the top, the water, and all feculent and extraneous particles subsiding to the bottom.*

sulphuric acid.

'degumming' today.

dregs

By this time it was known how to remove the offensive odour of whale and fish oil by means of an oxidising agent. One process was: *to 100 lb. of oil, add 25 oz. of roche alum, and mix, dissolved in 9 lb. of boiling water. After stirring it, about half an hour, add 15 oz. of nitric acid, still continuing to stir it. Let it stand for 48 hours, when the fine oil will swim on the surface, and then draw it off.*

Id., 299.

The smell of oils was a constant source of trouble before the days of alkali refining, fuller's earth treatment or steam deodorisation. To purify trotter oil (neat's foot oil) *put 1 quart of oil into a vessel containing a quart of rose-water, and set them over a fire till the oil melts and mixes with the rose-water, . . .* and so on.

Id., 300.

By 1825 alkali refining had been adopted. Thus, to refine spermaceti, *1 cwt. of unrefined spermaceti heated with 3 gallons of water. As soon as it begins to boil, add, from time to time, $\frac{1}{2}$ pint portions of the following liquor: 7 lb. potash + 2 gallons water, stood 24 hours; dip off the ley as required adding more water occasionally till the alkali is exhausted. The spermaceti is boiled for 4 hours, the scum being removed all the time.* The process is repeated three times.

Id., 300.

Some oils could be purified to some extent with bleaching powder, a process for which was devised by Charles Mackintosh and patented in 1799 by Charles Tennant, with whom he worked.

Archbald & Nan Clow, *The Chemical Revolution*, London: The Batchworth Press, (1952), p. 192.

The most far-reaching step in the history of oil bleaching before the introduction of fuller's earth was when Figuier in 1811 called attention to the bleaching effect of bone black. Accounts are available of some of the early methods of use of this in oil-bleaching.

Colin Mackenzie, *loc. cit.*, 302.

A method of purifying common fish oil, and rendering it equal to the best sperm oil, by the use of animal charcoal, has lately been discovered in Denmark. The description is very incomplete, but mentions that beef bones which have been boiled, are made into animal charcoal in a peculiar way.

The charcoal is mixed with the oil, and repeatedly agitated for two months, after which it is filtered through several strata of charcoal, and used as soon as made.

(After such a long process the user may have become impatient.)

Id., 301.

To sweeten, purify and refine Greenland whale and seal oil. The raw oil is filtered through bags 41 inches long, mouth held by wooden hoop, 15 inches diameter; there is an inner bag of flannel and an outer bag of jean, quilted, containing charcoal between them. The oil runs thence to a lead-lined tank, 8 feet by 4 feet by $4\frac{1}{2}$ feet deep, containing five or six inches of water in which are dissolved six ounces of blue vitriol, for the purpose of drawing down the glutinous and offensive particles of the oil, which have escaped through the charcoal; and freeing the oil of unpleasant smell. The oil passes on to another one or two sets of filter bags and tanks.

Jean, a stout, twilled cotton fabric, as used in trousers, from 'jean fustian' (Genoa fustian).

Though many other reagents were tried, including powdered chalk, slaked lime, tannin, and so on, active carbon remained the chief adsorbant in the oil industries until 1890.

Although, as already stated, alkali refining was known in 1825, Schmersahl took out a patent in 1842 for the refining of cottonseed oil with caustic alkali. This was important for the future use of fuller's earth, because unrefined cottonseed cannot be decolorised with the clay. Supplies of alkali became cheaper with the invention by Solvay in 1861 of the ammonia-soda process, adoption of which in 1872 in England at Brunner, Mond and Company's works reduced the price of soda. The Castner-Kellner electrolytic process for making caustic soda was introduced in 1885.

E. W. Blank, *Oil & Soap*, (1942), *19*, 110–113.

The use of fuller's earth for refining "paraffine" was patented by John Fordred, Frederick Lambe and Arthur Charles Sterry, paraffin refiners, in 1868. Their claims were for:

British Patent 610 of (24/2/1868).

1. The treatment of paraffine after being subjected to the sulphuric acid treatment in the manner specified with a solution of glue or other gelatinous or albuminous substance, or starch, or flour, or milk, in the manner and with the object described.

2. The treatment of paraffine after being subjected to the sulphuric acid treatment as described with fuller's earth or the substances named as its equivalent . . . (marl, clay, or other natural silicate of alumina, or silica or alumina.)

3. Treating paraffine (not previously chemically treated) with fuller's earth or the substances named as its equivalent as a distinct and substantive means of purification as specified; also the use of fuller's earth as a finishing agent in the place of animal charcoal as described.

Although these inventors recognised the superiority of fuller's earth over charcoal in paraffin refining and took out further patents the same year to cover its use for purifying hydrocarbon oil (No. 648) and for treating animal and vegetable oils (No. 2356), the process was not widely adopted by other oil refiners though it was not completely forgotten.*

Both of 1868.

In 1873 it was announced that

crude resin oil being generally turbid, Fuller's earth, as well as other varieties of clay, is sometimes used as a clarifier. The powdered clay is mixed with the oil, stirred, and, when it has subsided to the bottom of the tank, the oil is filtered and comes out quite clear. But this is not literally refined *oil. Crude resin oil contains a certain proportion of undecomposed resin, pyroligneous acid, liquid hydrocarbons of widely different boiling-points, also solid homologues of naphthalin. To refine this crude oil it is necessary to redistil it with slaked lime, and separate the successive products of the distillation. Having had considerable practical experience in the matter, I could give A. K. C. any further information he may require.*

Chem. News, (26/12/1873), *26*, 326.

*It is clear from Letters Patent (2.12.1873), No. 3942, taken out by Arthur Smith & Frederick Field, that the use of fuller's earth for decolorising hydrocarbons was known in 1873. See Recovery of oil from spent fuller's earth. (Recorded in F. Ullmann's *Encyclo. Tech. Chemistry*, (1914), *5*, 375.)

(The destructive distillation of naphtha had been studied by S. D. Hayes in 1871).

V. Ritter of Ofenheim, Vienna, used a magnesium silicate for the purification of ceresin, and he treated the spent adsorbant with steam to recover more wax. In a second patent in the same year (1879) he refers to the decolorisation of ozokerite and to the magnesium silicate as the *Entfärbungsmittel.* Later uses of this adsorbant are recorded by Kausch (1927).

D.R.P. 9291 of 21.8.1879.
D.R.P. 9981 of 24.10.1879.

Bleaching agent.

Das Kieselsäuregal und die Bleicherden, Oscar Kausch, Berlin: Julius Springer, (1927).

Nevertheless it was not until 1880 that fuller's earth came to be used for refining edible oils by a process which was adopted worldwide. According to David Wesson, N. K. Fairbanks and Company, of Chicago, news had reached them that in the Orient olive oil was sometimes improved in colour by agitating it with clay. *Various samples of clays were tried with cottonseed oil and the clay then used for fulling cloth in England gave much better results than any other clay that the company was able to obtain.* The process worked out by W. B. Albright and Henry Eckstein in the Fairbanks works in 1880 led to a rapid spread of the procedure throughout the United States and hence for a growing demand for English fuller's earth.

Mining & Engineering World, (1912) *37*, 667, cited by C. L. Parsons, *U.S. Bur. Mines Bull.* 71, (1913), 6.

At first the warm oil, after being stirred with fuller's earth, was originally recovered in bag filters as described by Colin Mackenzie in 1825, but by about 1900 these had been almost entirely replaced by filter-presses, enabling fuller's earth to be used on a much larger scale.

Albright's aim was to make a lard substitute: Eckstein developed the steam deodorising process in 1891, and a process for decolorising cottonseed oil in 1893. In 1900 David Wesson introduced the much more effective vacuum deodorising of this oil. The Sabatier-Normann-

Our Chemical Heritage.

OIL AND FAT
REFINING PLANT

FULLER'S-EARTH PROCESS

Fig. 100

S. H. Johnson & Co., Ltd.

Carpenters Road,

Stratford, LONDON, E. 15.

Telegrams : "Filtrum, Phone, London." Telephone : Maryland 3657.

 Fig. 84. Plant for refining oils and fats with fuller's earth; c 1930.

Kaiser process for hydrogenating vegetable oils was taken up by Proctor & Gamble in 1910, some 80 years after David R. Williams, ex-governor of South Carolina and "father of the cottonseed oil industry", operated a mill at Robbin's Neck, S.C., and fed the press cake to cows and pigs (1830).

In Britain oil refining plant was made by the leading oil mill engineers such as Rose, Downs and Thompson, Ltd., and Greenwood and Batley, Ltd. Both would normally have ordered the filter-presses required from S. H. Johnson & Co Ltd., of Stratford, London, the oldest makers in the country. Most of their records were destroyed, but they do have one record that they were designing oil refining plants as far back as 1893. The domestic sale of fuller's earth would then have increased.

In the USA a standard bleached colour of cottonseed oil of 25 Yellow, 2.5 Red (Lovibond) for bleachable oil *by the use of 6% of English fuller's earth, or other earth of equal quality, and best bleaching methods* was adopted in 1908 though without any definite method of analysis.

Extract from *Oil & Soap*, (1945), "Chemists' Committee of the National Cottonseed Products Association", p. 13.

In the course of time natural fuller's earth was used for bleaching oils or fats of low colour, such as groundnut and palm kernel oils, and its place was taken by the more active acid-activated earths for bleaching

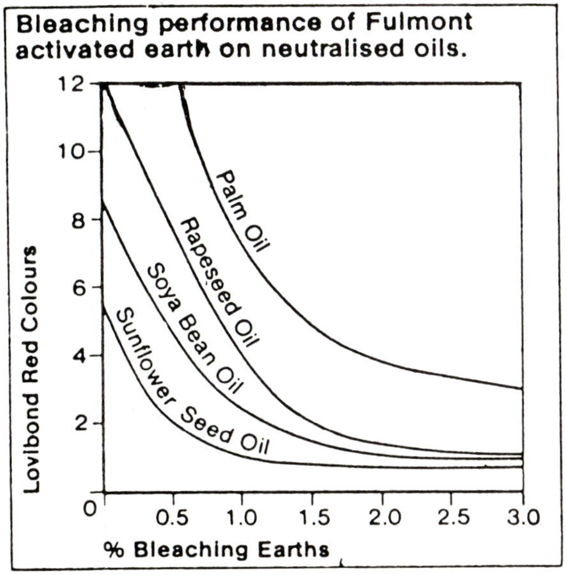

Fig. 85A

Laporte brochure

58i

The refining of animal fats, H. B. W. Patterson, *Chem. & Ind.*, (18.9.1976) 771–774.

Chem. & Ind., (16.9.1978), 692–703.

the more difficult oils. A mixture of 10–15 per cent of active carbon with 85–90 per cent of bleaching earth is sometimes used with more difficult fats, such as tallows.

Vernon Young reviewed the processing of oils and fats in 1978. He explained that in degumming (in which one or two per cent of activated earth may be used after phosphoric or citric acid) the impurities removed include phospholipids, sugars, resins, proteinaceous compounds, and trace metals. In the subsequent bleaching process pigments, oxidation products, trace metals, sulphurous compounds and traces of soap are removed. *Bleaching earth manufacturers have concentrated on increasing adsorptivity, improving filterability and reducing oil retention, and have tried with some success to enhance . . . trace metal removal. It is common practice for refiners to use of the order of 0.5 per cent of the natural clay initially to adsorb soap and, after 10–15 minutes, to add a similar quantity of the active earth.*

58ii

F. Krczil, *Untersuchung Verwertungtechnischer Adsorptionsstoffe*, (1931), Akademischer Verlag: Leipzig.

a. A. Markman & M. Kovalenka, *Masloboino Zhirovoe Delo*, (1929), (8), 10
b. T. Richert, *J. Industr. & Engg. Chem.*, (1917), *9*, 599
c. D. M. Guseinov, *Pochvovedenie*, (1950), 735–745
d. O. Eckart, *Seifensieder Zeitung*, (1931), *58*, 33 (bone fat)
e. O. Eckart & A. Wirzmüller, *Die Bleicherden*, (1929), Braunschweig.

Oil Retention

Table 14 Oil retention of various adsorbents, all in millilitres|per gram.

	Pfirschinger method	Eckart & Wirzümller	Others
Filter carbon	250–300	—	
Blood carbon	120–150	57	
Bone carbon	—	105	
Infusorial earth from Akhaltzikh	—	—	127 a
"Floridin" from Kutais	—	—	56 a
Isarit	—	74	
Tonsil	—	63	30–40 d
Alsil	—	59	
American natural earth (? Attapulgite)	—	—	40–48 b
German activated earth	—	—	19 b
Gumbrin	—	—	40 c
Franconit FC	60–65	36	
F	50–55	35	
CL	40–45	26	
S	30–35	24	
American "fuller's earth"	—	22	
English fuller's earth	—	18	8 b

THE oil retained in the filter-press cake has been a matter of concern since earths were first used for decolorising oils. In table 14 the oil content of spent earth has been taken from the older literature. 58ii

English natural fuller's earth has far the lowest oil retention of all these adsorbants.

Recovery of Oil from Spent Fuller's Earth

58iii

The spent and unconsidered earth
RUDYARD KIPLING. From the first poem in Puck of Pook's Hill.

AS early as 1873 Arthur Smith had spotted one of the disadvantages of using fuller's earth for refining hydrocarbons – the loss of valuable oil in the spent earth. He therefore made the first synthetic bleaching agents consisting of silicates of calcium, magnesium, iron or manganese or aluminate of magnesium. All these adsorbent gels could be decomposed by mineral acids and the oil released. The silica gel thus produced could be used for making more of the adsorbant.

Arthur Smith, of Pera Lodge, Loughborough Road and J. C. & J. Field, Upper Marsh, Lambeth, both in Surrey, Patent 3942 of 2.12.1873.

Charles L. Parsons showed that edible oil could be readily extracted from fuller's earth by naphtha, carbon tetrachloride, or other solvent, but that the colouring matter is left behind with the earth. *If this treatment is followed by a treatment with alcohol, the major part of the colouring matter is also removed from the earth, but the earth does not entirely regain its former bleaching power, even in the presence of alcohol. Accordingly, if to the alcohol, mixed with a certain proportion of an oil solvent like benzol, a small quantity of acid is added, the basic colors are converted to a salt of the acid, and in this form are then readily removed by the solvent. The whole process can be carried on at one operation by mixing the solvent for the oil, the alcohol, and the acid in one solution.* Parsons tried this many times with the same sample of fuller's earth, without any loss in bleaching power.

Fuller's Earth, *Bur. of Mines Bulletin* 71, Washington, Govt Printing Office, 1913.

58iv

The Regeneration of Spent Fuller's Earth

SINCE Parsons' clear description of how to regenerate fuller's earth used for refining edible oils, many attempts have been made to achieve this aim in practice. I suspect that one or more of the chemicals required was omitted in these failures, which will not be reviewed here. Less oil was lost when activated earths were introduced, and still less with 'superactivated' earths; and improvements in filtration technique have reduced the oil loss to an acceptable proportion. It is probable that edible oils recovered from spent adsorbants have been oxidised and so spoiled.

Vernon Young, *Chem. & Ind.* (16.9.1978), 692–703.

It is fair however to say that attapulgite, soon after its discovery in 1893, was used in granular form and the granules were regenerated by heating to a temperature below the sintering point. The same process applies to the clay mineral sepiolite. Parsons, again, gives an excellent description of this regeneration process. A spent acid-activated bleaching earth did not respond to Parsons' process, the result being worse than a natural fuller's earth.

58v

Lubricating Oil Regeneration

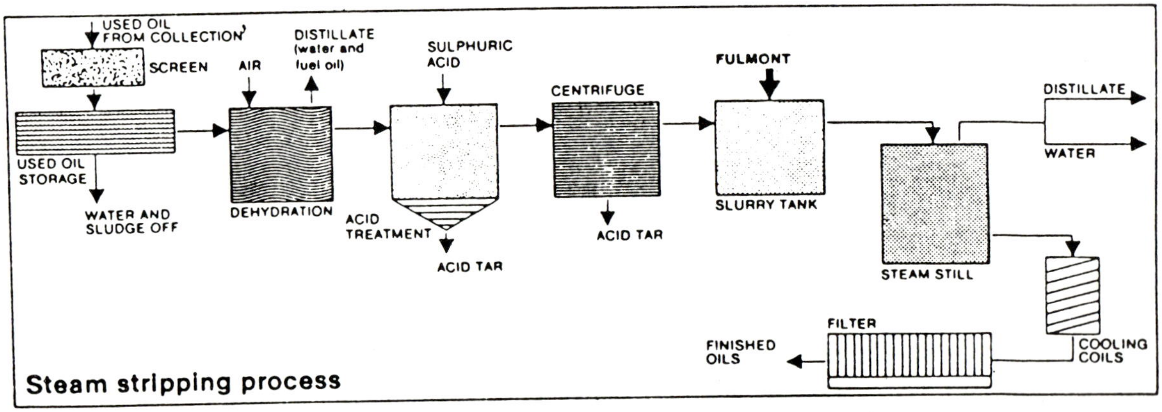

Fig. 85B. Steam stripping process for regenerating used lubricating oil.

Laporte brochure

CHARLES Ignatius Kelly, The FEU's Oil consultant in 1940 and author of Kelly's Viscosity Tables, introduced the company's acid-activated earths to most of the petroleum oil refineries in Britain. He also supervised the laboratory's work on oil regeneration. Two processes were tried, one a more or less conventional earth treatment and the other a high temperature earth treatment in the presence of superheated steam – "stripping" – to remove the volatiles. Fig. 85B.

The first to do practical trials was Bertie Bassett at his works near Redhill Station, where he replaced imported attapulgite granules by the locally activated earth.

More ambitiously The FEU asked the authorities for a sample of used aircraft lubricating oil. The oil, to their dismay, arrived in dirty containers which had previously contained a variety of liquids such as paint and ethylene glycol. By courtesy of Messrs S. H. Johnson & Co Ltd the low temperature process was carried out on their pilot plant. The filtered oil, returned in *clean* containers, passed *all* the tests for aircraft lube oil. The process was soon adopted by Dalton's of Belper.

For reclaiming lube oil used in buses the steam stripping method was applied at several centres throughout the country. An article written by Charles Kelly in an Australian technical journal led to the adoption of oil regeneration in Australasia and the use of plant designed by him and Messrs Highfield and Roger Smith, consulting engineers to The FEU.

59

WATER-SOFTENING PROPERTIES AND CATION EXCHANGE

There is a rivulet runneth close by the town, and another larger beyond Kilder's Well; this last serves for washing linnen, which it doth as well without soap, as other water does with it; of this we had experience, which was a confirmation of what had been reported to us concerning this water: we searched if in the brinks we could discover any fuller's earth, but found none; we discovered some pieces of iron-ore in several places of it; this rivulet drops from the mossy ground in the top of the hills.

MARTIN MARTIN, A Voyage to St Kilda, 1698

First published 1698; (1934 ed. Eneas Mackay Stirling, p. 415).

MARTIN Martin was commissioned by the Royal Society of London to make a journey to the Western Isles in 1695. He was an acute observer; he recorded several occurrences of fuller's earth (Ch. 30). [It was his book which made Dr. Johnson want to visit the area with James Boswell in 1773.]

Cation Exchange in Soils W. P. Kelley, Reinhold, New York (1948).

In his review of the history of cation exchange, Kelley also noted that the discovery that soils have the power of exchanging cations with solutions containing other cations was the outgrowth of observations dating back into the remote past.

Excursion to Woburn Sands and Sandy, 25 June 1892, *Proc. Geol. Assoc.* (1892), *12*, 395–403.

A. C. G. Cameron (1892) wrote that the water thrown out of the bed of fuller's earth at Aspley Heath near Woburn Sands, if not very abundant, is beautifully soft and pure, and acting on this the villagers place blocks of earth in some wells to purify the water. He was told that *in dry seasons dealers also carry the earth to the Fen district to clarify the peaty water; and the drier the season the brisker the demand. Filters of fullers' earth are set up and the water cleared and made palatable; the method adopted being somewhat after the old custom in country districts of placing perforated trays or "letches" containing wood ashes over vessels put out to catch the rain* [and dissolve the alkali]. *It seems therefore that in fullers' earth we have an agent that would be active in removing impurities from water and reducing the hardness, besides otherwise altering its character.*

Cameron also refers to the irony water that wells up through peat in the valley of the Flit, near Ampthill. The iron gives an amber colour to the water and an astringent and acidulated, but not unpleasant, taste. *If made to filter through fullers' earth, the liquid is bleached or paled and evidently contains a less quantity of iron.* He thought that the filtering effects were mechanical, but they may have been due to chemical action in a water such as that of Flitwick.

About the same time Cameron's colleague in the Geological Survey, H. B. Woodward, observed that the surface-water drained off the sediment pits at Midford, Somerset, was said to be very soft, pure and drinkable.

Proc. Geol. Assoc., (1893), *13*, 126.

George Dames, Director of the Fuller's Earth Mining Company at Aspley Heath, told Cameron that *those incrustations so destructive to boilers are carried down and disappear under the influence of fullers' earth. It is this carbonate of lime, as every housewife knows, that gathers as a crust over the interior of the family kettle with such serious effect, and it is well known to many that boiling a lump of fuller's earth softens this hard crust so that mere rubbing takes it off.* Another colloidal electrolyte, sodium alginate, was used for many years for the prevention of hard boiler incrustations according to H. W. Bannister, Director of The Liverpool Borax Co. Ltd. In his Patent of 1884 E.C.C. Stanford had written *I also use soluble alginates for completely precipitating iron and alumina from solution* and *Another useful application of soluble algin is for preventing boiler incrustation on account of its completely precipitating CaO, BaO and SrO from solution in a gelatinous form.* No doubt bentonite would behave in a similar way, but would not settle so well as fuller's earth.

Proc. Geol. Assoc., (1892), *12*, 396.

Chem. & Ind., (1944), p. 247.

British Patent No. 13433 of 1884.

In experiments begun in 1845, Thompson showed that when soils were mixed with ammonia and then leached with water a certain proportion of ammonia was retained. Way demonstrated that this cation adsorption – or exchange – was connected with the silicates in the soil in his study 'on the Power of Soils to Absorb Manure', and Forschamer showed that Ca and Mg ions were released from the soil when treated with seawater. All three published their findings in 1850.

On the Absorbent Power of soils; H. S. Thompson, *J. Roy. Agr. Soc. Eng.* (1850), *11*, 68–74.

J. F. Way, *id.*, (1850) *11*. 313–379 and (1852), *13*, 123–143.

G. Forschamer, reference in Wiklander, studies on Ionic Exchange, *Ann. Roy. Agr. Coll. Sweden*, (1946), *14*, 1–171.

It was not until 1912 that "base exchange capacity" of soils was first measured quantitavely by G. Wiegner. He studied zeolites and then clay minerals, developing an empirical equation derived from adsorption isotherms, and H. Jenny and P. Vageler expanded the studies, which are reviewed by C. Edmund Marshall (1949), who had introduced a method of calculating the ionic composition from the chemical analysis and the cationic exchange capacity of a montmorillonite in 1935.

J. Land. (1912), *60*, 111, 197.

The Colloid Chemistry of the Silicate Minerals, Academic Press: New York, chapters 11 and 6.

F. W. Bird helped to develop tablets which purified seawater; bentonite was used to cause the tablets to disintegrate rapidly so that the chemicals could react quickly.

60

CLAYS IN AGRICULTURE

Here lies Fuller's Earth
Memorial tablet to THOMAS FULLER, 1606–1661, divine, historian and wit, in Cranford Church.

Two miles east of Heathrow.

THE use of marl in agriculture was known in Antiquity, but Bernard Palissy, the potter and writer, was it seems the first to propose a theory of mineral nutrition of plants when he wrote "no vegetative thing could vegetate without the action of salts which is in seeds. . ."

B. Palissy, *'Discours Admirables'*, Paris: Martin le Ieune (1580) p. 164. *The Admirable Discourses of Bernard Palissy* by Aurièle la Rocque, Urbana: University of Illinois Press (1957), p. 19.

Palissy recognised that chalk if it could be finely milled would act as a fertiliser; calcareous clays or marls were in fact used because they broke down by weathering and became distributed in the soil. He also believed that some white clays might serve the same purpose. *I remember going from Partenay, as I went from Bresuyre in Poitou, and from Bresuyre to Thouars. . . the clays are very white. . . which leads me to believe that the clays of these regions could also serve as marl, and particularly the one used by the drapers to full and degrease cloth.*

id. (1580) p 324 *tr.* (1957) p 220.

The full passage is obscure, but he is making an independent and original attempt to recognise different species of earths and clays. As a potter, he objects to the use of the word fatty in describing clay *for if it were fatty it would be impossible to dissolve* [disperse] *it with water or frost: for all fats and oleaginous substances resist water and cannot have any affinity with it: but on the contrary, clay and marl remove all greasy, viscous and oleaginous stains: for these reasons the fullers use them to degrease cloth.*

id., (1580) p 326 *tr.*, (1957) p 221.

In a book by Sir Hugh Plat (1594) *Diverse new Sorts of Soyle not brought into any publique Use*, fuller's earth is mentioned. He also quotes Palissy; and Francis Bacon, Lord Verulam (1626), states that "bole arminick is the most Cold of them".

Part of *The jewell house of art and nature* (1594)
Sylva sylvarum: or a natural historie, (1626), para. 701.

Gervase (or Jervis) Markham (1568?–1637) has been cruelly dubbed "the first hack writer" and one cannot deny that in his first book of husbandry he "gathered his materials in a rather shameless manner". However, if his writings on angling are not original, he was a noted horse-breeder and is said to have imported the first Arab. *A Discourse of Horsemanshippe* (1593) and other popular works on horsemanship and farriery went through many editions, as did *A Farewell to Husbandry: or the Enrichment of all sorts of Barren and Sterile Grounds* (1620). Particular areas were examined as in *The Inrichment of the Weald of Kent*, (1625).

Encyclopaedia Britannica. 14th ed. (1929) *1*, 933.

As late as 1941, Stebbing sneered at the hardworking and observant Markham: *a final passage from the* Farewell *makes us realise that it was poss-*

Sub-soils in Farming, W. P. D. Stebbing, *Proc. Geol. Assoc.* (1941), **52**, 262–3.

ible for this class of writer to be as accommodating as the early pharmacist with his potions. Markham advised us that, "if any perfect and rich Marle was not to be obtained, that Earth which is called Fuller's Earth would do; where the one is not, commonly the other is, then you may use it in the same manner as you should marle, and it is found very near as profitable". Yet even today fuller's earth could be recommended for addition to very sandy soils to improve their ion exchange and water-holding capacities.

E.g. in glasshouses.

On the other hand Stebbing makes a more favourable comment on *Terra, a Philosophical Discourse of Earth* by John Evelyn (1676): *although Evelyn at first limits himself to treat of Mould or Earth* [by saying] *"we shall not . . . penetrate lower than is necessary for the planting and propaegation of Vegetables", he does list those* [sub soils] *which commonly succeed in* strata, *or layers, be it fat or lean,* Loam, Clay, Plastic, Figuline, *or* Smectic; *as* Chalk, Marle, Fullers-Earth, Sandy, Gravelly, Stone, Rock, Shelley, Coal, *or* Mineral. In this early attempt to classify rocks, fuller's earth is placed not only among the clays but in the rather natural series (chalk, marl, fuller's earth.) In 1686 Plot also makes the distinction between "the Smectic and figuline Earths", that is between two contrasting types of clay.

(1729 edition),

From L. *Figulus*, a potter.

Robert Plot, *The natural history of Staffordshire*, (1686).

In the next two centuries huge areas of soil were improved by claying or marling the land; thus Ward wrote in 1776: *clay put on to light and sandy grounds, which, unless clayed, will bear nothing but rye, with whatever other composts they are manured; but once clayed, they will produce oats, barley, pease, etc. . . . 100 loads to the acre – to last 40 years . . . Care should be taken not to use too much marle.* Sometimes 200–300 loads to the acre may be needed, depending on soil requirements.

Rev. Samuel Ward, *A Modern System of Natural History*, London: Newbury, (1776), *12*, 21–22.

An account was given of the history of the practice of clay marling by Middleton. In this century fairly pure clay minerals, kaolinite, illite, montmorillonite and others, have been used with silica sand to make synthetic soils by which the behaviour of the clay mineral could be compared. In the field soils containing even a high proportion of calcium montmorillonite can have a beautiful crumb structure when moist after rain and can be very fertile; but when sodium ions replace calcium ions through swamping by seawater or development of saline conditions the soil becomes impermeable and incapable of sustaining crops.

A. C. Middleton, *J. Minist. Agric.*, (1949), **56**, 80–84.

In the late 1930s.

Finally, there are those who think that on Mars an important component of the soil may be montmorillonite containing surface adsorbed iron compounds (1983).

A. Bain and L. Margulies. *Nature*, (6/10/1983), *305*, 523–5

61

SEALANTS AND GROUTS

A. J. and G. Hubbard, *Neolithic Dewponds*, London, (1904, 1907).

IN the construction of dew ponds on the Chalk Downs of the south of England and elsewhere neolithic man must have used clays to make these artificial basins watertight. Once the impervious basin was filled artificially with water, it would begin to function as a means of providing a water supply for cattle.

The role of clays in irrigation channels may have been important on porous soils, but it was not until canals were cut as major routes for transporting merchandise that water impedence became of vital importance. The development and use of puddling clays is dealt with in Chapter 26.

Science News Letter (13 March 1937)

Already by 1937 bentonite was being used for sealing dams and the walls of reservoirs; and engineers were looking into the possibility of using it in levels along the Mississippi and Ohio rivers. The method used was injection of a slurry under pressure. *When drying occurs the bentonite shrinks and clings tightly to the sand or gravel. But as soon as leakage occurs the bentonite. . . swells again and reseals the entire mass.*

See ch. 64.

Bentonite behaves as a sealant in oil well drilling, though it has other functions; and in nuclear waste containment it not only reduces the permeability of the soil, but adsorbs some of the most dangerous fission products.

C. D. Weaver, *Proc. Soil Sci. Soc. Am.*, (1946), *11*, 196–254.

Bentonite is not only used as a grout; it is also used in a "placement" method, by which granular bentonite is allowed to fall to the bottom

E.S.P. = exchangeable sodium percentage.

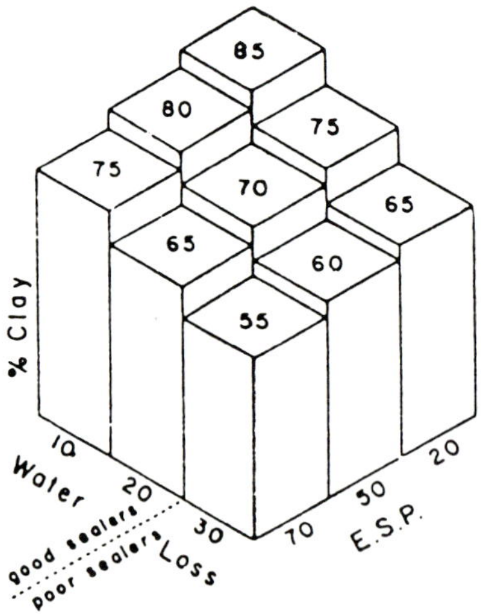

Fig. 86. Predicted clay content necessary to hold water loss at various clay and exchangeable sodium ion contents.

M. B. Rollins (1969)

of a pond or lake, forming a blanket there, perhaps 3 to 4 in. thick. When mixed with the soil, about 1 part of bentonite to 5 parts of soil is adequate – a technique not unlike puddling mentioned above.

75–100 mm.

Rollins showed that water impedance, as measured by water-loss in the filter-press test, correlates highly with montmorillonite content and exchangeable sodium percentage, but to a less extent with swelling, viscosity and gelation.

M. B. Rollins, *Clays & Clay Minerals*, (1969), *16* (6), 415–443.

Bentonite mixed with sodium silicate with or without sodium silicofluoride is also the basis of sealing systems for controlling airflow in coal mines.

For other uses of smectites in civil engineering see Chapter 65.

For further information see R. E. Grim, *Applied Clay Mineralogy*, (1962), McGraw Hill Book Co. Inc., pp. 404–5.

62

FOUNDRY BONDING CLAY

Ind. Min. (126), 57.

Berk in the Foundry, Foundry Bonding materials, marking 25 years of Berk service to the foundry industry from 1931 to 1956, F. W. Berk & Co. Ltd., Commonwealth House, 1–19 New Oxford St., London, W.C.1. [1956].

THE sand moulding process was invented in the 14th century. Before that the lost wax process was used.

Molten metal is poured into a mould made of sand, clay and water in greensand practice or sand bonded by clay, after the water has been dried out in drysand practice. For centuries natural sands were used. These usually contained a high proportion of clay, often illite or kaolinite, which gave a good finish but frequent faults in the casting, due to the low permeability of the moulding sand.

Synthetic moulding sands were first made in the USA from a mixture of pure silica sand and Wyoming bentonite, still the standard bonding clay in the industry, in the 1920s.

F. W. Berk & Co Ltd were the first firm to import Wyoming bentonite to the British Isles, the pioneers in its foundry use being the Ford Motor Company in Cork in 1931 and then from 1932 onwards at Dagenham. From 1933 to 1940 the proportion of foundries which had tested the bentonite and then became regular users increased every year as more and more foundries became assured of the value of bonding materials.

Table 15 Bentonite first used in the UK

Year	*Number of foundries who had tested Bentonite supplied by Berk.*	*Number of foundries who were regularly using Bentonite*	*Regular users as %age of foundries who had tested Bentonite*
1932	*20*	*2*	*10%*
1933	*50*	*3*	*6%*
1934	*77*	*5*	*6.5%*
1935	*89*	*7*	*7.9%*
1936	*97*	*10*	*10.3%*
1937	*125*	*16*	*12.8%*
1938	*200*	*27*	*13.5%*
1939	*267*	*37*	*13.9%*
1940	*332*	*62*	*18.7%*

Meanwhile a Mr. Roland was sent by The Fullers' Earth Union Ltd a sample of No. 2 Special – from about 2 feet below the 'Black Jack', near the top of the All Top seam at Combe Hay. On 21.8.1933 he took a consignment which was definitely stated to be for foundry use. (In 1928 the Widnes Foundry (1925) Ltd took 2 cwts of 200 mesh Fuller's Earth, but this was used for oil refining).

It was not until Algernon Lewin Curtis of Westmoor Laboratory,

Chatteris, Cambridgeshire, called on 27.6.1934 at The FEU Copyhold Laboratory that the foundry application was seriously considered. Mr. Curtis said he was sure that a very permeable moulding sand with high green strength could be made with Redhill fuller's earth. He thought a suitable composition would be 84% pure sand, 15% fuller's earth and 1% bentonite and that a cheap ferruginous fuller's earth, possibly a waste product would be most economic.

He suggested the trade name FULBOND.

The visit of A. L. Curtis (Editor of *Sands, Clays and Minerals* from April 1932) not only led to the production of the first bonding clays made from English fuller's earth but led to the testing of the "Bunkum" seams of clay which had hitherto been thrown away. The thickest of these seams turned out to be of excellent quality.

A precursor of *Industrial Minerals*.

By 1936 the Laboratory felt confident enough to write a memorandum 'On the Use of Fuller's Earth in Moulding Sands'. Copies were sent 21.7.1936 to the Secretary of the Iron and Steel Institute, Mr. T. R. Walker of The English Steel Corporation, Vickers Works, Sheffield, at the suggestion of Sir William Larke, FRS, a friend of Sir Robert Robertson, FRS, the Government Chemist.

Report K3/36 of the Moulding Materials Sub-Committee of the Steel Castings Committee of the Iron & Steel Institute, by R. H. S. Robertson.

Also in the 1930s QUESTAL KWK bentonite was imported by Quest Products, later Colin Stewart Ltd., whose owner, James N. Wilson, wrote a useful article on Bentonite in 1937.

Bentonite – its Uses in Industry, *Sands, Clays & Minerals*, Sept. 1937, p. 149 with advertisement opposite.

F. W. Berk & Co Ltd, having then no deposits of fuller's earth, introduced in 1940, a foundry bonding clay containing illite and some montmorillonite, called DISTRIBOND. This consisted of London clay from excavations made when the District Underground line was being extended. It had a useful 'flowability' and was used mainly in backing sands, but occasionally in facing sands.

Berks' eight years of pioneering effort were an important factor in enabling the foundry industry to do the magnificent job it did in war production. Indeed the outbreak of war gave a great impetus to the adoption of synthetic sand practices. The ever widening use of bonding materials speeded up mechanisation and productivity in the foundries. Output and quality increased sharply; but also, it is estimated, over 50 million ton-miles of war-time transport was saved in moving sand around the country.

Shipping restrictions were soon imposed on the import of Wyoming bentonite at a time when more of it than ever was wanted. The Moulding Materials Sub-Committee invited me to their meetings from 27.5.1941 and agreed to have FULBOND moulding clays tested by English Steel Corporation, Hadfield's and Jessop's, Dr. R. J. Sarjant having reported that 5 cwts used in the foundry had given very good results. At this meeting The FEU calcium montmorillonite was called FULBOND No. 1, and the clay treated with 4% of sodium montmorillonite FULBOND No. 2.

M.M. Sub. C. Report K7/41 Comparative properties of Fulbond and Bentonite.

S. George Henderson, who had carried out the lab work on preparing the FULBONDS, was also invited to the Sub-Committee Meetings

Committee Reports
K5/42 RHSR
K12/42 RHSR
K13/42 A. Wheeler.

from 16.4.1942. FULBOND 1 and 1a were untreated, and 2 and 2a were alkali-treated, 2a being a variety which developed its bond quickly with only light milling, an essential requirement for Ford's, Dagenham. FULBOND No. 3 was introduced by letter to the Sub-Committee 31.3.1942.

Later FULBOND No. 4a was introduced with a wide range of mouldable moisture contents and high greenstrength – higher than No. 3A.

By 1942 Henderson was not only studying all the different foundry practices and tailoring these different grades to suit them, but had also developed 'A New Method of Measuring the Plasticity of Foundry Moulding Sands' – a method which was approved by the rheologist G. W. Scott Blair.

K11/42 and K21/42 RHSR and S. G. Henderson. See p. 396.

K16B/42 RHSR and B.S. Emődi K16c/42.

Nature (Lond.), 1943, *152*, 728.

Dr. Barbara S. Emődi, later Neumann, measured the Specific Surface of three Silica Sands and of Southport Sand. This work led to the measurement of Rugosity [later Angularity] of Granular Solids, defined as the ratio of the measured specific surface to the hypothetical surface area which the particles would have if they were all spherical. For sand grains the method used air permeability and sieving techniques. It was adopted by Dr. W. Davies and Professor W. J. Rees, the Chairman of the Moulding Materials Sub-committee, in their study of the effect of grain shape on the moulding properties of synthetic moulding sands, and they used it also for measuring the grain-shape of glass-making sands. After the war Professor Rees was awarded the CBE for his work on foundry sands and moulding materials and Freddie W. Bird, superintendent of the Redhill laboratories of The FEU was awarded the MBE.

J. Iron Steel Inst., 1944, II, 19 p–47 p *and Refractories J.*, 1945, *21*, 98.

Trans. Soc. Glass Techn. 1945, *29*, 279–288.

K6/44 SGH

K14/45 SGH

Atherfield clay operated by the Leeds Fuller's Earth Company, until they were acquired by The Fullers' Earth Union, turned out to have some useful properties in a moulding sand, behaving rather like a mixture of Wyoming bentonite and fireclay then in vogue in the USA, (FULBOND K).

First AGM of The FEU after the war.

In June 1945, Mr. J. S. Highfield said that these developments would also be of value in peace. The Bath Works had doubled in size when it was equipped with modern plant, which gave excellent results in the foundries where it was used.

It was not until after the war that other montmorillonites were imported, such as North African (1947), Italian, from Ponza (1950), and Cyprian (1954) by Berks.

In 1953 Berks extracted their own fuller's earth from Aspley Guise in Bedfordshire, and made BERKBOND No. 1 by drying and grinding and BERKBOND No.2 by the addition of sodium carbonate at their works at London Colney, near St Albans.

Sir Hugh Plat, *The jewell house of art and nature* already referred to in ch. 60.

I have not attempted to write the history of metal casting but an interesting early work, by Sir Hugh Plat (1594), includes a section on 'The art of molding and casting.'

SUSPENSION AND THIXOTROPY

63

THESE properties are virtually inseparable. The ability to keep solid particles in suspension is of importance in many industrial fields. Finely divided barium sulphate and calamine are held in suspension in medical and pharmaceutical preparations by bentonite or the synthetic smectite LAPONITE. Lens-grinding abrasives are similarly held in suspension.

Thixotropy is the reversible sol-gel-sol transformation produced by shaking. The first books on the subject were by Freundlich (1935), summarising his own experimental work, and by E. A. Hauser (1930) and others. Thixotropy is important in soil science, as in the regain of strength in remolded soils; Lambe (1958) held that this effect was due to a crystallisation of the water between particles. This phenomenon is met in grouting and in diaphragm wall construction (chapter 65); as well, of course, as in drilling mud (ch. 64).

> Evans and Reid (1936) *have used the Scott Blair-Crowther plastometer to make a very complete study of the flow properties of drilling muds. Thixotropy can be both advantageous and disadvantageous in this material. If the mud is not sufficiently thixotropic, settling takes place as soon as agitation of the mud ceases, and this gives an uneven concentration when the agitation is resumed; but too much thixotropy is disadvantageous. Turbulence comes in at a surprisingly low rate of shear, in fact plug flow conditions and turbulent flow conditions almost overlap.*

cf. A. Hauser. *Proc. Chem. Eng. Group, Soc. Chem. Ind.*, (1930), *12*, 137–141

H. Findlay Freundlich "*Thixotropy*", Paris: Hermann et Cie, (1935).

T. W. Lambe, Papers 1654 and 1655 *J. Soil. Mech. & Foundation Div., Am. Soc. Civil Engrs*, (1958).

P. Evans & A. Reid, *J. Inst. Petr. Techn.* (1935), *21*, 677 and later papers. See also G. W. Scott Blair, *An Introduction to Industrial Rheology*, London, J & J. Churchill Ltd., (1938), 89.

For more modern work see R. E. Grim, *Applied Clay Mineralogy*, (1962).

64 OIL WELL DRILLING

Editorial article on bentonite, etc., *Industrial Minerals* (1978), (126), 57.

IT IS probable that the advantage of having a mud present in the drilling fluid was discovered by accident. In 1894, oil was encountered in a well being drilled for water at Corsicana, Texas, the first time that a rotary water flush drilling equipment had been used to produce oil in the USA. At other sites, a natural water/clay suspension was often formed when the drilling bit passed through shale formations. One of the first recorded accounts of a clay-water mixture being introduced into the annulus by the drill crew was in 1900, during the drilling of the famous Spindletop well, near Beaumont in south-east Texas. Soon after the well was spudded in, quicksand was penetrated, resulting in extensive fluid loss. It was then remembered that clay contamination of water drilling fluids helped line the walls of the hole, and several nearby cattle were herded into the shallow pit holding the water supply, producing a muddy fluid which immediately sealed off the permeable sands. The crew continuted to use the muddy fluid to aid in the retrieval of cuttings and to counterbalance gas pressures.

G. & W. Base & Industrial Minerals Bulletin No. 8 (May 1963).

This experience led Captain Lucas, who was using a light rotary drill at the same site in 1901, to make up a thick mud suspension using bentonite in an attempt to prevent the hole he was drilling from caving in. From this famous well at Spindletop oil poured forth from a depth of 1000 feet at a rate of 100,000 barrels a day. Since that historic day the technique of drilling with bentonite muds has advanced to such an extent that major drilling organisations employ mud chemists (and a host of instruments) to maintain mud circuits in optimal condition for each drilling environment.

Nevertheless, it was the use of weighting agents – in particular barytes – that led to the universal necessity for an efficient, low-cost suspending agent, and the mineral bentonite was first introduced as a drilling mud additive in 1929, when a patent application was filed by P. E. Harth of National Pigments and Chemical Company for its use as a suspending and gelling agent for muds – many years after this practice had been established.

B.P's First Quarter Century of Crude Oil **Production in Britain**, R. G. W. Brunstrom, *B.P. Magazine*, (Autumn 1964), no. 13, 3–7.

In Britain the first oilfield was discovered in 1919 at Hardstoft, Derbyshire, in a well drilled for the Ministry of Munitions. The Petroleum (Production) Act, 1934, vested in the Crown the ownership of all mineral oil not discovered up till then, and provided for attractive operating conditions. The Anglo-Iranian Oil Company (now British Petroleum Co Ltd) were quick to move but their D'Arcy Exploration had at first no luck in the south of England in a drilling programme beginning on 30th March 1936. However they found gas at Cousland, Midlothian, in March 1938, and at Eskdale, Yorkshire, in January 1939, the first two commercial gas fields in Britain, and in the summer of 1939 oil was struck in the Eakring anticline, and further discoveries were made at Kelham Hills, January 1941, Dukes Wood, July 1941, Caunton, April 1943 and Nocton, south of Lincoln in December 1943 – the last yielding only 40 tons of oil.

During the war production increased rapidly and was maintained for many years by "secondary recovery", achieved by driving the oil towards the production wells by a "water flood".

The man chiefly responsible for this successful programme was G. M. Lees, Chief Geologist of the British Petroleum Company (and its predecessor) from November 1930 to August 1953.

Table 16 Production of Crude Oil/English Oilfields (tons)

1939	3,145	1944	94,414	1949	45,330	1954	59,000	1959	82,644
1940	16,689	1945	71,427	1950	45,468	1955	53,482	1960	85,281
1941	29,991	1946	55,240	1951	45,079	1956	65,996	1961	105,770
1942	81,298	1947	47,092	1952	55,358	1957	81,926	1962	110,872
1943	112,622	1948	43,325	1953	54,636	1958	79,819	1963	122,764

In 1937 D'Arcy Exploration, worried by the importation of bentonite for oil well drilling, wanted to know if fuller's earth (calcium montmorillonite) could be converted into sodium montmorillonite, or bentonite. Dr. S. H. Bastow, a Tasmanian physicist who had worked under Professor Charles Ellis at Cambridge University, visited the laboratory of The FEU at Redhill. Of course it was theoretically possible but it had not been achieved.

25 Nov. 1937 "discussed drilling muds with D'Arcy Exploration People" RHSR's diary. 1 Dec. 1937. Dr. Stewart H. Bastow's visit.

The lab had at that time, in seven pound tins, reference samples of about fifteen different grades of natural fuller's earth. The same proportion of powdered soda ash was mixed with all the samples and water was added to allow the reaction to proceed. Only one sample produced a swelling clay: this was a 200 mesh pharmaceutical grade made from Chart Quarry (yellow) fuller's earth. All 26 hundredweight sacks of this grade in store were ordered for Eakring, though the destination was not allowed to be written on any of the Company's shipping documents. Secret delivery instructions were handled by the Railway Company.

Na_2CO_3

The only successful sample was moister than all the others. Research during that winter confirmed that for bleaching vegetable oils a natural fuller's earth should be dried down to about 4 or 5 per cent moisture, but for reacting with sodium carbonate, the wet clay must not be dried below about 14 per cent moisture.

Control of moisture content was not easy in the old kilns; in these, hot combustion gases were drawn through clods of fuller's earth held on perforated steel plates. The outside of a large lump became "over-dried" and the inside could still be too moist. Experiments showed that to obtain even drying one had to dry evenly size-graded pellets in an atmosphere almost saturated with water vapour. Drying timber came to mind. The outcome of the research was the introduction of the Dunford & Elliott Rotary Louvre Dryer and the resumption of supplies of clay for oil well drilling from 1941.

The sodium-exchanged swelling clay, the first commercial synthetic bentonite in Britain, was at first called "UNION BENTONITE No 1", but the trade name FULBENT was later adopted.

Seismic refraction had so far found three oil-fields but had reached its limits of usefulness. The more precise seismic reflection technique was introduced in 1949, and new oilfields were discovered at Plungar, Leicestershire in the autumn of 1953 and at Egmanton in the summer of 1955.

The probability of finding oil under the North Sea spurred F. W. Berk & Co Ltd also to make a synthetic bentonite. In August 1963 they had made a drilling mud grade of BERKBENT on their plant at London Colney, a year before the list of applications to drill had been closed by the Ministry of Power. They were eventually to become the leading British suppliers to this area.

OTHER CIVIL ENGINEERING APPLICATIONS

65

χρῶνται γὰρ πρός τε τὰ οἰκοδομήματα
αὐτῷ τῷ λίθῳ περιχέοντες κ'ἄν τι
ἄλλο βούλωνται τοιοῦτο κολλῆσαι

It is used in building, being poured round the stones themselves, and is also applied to any similar material which requires bonding.

THEOPHRASTUS, De Lapidibus, translated by D. E. EICHHOLZ, Section 65.

Oxford at the Clarendon Press, (1965), pp 82, 83 and 132.

HERE grouting (περίχευμα) is by plaster of Paris and not by lime mortar as given in Caley and Richards (1956).

E. R. Caley & J. C. Richards, *Theophrastus on Stones*, Columbus: Ohio State University (1956), 215.

A short history of grouting was given by R. Glossop in his opening address to the Symposium on Grouts and Drilling Muds in Engineering Practice (1963). The process of *injecting* a suitable fluid under pressure was invented by Charles Berigny (1772–1842), who in 1802 repaired a scouring sluice at Dieppe by injecting grouts of clay and of hydraulic lime beneath it.

London: Butterworths (1963).

The scientific study of drilling muds and the use of additives appear to date from about 1921, and bentonite was first used for this purpose in 1929. In 1931 the Størmer viscometer was applied to the measurement of the properties of drilling muds, and in the same year Marsh invented his funnel viscometer which is still generally used in the field.

C. Veder, *Habilitationsschrift, Techn. Hochsch.*, Graz, (1950); *Comptes rendus. III Congr. Int. Mécanique des Sols.* Zürich, 16–27/8/1953, Vol. II, Session 5, no. 18.

The continuous concrete diaphragm formed in a trench supported by means of a bentonite drilling mud originated in the invention of Dr. Christian Veder of the Technische Hochschule, Graz (1950). His process was immediately taken up by Impresa Costruzioni Opere Specializzate of Milan (I.C.O.S.), and similar processes were used by E.L.S.E – Milan and Solétange – France. See Fig. 87. These processes which use large quantities of bentonite enable concrete walls to be built close up to existing structures. The range of applications is already quite wide. H. Lorenz invented the slip caisson method and once made a circular caisson 60 metres in diameter supporting a building in Geneva.

The I.C.O.S. Processes, monographs of characteristic works, by I.C.O.S., Milan, (1959).

See Lorenz's contribution to the Symposium (1963) p. 20.

In the same Symposium G. K. Jones of The Fullers' Earth Union Ltd

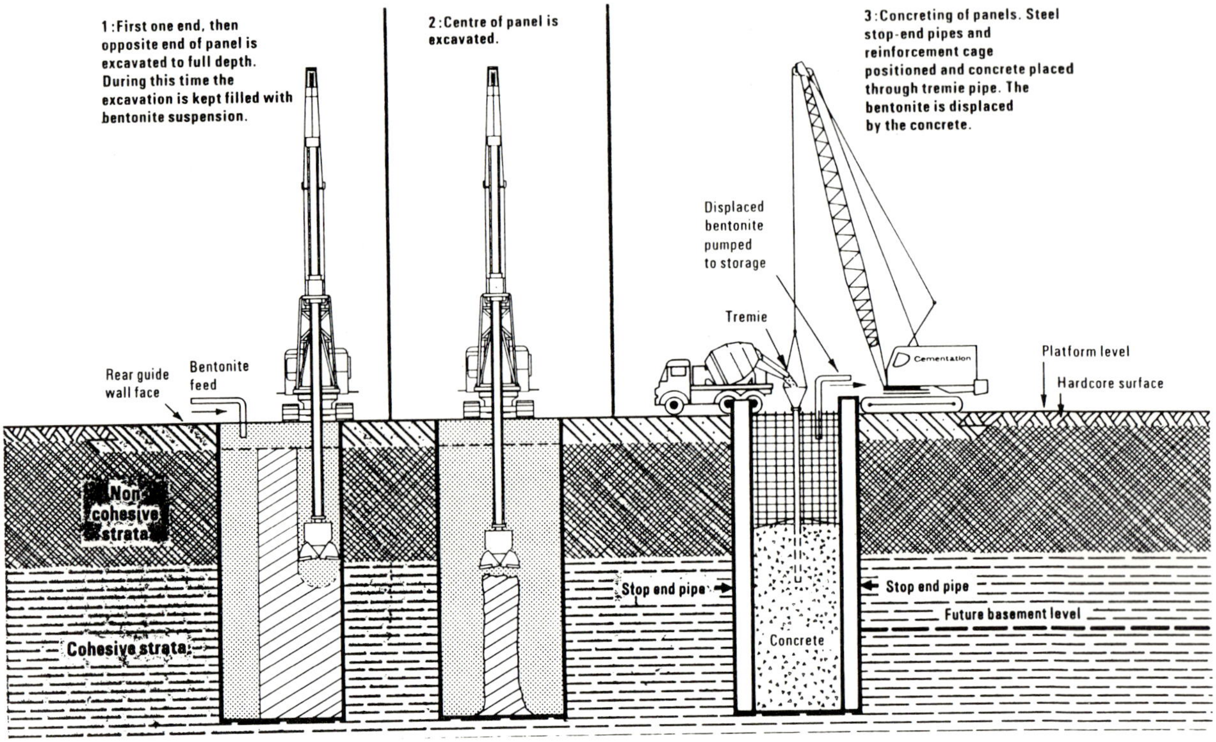

Fig. 87. Diaphragm walling: sequence of operations.

Cementation Piling and Foundations Ltd.

describes the chemisty and flow properties of bentonite grouts, and with Professor J. K. T. L. Nash the support of trenches using fluid mud. Veder himself illustrates the use of diaphragms in the Hyde Park Corner Underpass (1960). The FEU were awarded the contract to supply FULBENT for this, the first major contract of I.C.O.S. in Britain. F. W. Berk & Co. Ltd were not slow to make BERKBENT C.E. for this and other civil engineering work.

id., p. 22

id., p. 177

Chemical Age, (8 Oct. 1960) 582

In 1971 Laporte Industries Ltd patented a clay grouting composition consisting typically of cement (30), fly ash (70), synthetic clay, LAPONITE, (0.12), and tetrasodium pyrophosphate (0.0024 parts by weight). The set grout has a final strength of 200–800 lbf/in^2 (1.38–5.52 million pascals) with little shrinkage or separation of water or solids during setting.

Brit. Pat. 1 228 155 (15/4/71).

66 PELLETISATION

R. E. Grim, *Applied Clay Mineralogy*, (1962), McGraw-Hill Book Co. Ltd., pp. 388–389.

GRIM provides us with an adequate history of pelletising powdered ores with bentonite. I need only give a short summary here.

Pelletising powdered ores with bentonite

Year	Pelletising	Bonding agents Bentonite + Additive	Authors	Reference
1940	Powdered fuel	~ 5% + quebracho extract	G. G. Rick C. E. Loetel	USP 2 217 994
1940	Ground fluorite	~ 1% + sulphite liquor	F. C. Abbot C. O. Anderson	USP 2 220 385
1946	Ore Slimes	~ 2% + corn flour or starch	T. A. Klinefelter	U.S. Bur. Mines Rept. Invest. 3846
1952	Taconite iron ore	~$\frac{1}{2}$% (– 1%) ± lime	G. Komarek W. J. Chapman	USP 2 582 386

loc. cit., pp 367–369.

Other patents mentioned by Grim describe improvements to the process of making pellets, typically 25 mm in diameter, of clay-bonded iron ore, first dried and then calcined.

A related field is the bonding of animal feed pellets. Bentonite has several functions: it bonds rather dry ingredients; it acts as a lubricant; and it may help animals to put on weight faster. Bentonite was introduced into this field (or byre!) between 1953 and 1960.

See *Ind. Min.* (March 1978), (126), 61.

Ore pelletising by 1978 accounted for perhaps 50% of the total demand for bentonite. A high proportion of Greek bentonite is sold to Canada for pelletising.

PESTICIDES

67

NICOTINE-fuller's earth dust and a sulphur-fuller's earth dust, prepared by melting the sulphur in and milling, were tested in an orchard in Worcestershire in 1939. The dusts were easy to apply and a good coating remained after a heavy downpour.

Thanks to E. N. G. Crane, Oakhampton, Stourport-on-Severn.

Grim has reviewed the role of clays in pesticides. They may be used either as carriers in a master-batch or as diluents.

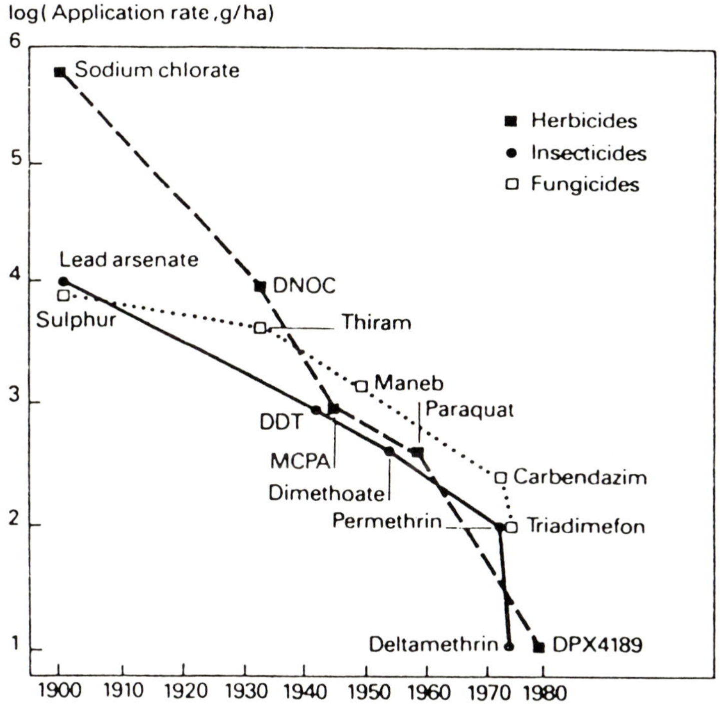

Fig. 88. (left). The progressive increase in activity of pesticidal chemicals during the 20th century.

Phil. Trans. Roy. Soc., Series B, No. 1076, 23 Sept. 1981. Fig. thanks to Sir Leslie Fowden FRS; Dr. I. J. Graham-Bryce. D. Phil., FRSC, and the Royal Society.

As the activity of the toxicant increases so does the proportion of the diluent. There have been hundreds of formulations of pesticides containing montmorillonite, kaolinite or attapulgite and sepiolite. Clays give uniform dispersion of the toxicant, help the plant retain the pesticide in wind and rain, and often preserve the toxicity of the product. The bentonites stick well to leaves, but detoxify certain chemicals such as paraquat.

R. E. Grim, *Applied Clay Mineralogy*, McGraw-Hill Book Co Inc (1962) should be consulted.

In 1951 The Fullers' Earth Union introduced agricultural granules for the accurate emplacement of toxicants in pesticides.

68

CATALYSIS

J. H. de Boer, *Chem. & Ind.* (25 July 1959), 934–941; see also John H. Sinfelt, *id.*, (4 June 1984), 403–406.

SOME events in the history of catalysis and the use of clay, especially montmorillonite products, are given in the following table (No. 17)

Table 17 History of catalysis

1831	P. Phillips, patent. Oxidation of SO_2 on Pt surface.
1835	J. J. Berzelius suggested the word 'catalysis'.
<1875	Manufacture of oleum, not important till late '90s.
1897	Paul Sabatier, H_2 reacts with ethylene on Pt or Ni.
1901	K. P. W. T. Normann hydrogenation of glyceride oils.
1909	Crossfields of Warrington hardening 100t/week using Normann's powdered Ni process.
1915	L. Gurvich, attapulgite, also natural and activated montmorillonite, polymerise pinene.
1916	L. Gurvich, theory of heterogeneous catalysis.
1923	K. Kobayashi *et al.*, many catalytic reactions (ch. 39).
1928	E. Houdry, silica-alumina to make anti-knock petrol.
1929	Egloff, V. Ipatieff, and later H. Tropsch joined UOP to develop catalytic processes.
1931	Catalytic cracking began. 1937 firmly established.
1939	Fluid-flow method of catalysis.
1942	First enquiries to FEU for catalysts.
<1948	FILTROL and RETROL used for catalysis in USA.
1950	G. A. Mills, J. Holmes & E. G. Cornelius, a substantial proportion of the montmorillonite structure must be retained in acid-activated catalysts. *J. Phys. & Coll. Chem. 54*, 1170–1185.
1950s	FEU selling solid montmorillonite catalysts, as advertised *Chem. Age*, (21 July 1962).

L. Gurvich, *J. Russ Chem. Soc.* (1915), *47*, 827; *id* (1916), *48*, 837.

C. Fuchs, Bleaching earths as catalysts, *Chem. Trade J.*, (1938), 295.

R. H. S. Robertson, Clay minerals as catalysts, *Clay Min. Bull.* (1948), *1*, 47–54.

R. E. Grim, *Applied Clay Mineralogy*, (1962), 315–323.

Catalysts for different duties are made by varying the amount and strength of the acid, temperature and time of the reaction and other parameters.

In 1964 came J. W. Jurg and E. Eisma's practical demonstration that montmorillonite converts fatty oils to hydrocarbons; this is one possible method for the natural formation of petroleum oil. Catalysis not only formed oil, it seems, but it allows one to transform oil into desirable products.

GEOPHAGY AND FOLK MEDICINE

69

Geophagy

69i

> *It is a very curious fact that clay was largely eaten at one time in Europe, as it is still eaten by some races in other parts of the world.*
> **SIR WILLIAM BRAGG, Old Trades and New Knowledge, (1926)**

London: G. Bell & Sons, Ltd., (1926), p. 190.

THE practice of eating earth or clay is or was so widespread throughout the world that it is surprising how seldom the scattered records have been brought together. In an earlier article (1947) I arranged the material roughly in order of increasing discrimination in the choice and preparation of earth for ingestion and of increasing medical knowledge by those who practise geophagy.

R. H. S. Robertson, Geophagy, or Earth Eating, *Discovery* (1947), *8*, 213–215.

In the condition known as pica, the chances of fuller's earth being eaten accidentally are remote; but in communities where "dirt eating" was prevalent, people would walk miles for a spoonful of earth from a favourite bank of clay. In Mexico addicts became pot bellied but otherwise thin and weak. No one objected to admitting he was a drunkard, but dirt-eating was unmentionably shameful. In 1963 a group of poor, rural Iranian children, given to earth eating, became the first examples of overt human zinc deficiency known to medicine: they suffered terribly stunted growth.

Latin for *magpie*.

Prevention "You're sunk without zinc," (Aug. 1982), 52–55 (54).

Yet carefully chosen and prepared comestible clays have been on sale in bazaars for centuries. Dr. Leonhart Ranwolf (mid 16th century) said that in Tripoli an ash-coloured earth called *malun* was used for washing the hair and another earth called *jusabar* was eaten by women.

Very similar practices were recorded in India by Sir George Watt (1908) – a pale yellow mud was eaten medicinally, and sold under the name of *multani mitti*, earth from Multan.

The Commercial Products of India, (1908), 329.

According to Lawson and Moon (1938), primitive descendants of the Incas on the Capachica Peninsula near Puno on Lake Titicaca, Peru, dip their potatoes into an aqueous suspension of a clay, called *chacco*, which may have been kaolin. It is not known whether the clay was used as a protective remedy against bacterial infection of the intestinal tract or whether it was used to adsorb solanine from green-

Alexander Lawson and H. P. Moon, A clay adjunct to a potato dietary, *Nature, London*, (1938) *141*, 40. The Quetchua Indians said the clay prevented "souring of the stomach".

skinned varieties. G. ap Griffith wrote a few weeks after that a clay is also eaten by the Baganda, whose diet is dominated by plantains and sweet potatoes, but here, as frequently in India, this use of clay is confined to child-bearing women and is eaten after work almost as a sweetmeat, the clay having first been smoked in the fire.

British Borneo Geological Survey Memoir, *6*, 28–31.

In North Borneo montmorillonitic clay is eaten by almost everyone, male and female, and held to be a remedy for stomach complaints, an excellent tonic for nursing mothers or even as a sustaining "food" for long journeys. It is shaped into small cubes and baked. The best quality is white and smooth in texture and devoid of grit. The texture is described as being like caramel. Other types are yellow or even reddish, but these are inferior and cheaper.

(1) (1790), *2*, 213 (2) *Natuurkundig Tijdschrift voor Nederlandsch Indie,* Batavia (1857), *13*, 83 (3) Brussels, (1892), 163 (4) *Mem. Asiatic Soc Bengal*, (1906), *1*, 259.

Dr. Nathaniel Mutch (1936) gave examples of comestible earths in the past: (1) J. Labillardière, Voyage in search of La Pérouse (1790), (2) J. J. Altheer in the Netherlands East Indies (1857), (3) A. A. Fauvel, La Province Chinoise du Chantoung, (1892) and (4) D. Hooper and H. H. Mann, Bengal, (1906).

Vol 5.

In the *Neue Brockhaus Enzyclopädie* of 1968, support is given to the idea that valuable mineral substances [iron for pregnant women] and trace elements are probably yielded by these edible earths, but Professor John Yudkin, in his review of "Consuming Passions": *The Anthropology of Eating* by Peter Farb and George Armelagos, wrote *The plausible suggestion that pica usually has a nutritional significance – for example, that eating clay provides minerals lacking in the diet – has been repeatedly disproven.* On the other hand the eating of earth is often not a mindless form of *pica.*

Nature, (22/1/81), *289*, 334; Review of Farb and Armelagos (1980); Houghton Mifflin.

In the First World War British medical officers in the Near East noticed that the French troops suffered from dysentery far less than the British troops. It became known that a French medical officer had ordered fuller's earth to be mixed with some food and given to the men. Up to the early 1940s, when large quantities of fuller's earth were regularly exported to America, sailors at Angerstein's Wharf in London used to ask for fuller's earth which had spilled from the sacks, so that they could use it medicinally against dysentery and stomach troubles on their voyages at sea.

Oral tradition ~ 1934.

In nature animals often get their salts from efflorescences on clay pans. The okapi at the London Zoo had a special salt clay-lick made for it with Surrey fuller's earth; but some clay-licks resorted to by elephant and other species in India are not notably salt, but mainly composed of clay. However mahouts do not allow their charges to eat the clay as they believe it is not good for them. I should say that the medical aspects of this habit are far from being fully understood.

Brockhaus gives further references to geophagy, including one book:
(1) E. Mayerhofen & C. Pirguet, a general account.
(2) B. Laufer, *Geophagy* (1930), a book describing the practice in Senegal, Iran, Peru, Australia, New Mecklenburg, New Caledonia and Papua.

(1) *Lexikon der Ernährungskunde*, in 5 volumes, (1923–1925)
(2) Anthropological Series, Chicago, (1930)
(3) *Handbuch der Bodenlehre*, (1932), *10*,
(4) Uppsala, (1958).

(3) H. Plischke (1932), a general account.
(4) B. Anell & S. Lagercrantz, geophagical customs in Africa, with bibliography.

69i

An analysis of thirty-three of the various earths still sold in the bazaars of India and eaten mainly to relieve acidity of the stomach or women's ailments, and to improve the complexion showed most of them to be montmorillonitic albeit often siliceous.

We have seen how Professor J. Plesch tried, from 1928, to find a way of making an acceptable laxative containing bentonite, but could not compete with NORMACOL. More recently interest has shifted to the use of a bentonite product in combating diarrhoea in both adults and in babies. The work of the Institut Henry Beaufour is outstanding in this connection.

Chapter 37.

Institut Henri Beaufour, 17 Ave. Descartes, 92350, Le Plessis Robinson, France.

Before we leave the subject of geophagy, we should not forget to mention, with honour, the greatest geophagist in the world – the worm.

Folk Medicine

69ii

Se vautrer dans la fange,
to wallow in mud.

LA FANGE, the French word for mud, dates from the 12th century and came from Corsican *fangu* or fanga, derived from the Gothic word *fani*. In 1952 the therapeutic application of hot mud was called *fangothérapie* and the mud itself *le Fango*.

"L'argile, Terre curative"

Paris: Association Amour et Vie, 36 rue de Lancy, 10e (1956), 11th ed.

Si nous considerons le soleil comme le Père de la Vie manifestée, nous pouvons en admettre que la Terre en est la Mère.

So begins a remarkable little book (Fig. 89) extolling the virtues of clay "pour guérir toutes les maladies," and recording thirty years of happy trials and success with curative earth. Many suggestions are given for both internal and external application to the members of the Association AMOUR et VIE. Some of the clay bought by this society came from the deposits of calcium montmorillonite at Apt and Mormoiron.

Interest in France does not seem to be waning, judging by the publication of two books *La Guérison par l'Argile* – healing with clay – and *L'Argile pour votre santé* – clay for your health. The former describes the use of unspecified types of clay in a very wide

Gérand Coutaret, Paris: La Diffusion Scientifique, n.d.

Fig. 89. Jesus presenting samples of curative earth to the sick.

Association Amour et Vie

range of maladies; the latter lists the different sorts of clay used in medicine, rather oddly since bentonite is a smectite: smectites, refractory clays, bentonites and kaolins. Fuller's earth is said to be "less pure than kaolin". Clay is sold as pills, as dried lumps or as powder. For external application; a mud is made into a poultice (*cataplasme*) and covered with a bandage. For internal use a weaker suspension is allowed to settle overnight; only the supernatant 'milk' is swallowed.

Dr. André Passebecq, St. Jean de Braye; Editions 5 Dangles (1978).

Argiletz S.A. supply a wide range of clays and argilliferous medicines and cosmetics, as well as books by Amour et Vie. Their powdered green clay, available in some health food shops in England, consists mostly of a mica clay mineral, probably glauconitic illite, quartz and small amounts of gypsum, calcite and kaolinite. One could liken this clay to some of the terrae sigillatae of the past. Liz Taylor, a journalist, found that Green Clay, taken internally, improved her skin and made her feel better.

203 Allée de Montfermeil, 93390 Clichy-sous-Bois.

Distributors: Larkhall Laboratories, 225 Putney Bridge Rd. London SW15 2 PY.

XRD Thanks to C. V. Jeans.

Scotsman, 23.8.1984.

On the other hand, S.A.R.L. Argiles de Haute Provence, Hyères, supply for medical purposes a highly smectitic clay, containing about 10 per cent of illite and about 2 per cent of non-clay minerals. Its therapeutic value is ascribed to its high cation exchange capacity; four grams of it, for example, adsorb one gram of strychnine.

Monsieur J.P. Guénot, head of this firm. (B.P. 598, Hyères, 83411, France) has published a book *Comprendre l'argile: terre naturelle et mystérieuse.*

Local muds of varied composition are used therapeutically at Saint-Amand (Nord), Balaruc-les-Bains (Hérault), Barbotan-les-Thermes (Gers) and elsewhere; but "bollboue" is imported from Germany. For folk medicine grit-free clays should be supplied, preferably to the standards of the country's Pharmacopoeia.

Bolus

In Germany a montmorillonitic clay is sold in a more conventional way as Adolf Just's LUVOS Heilerde, "damit Sie auf dem Posten bleiben". Internally it is taken for absorbing poisons, controlling the acidity of the stomach, stomach-ache, fermenting and putrifying conditions as well as diarrhoea. It is recommended in mouth care by rinsing or gargling. Externally it is used, as in French practice, for treating discharges, sprains and strains, inflammatory rheumatic and neuritic diseases, burns and insect bites. In these and many other treatments it is used either as a paste for instance in a face mask, or as a powder, especially for babies.

Heilerde-Gesellschaft Luvos Just KG, Bad Homburg V. D. Höhe.
(To be on one's toes, in good condition, or even "full of beans.")

Grim's *Applied Clay Mineralogy* (1962) may also be consulted for the use of clays in medicines, pharmaceuticals and cosmetics.

R. E. Grim, McGraw-Hill, (1962), 372–373.

Fig. 90. "Luvos" curative earth to keep you in good condition. A German advertisement.

APPENDIXES

THE APOSTROPHE

70i

UP till the end of the 18th century Fullers-earth with no apostrophe was the common usage. By 1825 when the apostrophe had been commonly adopted, its position was inconsistent; and so it has remained, though fuller's earth is much commoner than fullers' earth. Both forms are right, the New English Dictionary giving the former and the Century Dictionary the latter. In 1935 all three forms were found in telephone directories, The Fullers' Earth Union Ltd., Baynards Fuller's Earth and the Leeds Fullers Earth Co Ltd.

Five Thousand Receipts, Colin Mackenzie (1825), 5th ed.

Baynards and Leeds were not smectitic.

The form fuller's earth is used in this book not only because it is the commoner form but because it is the more logical. There are many phrases which describe *a material used by a craftsman*; such as baker's yeast, brewer's yeast, brewer's gypsum, carpenter's glue, cobbler's wax, glazier's putty, goldbeater's skin, potter's clay, potter's flint, printer's ink, and tinman's solder.

Admittedly one can find these phrases in the form brewers' yeast fairly frequently, but it is not necessary to invoke more than one craftsman to give a clear indication of what is meant.

These phrases should not be confused with names of *products* including the waste products of craftsmen. Here the picture is that of a number of craftsmen producing a material such as brewers' spent grains, distillers' slop, plumbers' tinny sweepings, refiners' blue verditer, soapers' leys, salts or waste, or soap boilers' ashes.

Verditer = blue carbonate of copper.

However, in *animal products* the animal's name is singular in such phrases as bee's wax, cow's, ewe's and goat's milk, hen's egg protein, Egyptian cow's butter, hog's lard, neat's foot oil and a spider's web; but when the product of these animals is plural, the animal may also be given in the plural genitive, as in ducks' and hens' eggs and spiders' webs. Here the usage is similar to one ship's bottom and several ships' bottoms – what might be called the uncomplicated genitive.

Where the emphasis is on *consumption*, the plural genitive is correct in, say, infants' milk; where on *production*, the singular form is clearer as in mother's milk. The plural form is also logical in druggists' sundries, apothecaries' wares or cordial makers' drugs; or indeed in welders' disease or housemaids' knee, which are conditions affecting these people.

70i

English Grammar, 1876 ed., p. 29.

cf. The Australian Poem. "A sunburned bloody stockman stood, and in a dismal bloody mood, *apostrophised* his bloody cuddy, this bloody moke's no bloody good, bloody, bloody, bloody."

NOTE. I should perhaps add that the apostrophe was originally placed where a letter or letters had been elided, as in o'er, yestre'en, or, in Scots, fa'. Charles P. Mason pointed out in 1876 that is is "meaningless" to put the apostrophe after possessive plurals because no vowel has been dropped there. Nevertheless the desire to indicate genitives in this way came gradually to be felt. What we have tried to do here is to bring order to present chaos. On the other hand we shall not start a campaign to pronounce the word apostrophe with only three syllables, as in French, even though the commonly accepted pronunciation has arisen by ignorant confusion with the four syllabled word which means a dramatic turning aside to address someone.

70ii

On the Naming of the Essential Clay Mineral of Fuller's Earth

IN Classical Greece and Rome, earths were classified usually according to their uses or place of origin, such as terra fullonia or kimolian earth, but they were also given a trade name, such as ASTER – the star-grade of kaolin or a descriptive name – SAXUM – or lump fuller's earth.

De Natura Lutorum, R. C. Mackenzie, *Clays & Clay Minerals*, (1963) **11**, 11–28.

The earths of Theophrastus, R. H. S. Robertson, *Classical Rev.*, **72**, 222–223.

As Mackenzie has pointed out (1963) such nomenclature is of relatively little use now since it is doubtful whether most of the *earths* were mono-mineralic and one name could be used in two contexts, e.g. Melian earth could be either a kaolin or a fairly pure, very fine silica, since deposits of both these types are found in Melos (Robertson, 1958). Agricola (1546) accepted the common practice of giving locality names to earths, but believed that earths from one locality may be worthless, while similar ones from another may be valuable. That varietal differences were often due to impurities could not be proved before the 20th century.

The Irish scientist, Richard Kirwan (1733–1812), in writing his *Elements of Mineralogy*, the first work on that subject in English, (1794), wrote *As earths originally received their denominations from external and not from internal characters, they cannot now, without creating much confusion, be arranged according to their real specific properties.* Not only did he record such properties as fracture, friability, hardness, unctuousness, adhesion to the tongue, colour, opacity and so on, but he introduced the measurement of specific gravity, which, with the chemical analysis, we now know are related by a formula defining the structural cell contents.

Edward Phillips *The new world of English words: or a general dictionary*; ed. 6 by J. Kersey 1706.

The first 'scientific' name given to fuller's earth was *smectis*: thus it appears in Phillips' dictionary in its 1706 edition. John Hyacinthe de Magellan used the same word in his translation of Axel Cronstedt's *Mineralogi* (Stockholm, 1758), as well as "Walkerde" and fuller's earth. In the Royal Society's *Philosophical Transactions*, (1783), *73*, 227, one reads *it feels like hard soap or rather like that kind of stone which the mineralogists call* smectis; and Richard Sullivan (1794) asked "How is a clay to be distinguished from the smectis or soap rock?"

Richard Sullivan, A view of nature, (1794), **2**.

Hippocrates had used the word σμηκτίς [γῆ] for fuller's earth, but later writers preferred the form σμηκτρίς to mean a kind of fuller's earth for cleaning cloth. The etymology of these words has been discussed in chapter 7i.

cf. χνόον ἐκ κεφαλῆς ἔσμηχεν

Od. *6*, 226 to wash my skin (salt free) from the head down.

Kirwan (1794) following Cronstedt had used the terms *smectis* and *Walkerde* in addition to fuller's earth, but the Abbé R. J. Haüy (1801) used the word smectite as an adjective in the phrase 'smectite clays and fuller's earth.' Certainly by 1811 smectite had come into use as the modern-style mineral name, John Pinkerton specifically referring to the smectite of Cimolus though J. C. De Lamétherie (1812) did not improve matters by using smectite as an inclusive term covering mica, talc and steatite.

Traité de minéralogie, Paris, **4**, 444.

Petrology, a treatise on rocks, (1811), *1*, 278.

Leçons de Minéralogie, (1812), *1*, 395 & *2*, 186.

August Breithaupt (1841) had the correct impression that smectite was a hydrous aluminium silicate containing magnesium, calcium, and other constituents, but by 1868 the name was being applied to unrelated mineral species. Thus Henry Watts (1868) refers to *Smectite, a term . . . applied . . . more particularly to an argillaceous mineral from Cilly in Lower Styria*, . . . and another from Condé in France, the latter now known to contain halloysite. Unfortunately E. S. Dana (1892) applied the name to the mineral from Condé near Houdon, which he describes as greenish, and in certain states of humidity, transparent and gelatinous.

Vollständiges Handbuch der Mineralogie, Dresden & Leipzig, (1841) *2*, 344.

A dictionary of chemistry 1859–1868. For *cimolite* (1848), see p. 318.

A system of mineralogy, 6th ed., John Wiley & Sons, Inc., New York (1892), 688, 695.

In view of this confusion it was fortunate that a new name *montmorillonite* came in and was eventually universally adopted. Mackenzie has unravelled this tangled skein of nomenclature and may be quoted in full.

op. cit.

The original material from Montmorillon, France, was named "montmorillonniste" by Mauduyt (1847), who simply stated that it contained silica, alumina and water. This material was analysed in the same year by Damour and Salvétat (1847), who, on comparison with the analysis of a reputed "halloysite" from Confolens described by Berthier, renamed the mineral "halloysite". The name "montmorillonite", or rather the German form "Montmorillonit", appears to have been used first by Naumann in 1850, but despite this Salvétat in 1851 described as "halloysites" some other minerals which from their properties and analyses† appear to be essentially montmorillonites; he also names a mineral from Condé, which from its analysis appears to be halloysite, "smectite". This is the only instance the author has been able to find of the application of the name smectite to a mineral not of the montmorillonite group, for which, as fuller's earths, the term "smectis" has been used since early times. This error would undoubtedly have been avoided had Damour and Salvétat (1847) appreciated that their type material from Confolens – later called confolensite by Dufrénoy (1856) –was*

– Mauduyt. *Bull. Soc. géol. Fr.*, (1847), **4**, 168–170.

A. D. Damour & L. A. Salvétat, *Ann. Chim. (Phys.)* (1847), *21*, 376–383.

*C. F. Naumann *Elemente der Mineralogie*, (2nd ed.); Leipzig (1850), p. 307.

†L. A. Salvétat, *Ann. Chim. (Phys.) 31*, 102–116.

not identical with the original halloysite of Berthier (1826) from Angleur, but that Mauduyt's (1847) identification of the new mineral "montmorillonniste" was correct and that their attribution of names ought to be reversed. The fact that montmorillonite and confolensite were distinct from halloysite was not, however, appreciated for many years to come.

Indeed it was not until 1932 that P. F. Kerr showed that the smectite from Cilly, Lower Styria, (see chapter 51), and the fuller's earth from Woburn in England are apparently bentonites derived from volcanic ash, and are composed of the mineral that has generally been known as montmorillonite. He concluded that *In view of the large amount of modern literature on montmorillonite, it seems in the best interest of science to continue the use of montmorillonite and to drop that of smectite.*

70ii

A. Dufrénoy, *Traité de Minéralogie* (2nd ed.) (1856), Dalmont, Paris, **3**.

P. Berthier, *Ann. Chim. (Phys.)* (1826), **32**, 332–334.

Amer. Mineral., (1932), **17**, 192–198.

Also P. F. Kerr and P. K. Hamilton, Glossary of clay mineral names, *Amer. Petrm. Inst. Project 49, Preliminary Report* No. 1 (1949), p. 59.

70iii

How Smectite came to be used as a Clay Mineral Group Name

Men ever had, and ever will have, leave
To coin new words well suited to the age.
Use may revive the obsoletest words,

And banish those that now are most in vogue;
Use is the judge, and law, and rule of speech.

HORACE, De Arte Poetica, translated by the Earl of Roscommon, 1860, in The Complete Works of Horace, J.M. Dent & Sons (Everyman Edition) 1927, p. 134.

Thanks to R. C. Mackenzie for these quotations.

JOSEPH Henry Collins (1841–1916) was the first to show the family resemblance of the "minerals" chloropal, nontronite, gramenite, keffekilite, bole and montmorillonite, calling them the chloropal group. He thought that these highly hydrated ferric-aluminous silicates pass into one another in many instances by insensible gradations, but the extremes of the series (nontronite = chloropal for the ferric end and montmorillonite for the aluminous) are too far apart to accept their disposition as a single species. It is pleasant to recall that a Mr. Letcher, a student, expressed his gratitude for the convenient ordering so proposed.

Mineral. Mag., (1878), **1**, 67.
and
R. H. S. Robertson, History of Clay Research in Great Britain, *Clay Minerals Group:* In the press.

Collins' group name was forgotten, and when the true relationship of all the clay minerals was revealed by x-ray structure analysis, the need for group names was generally felt. Correns suggested montmorins to parallel kaolins, and MacEwan used the name montmorillonoids; but George Brown proposed *smectites* as the name for the group of minerals composed of 2:1 or triphormic layers, which, when the readily exchangeable cations are replaced by Na^+ and the material is saturated with glycerol, give a basal spacing of 18Å approximately. Such minerals generally have a cation exchange capacity of 70–120 m.e./100 g. for readily exchangeable cations.

70iii

Clay. Min. Bull., (1955), *2*, 296.

Robertson pointed out (1962) that whereas mineral *names* were subject to the international rules of nomenclatural good manners, group names were *words* which should also conform to semantics and communications theory. There have been very few studies of the acceptability of made-up words: the pioneer in this work was Sir Richard Paget (1935) who wrote that:

R. H. S. Robertson, Some notes on the acceptability of mineral group names, *Clay. Min. Bull.*, (1962), *5*, 41–43.

This English, Kegan Paul, London, (1935) p. 88.

Though a word may have come into use by some other means than instinctive gesture, its original selection and its survival value are largely determined (though unconsciously) by whether gestures of articulation which produce it are pantomimically appropriate or not.

Thus kandites is an excellent word gesturally in spite of its artificial invention. It means a flat object with some peripheral points or angles, as in hand *and* strand. *It is in shape something between a card and candy. (By the same reasoning* boycott *is cognate with* picket *but had the gentleman been called Smith he would not have thus been immortalised).*

Robertson, *loc. cit.*

Smectites *is gesturally and philologically related to* smear *which is the formation, away from one, of a thin film of undefined outline.*

Illites *are small, rather round-edged things. Preceding consonants give the word a third dimension, so a* prill *is a small sphere and a* frill *is a repetition of disc-shaped objects. The harder consonants give a more precise shape, as in* till *or* sill, *which is of course longer than* till. Fill *denotes change in volume.* Mill *depicts rotatory movement.*

The Greek word for "a chain or row of similar things hanging from one another" commonly in a straight line, such as a string of onions, is hormathos (ὁρμαθός *); in order to match the other disyllables I suggested* hormites *from hormos (* ὅρμος *), a chain of the usual catenary or looped sort. Nevertheless, the shorter word is easy to remember and to use and may be almost as acceptable as the gesturally more perfect* hormathite.

Unfortunately, although kandites, illites and smectites are now in common usage, hormites was rejected by the Commission on New Minerals and Mineral Names of the International Mineralogical Association. They offered no alternative.

Modern Research

70iv

DETERMINATION of the structure of phyllosilicates by Linus Pauling, of Japanse acid clay by Kameyama and Oka and of the kaolinite, montmorillonite and hydromica groups (kandites, smectites and illites) was made in the year 1930 which marked the transition from the classical to the modern period.

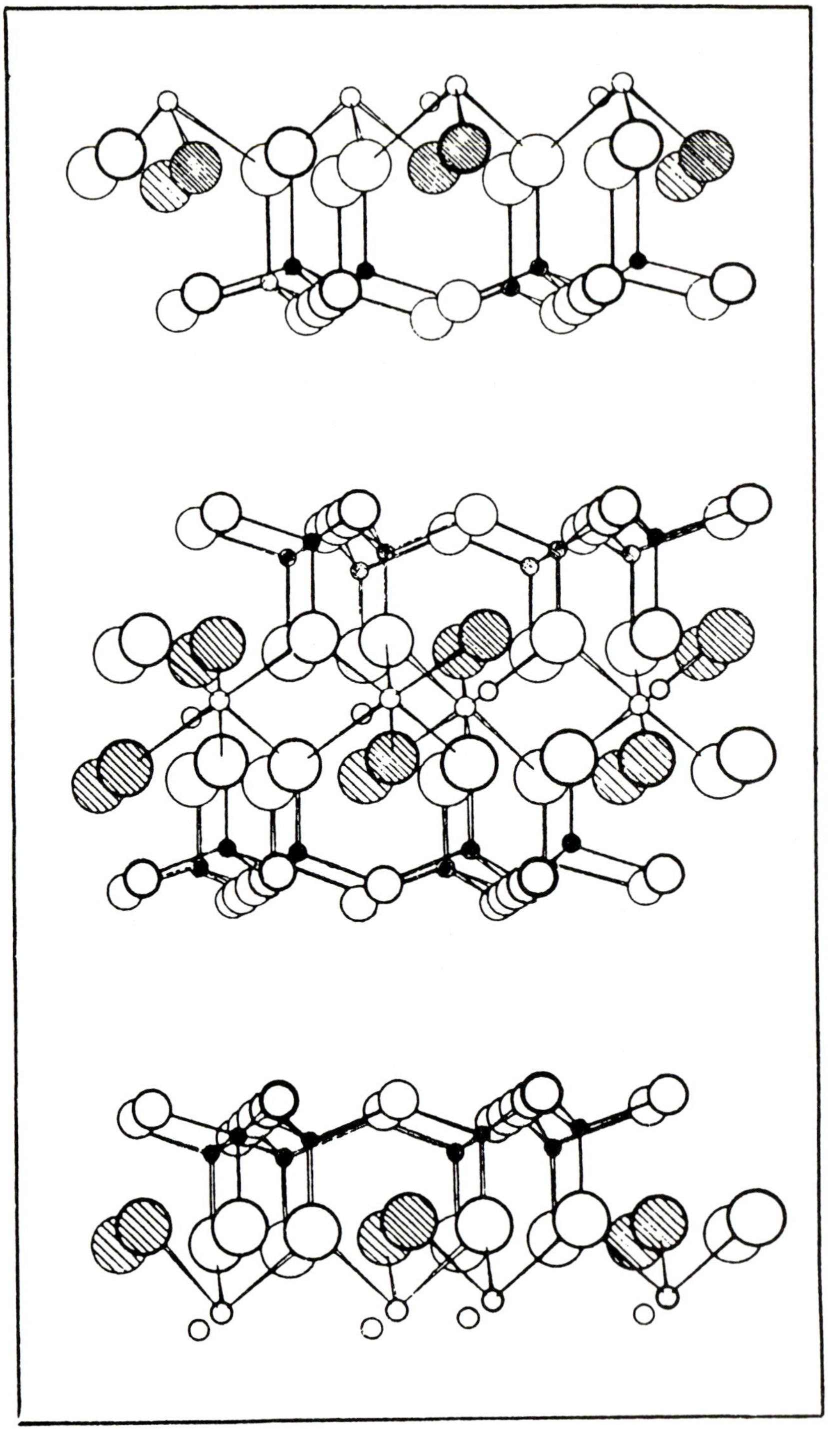

Fig. 91. The structure of montmorillonite. From MacEwan, Chap. 4 *X-Ray Identification and Crystal Structures of Clay Minerals* (Min. Soc.) 2nd. edition.

- ● silicon
- ○ aluminium
- ◯ oxygen
- ⊘ hydroxyl

The history of clay research in Great Britain was presented to the Centenary Celebrations of the Mineralogical Society of Great Britain and Ireland in 1976. **70iv**

However most people coming to the study of fuller's earth scientifically will begin by reading Grim's *Clay Mineralogy* and his *Applied Clay Mineralogy* (1953, 1962). I am indebted to Ralph for many pieces of information in this book.

R. E. Grim, McGraw-Hill Publishing Co. Ltd. New York, London and Toronto (1953 and 1962).

The "higher hydrates" of Calcium Montmorillonite

70v

WELL might one ask if there is any unifying factor which explains how calcium montmorillonite performs so many different tasks. The answer is water. To arrive at this conclusion a number of physical properties were measured at a range of moisture and water contents, which were all expressed in relation to the double unit cell (O_{20}) as H_2O contents or better as nominal water layers. The word nominal is used because no attempt is made here to find out how much of the water is neatly emplaced between the layers and how much is loosely held.

Molecular Weight

TO set the scene we shall relate the chemical analysis, ionic composition of the unit cell and the molecular weight contributions of the various oxides. In the table are shown figures obtained for the 'ideal' composition of this mineral (1); for Chart Lodge fuller's earth from Redhill (2) and for Baulking fuller's earth from Berkshire (3).

Atomic composition and molecular wts. of fuller's earth montmorillonites

	(1) Si_8	$Al_{3.33}$				$Mg_{0.66}$	$(OH)_4$.	$8H_2O$.	$\frac{Ca}{2}0.66$			
	$\equiv 8SiO_2$	1.66 Al_2O_3				0.66 MgO	$2H_2O$	$8H_2O$	0.33 CaO			
M.W.	480.70	169.44				28.86	36.03	144.13	18.69			
%	54.85	19.39				3.06	4.11	16.45	2.13			
	(2) $Si_{7.98}$	$Al_{2.61}$		$Fe^{3+}_{0.61}$	$Fe^{2+}_{0.10}$	$Mg_{0.7}$	$(OH)_4$.	$8H_2O$.	$\frac{Ca}{2}0.67$	$\frac{Mg}{2}0.15$		
	$\equiv SiO_2$	Al_2O_3		Fe_2O_3	FeO	MgO	H_2O	H_2O	CaO	MgO		
M.W.	479.49	133.06		48.71	7.18	28.21	36.03	144.13	18.78	3.02		
%	53.36	14.81		5.42	0.80	3.14	4.01	16.04	2.09	0.34		
	(3) $Si_{7.98}$	$Al_{2.52}$	$Ti_{0.11}$	$Fe^{3+}_{0.61}$	$Fe^{2+}_{0.18}$	$Mg_{0.6}$	$(OH)_4$.	$8H_2O$.	$\frac{Ca}{2}0.36$	$\frac{Mg}{2}0.24$	$Na_{0.04}$	$K_{0.02}$
	$\equiv SiO_2$	Al_2O_3	Ti_2O_3	Fe_2O_3	FeO	MgO	H_2O	H_2O	CaO	MgO	Na_2O	K_2O
M.W.	479.49	128.47	4.39	48.71	12.93	24.18	36.03	144.13	10.09	4.84	2.48	1.88
%	53.42	14.31	0.49	5.43	1.44	2.69	4.01	16.06	1.12	0.54	0.28	0.21

The molecule weights are (1) 876, (2) 899, (3) 898

70v

D. M. C. MacEwan, Montmorillonite Minerals, ch. IV, *The X-ray Identification and Crystal Structures of Clay Minerals*, Mineralogical Society: London, (1961), p. 167.

*Ti has been included with the Fe^{3+}

Moisture adsorption and specific surface area, P. S. Keeling, E. C. Kirby and R. H. S. Robertson, *J. Brit. Ceram. Soc.*, (1980), *79(2)*, 36–40.

(EGME). D. L. Carter, M. D. Heilman & F. L. Gonzalez, *Soil Sci.*, (1965), *100*, 356 & 409.

E. C. Ormerod & A. C. D. Newman, *Clay Min.*, (1983) *18*, 289–300.

The surface area of Calcium Montmorillonite

(a) THE cell dimensions of dioctahedral smectites can now be calculated by using MacEwan's formula $b(\text{Å}) = 8.92 + 0.06x + 0.09q + 0.18r + 0.27s$ where p, q, r, s are the numbers of Al^{3+}, Fe^{2+}, Mg^{2+} and Fe^{2+} per *half*-unit-cell-layer in octahedral coordination, and x the number of Al^{3+} ions in tetrahedral coordination. As this is an orthohexagonal cell the a axis (Å) is $\sqrt{3}b$, or 0.577 b. We thus obtain for the two fuller's earths the following dimensions.

	b	*a*
(2) Chart Lodge	9.030	5.210
(3) Baulking	9.076	5.211

The specific surface can now be calculated by the formula $(2\,N.a.b)/M$, where N is Avogadro's number 6.022×10^{23} and a and b are the cell dimensions in Ångström units. Since the molecular weight of the anhydrous material was 754 and 753, respectively, the specific surface is found to be 751 and 756 $m^2 g^{-1}$.

(b) The moisture adsorption of undried, impure Baulking fuller's earth, containing 12.7 per cent of impurities and equilibrated at a relative humidity of 75 (that is, over concentrated sodium chloride solution), was found to be 19.9 per cent. The specific surface estimated from this moisture adsorption is $N/(100 \times 18.016) \times 10.5 \times 10^{-20} \times 19.9$, where N is Avogadro's number and 10.5 Å^2 is the surface area occupied by one water molecule. The area arrived at in this way is 698 $m^2 g^{-1}$. If we deduct 7.5 $m^2 g^{-1}$ for the illite in this clay and 17.2 $m^2 g^{-1}$ for the zeolite, which sorbs a lot of water internally, and divide the result by 0.873, the montmorillonite content, we arrive at a figure of 771, which is higher than that given by MacEwan's method, presumably because the edge area was also partially measured.

(c) The specific surface obtained by the adsorption of ethylene glycol monoethyl ether was 660 $m^2 g^{-1}$ for the Baulking earth. Assuming again that the montmorillonite content is 87.3 per cent, the EGME surface area is 756 $m^2 g^{-1}$, which is the same as that calculated in **(a)**.

Ormerod and Newman (1983) in a more detailed study calculated 740 $m^2 g^{-1}$ + 60 $m^2 g^{-1}$ for the flat surfaces and edges, respectively, from crystallographic parameters, and found experimentally 799 $m^2 g^{-1}$ from moisture adsorption using a desorption curve.

Moisture content

MOISTURE content was traditionally expressed as the proportion of water driven off in an oven at 105° or 110°C; when it was suspected that some adsorbed water survived this treatment, higher temperatures were used. The water released was expressed as a proportion of the moist clay. It is really more informative if the moisture removed is expressed as a proportion of the dry clay, as in the timber industry. Even more information is given by relating the moisture or water content to the ionic composition. Barshad (1949) showed that at low relative humidity six and at high relative humidity eight molecules of water were associated with each (double) unit cell containing nearly eight silicon atoms. In this chapter all moisture or water contents will be calculated as nominal water layers – the 8 H_2O of the unit cell amounting to two layers.

I. Barshad, *Amer. Mineral.*, (1949), *34*, 675–684.

Relating physical properties to moisture content

THERE is no doubt that most of the important uses of calcium montmorillonite are determined by the water content. Grim reviewed the theoretical aspects of what appeared to be higher hydrates of both calcium and sodium montmorillonite; and, with Cuthbert (1945, 1946) he related the greenstrength and test piece density to the moisture content expressed as water layers. This concluding chapter is a continuation of their method of study, which is also reproduced in Grim's *Applied Clay Mineralogy* (1962).

R. E. Grim, *Clay Mineralogy*, McGraw-Hill Publishing Co. (1953), ch. 8.

R. E. Grim & F. Leicester Cuthbert, The Bonding Action of Clays, I, II, Engineering Experiment Station of the University of Ilinois, Bulletins 357, 362. Urbana, Illinois.

Decolorising glyceride oils

AS may be seen in Fig. 93 the decolorising of glyceride oils is best done when the natural earth has a moisture content of 4 to 6 per cent. Part of the colouring matter of these oils appears to be attracted to the clay by exchange or chemisorptive bonding. It is thought that some moisture is necessary to allow the exchange reaction to proceed. If the clay contains 80 milli-equivalents of Ca ions per 100 g and the Ca ions are surrounded by a shell of six H_2O in octahedral arrangements, the water of hydration of the ions would be 4.324 per cent; if the state of hydration is $Ca(H_2O)_8$ the amount would be 5.765 per cent. Expressed as H_2O in the double unit cell, these hydration states are equivalent to 2.2 to 3 H_2O.

The bleaching power of the earth is definitely impaired when the moisture reaches or exceeds 8 H_2O in the formula; yet it is strange that glyceride oils are sometimes better bleached when $\frac{1}{2}$ to 1 per cent of water is added to the hot oil.

70v

E. C. Kirby

Decolorising mineral oils

ENGLISH fuller's earth does not decolorise mineral oils, such as white spirit, spindle oil, lubricating oil and so on, as well as lightly calcined attapulgite or acid-activated montmorillonite. Adsorption of colouring matter from non-polar solvents is more a matter of a large active surface than of ionic reactions; therefore the optimal conditions of drying are those which give the highest available surface area *and opportunity for bonding to π electron systems*. Certainly acid-activated earths work best when they are dry, and their decolorising power falls off more steeply than with glyceride oils.

Dehydration studies

AN attempt was made to find the optimal moisture content for various purposes by air-drying some moist Surrey fuller's earth, milling it and proceeding with various tests at lower moisture contents.

Density

BETWEEN about 3 H_2O and 7 H_2O in the formula, the density measurements lie on a straight line pointing to 2.53 as the density of the anhydrous clay, the actual density of which was found to be 2.64. The density of clays at 8 H_2O or above point to 2.38 for the dry clay. There is no evidence here that the inter-layer water has a density higher than 1g/ml, but the figures may not be wholly reliable if the montmorillonite contained pores or spaces which were too small for the paraffin to penetrate.

Swelling index

SAMPLES of Surrey fuller's earth with decreasing moisture were converted substantially to Na-montmorillonite by the addition of sodium carbonate. The swelling index rose slowly to a peak at 8 H_2O in the formula. The slow rise is consistent with the presence of adsorbed water acting as a diluent; but if any of the 8 H_2O were removed from the unit cell, the swelling power decreased dramatically.

The octahydrate of the pure clay mineral contains about 16 per cent of moisture. If a sample contained, for example, 10 per cent of impurities, it is obvious that no single particle should be dried below 14.4% of moisture. In practice it is impossible to dry the clay down so that no particles are overdried while others are underdried, but some driers are more capable than others of drying fairly evenly. Consequently, to avoid overdrying, the moisture content of such a product should not be less than about 15 per cent.

Greenstrength of foundry sands

GRIM and Cuthbert (1945) were the first to show that sodium montmorillonite required 3 water layers at its maximum greenstrength, whereas the calcium clay required 4. One can infer from these figures that more water is held in a rigid condition by the calcium than by the sodium clay and therefore the former has the higher greenstrength.

op. cit.

Henderson (1942) measured the greenstrength of 95:5 per cent sand: clay mixtures starting with 2 to 3 per cent of water. Allowing the mixture to dry, he measured the compression strength at decreasing water contents. The greenstrength rose to a maximum – the "peak greenstrength" – and then fell sharply. His peak was at 3.6 layers.

In R. H. S. Robertson, *Tonindustrie Zeitung*, (1951), *75*, 2–6.

Franz Hofmann found a peak at 1.63% moisture content equivalent to 4.1 layers.

F. Hofmann, *Giesserei*, (1956), *16*, 857–864.

Henderson's curves show a slight inflection at 6 water layers.

Bulk density

GRIM & Cuthbert (1945) found a minimum bulk density in test pieces containing 5 molecules (water layers) in the calcium montmorillonite. At a lower moisture there was a maximum bulk density at about 4 water layers.

Grim's Figs 7 and 20.

Natural moisture content of fuller's earth in the ground

Nutfield, Kent and Bath. The moisture content of the natural fuller's earth 'untouched by human hand' is 34–36% on a calcite-free basis. This seems to be valid for all the Nutfield, Bath and Kent earths, mainly because the content of other minerals which do not adsorb substantial amounts of water is similar in all these occurrences, at 6 to 10 per cent. A sample containing 8 per cent of impurities had a moisture content of 35 per cent, so the moisture content on the basis of a pure clay was 38 per cent, associated with 62 per cent dry clay, which in Chart Lodge fuller's earth has a molecular weight of 898.61 – 144.13 = 754.48. The pit-moisture therefore has a 'molecular weight' of 754.48 × 38.04/61.96 = 398.89 or 25.7 H_2O in the formula = *6.43* 'water layers'.

Dr. B. S. Neumann, Laporte Industries Ltd.

(The decimal fractions have no meaning but are used so as not to increase error where figures are multiplied).

Baulking. The moisture content of the pit-moist clay was 35.49 ± 0.13 per cent. As the impurities of this clay amounted to

9.76% of the pit-moist clay, and these impurities contain about 0.37 per cent of moisture, we have 35.12 parts of moisture associated with 74.51 − 9.76 = 64.75 parts of dry clay. The molecular weight of the moisture is therefore 753.49 × 35.12/64.75 = 408.69, equivalent to 22.68 H_2O or 5.67 water layers.

Steetley Minerals.

Woburn. Three samples taken from the stockpile and one from the pit (No.3) were tested by B. B. Dixon for moisture and loss on ignition.

Figures in italics are the soundest, for the reason given below in the text.

Moisture and loss on ignition of undried fuller's earth

	1.	2.	3.	4.
Position in quarry	top	near top	below middle	near bottom
Moisture content	41.40	35.80	36.14	41.47
Dry solids (105°C)	58.60	64.20	63.86	58.53
LOI (1000°C)	7.06	5.34	6.04	6.70
H_2O in formula	29.59	23.35	23.70	29.67
Water layers	7.40	*5.84*	*5.93*	7.42

It would appear that the middle and upper parts of the seam are much like the fuller's earths of the other four occurrences, but the bottom, accessible to oxidising water flowing in the Woburn Sands, has now a proportion of ferrialuminous interlayer material, confirmed by infrared spectroscopy. This would increase the porosity of the clay.

Dr. V. C. Farmer, personal communication.

As it is very difficult to know the impurity content of a sample taken for moisture determination, I suspect that the moisture content of most fuller's earths in the pit is about *6* water layers.

Plasticity. In 1942 Henderson converted a British Cast Iron Research Association standard compression test apparatus to measure plasticity by the simple expedient of putting the dynamometer out of action by tying up the spring assembly with wire. Test piece cylinders of sand, clay and water, of standard dimensions, rammed to a bulk density of 1.5, were placed between the plates of the apparatus and squeezed at a standard speed by turning the handle. The end-point was taken as the percentage reduction in distance between the plates when a lump of sand fell off the test piece. Scott Blair assured us that what was measured conformed to the definition of plasticity. Fig. 92.

Made by Ridsdale & Co. Ltd. 5 Wilson St., Middlesborough.

G. W. Scott Blair, author of *An Introduction to industrial rheology* (1938). and many other works on rheology.

Two silica sands were used, one more round-grained (angularity 1.08) than the other (angularity 1.11). The Figure (92) shows that the round sand mix had no plasticity below 4 layers of water, when the mix had maximum greenstrength. The plasticity is very rapidly developed as more water is added until 4.8 layers of water are present with the more angular or 6.2 layers with the rounder sand. Above these water contents plasticity is developed much more slowly than at first. The rounder sand is too plastic to handle at 18 water layers but the angular sand can be handled up to 20 water layers.

PLASTICITY NUMBER (Henderson) against Layers of Water

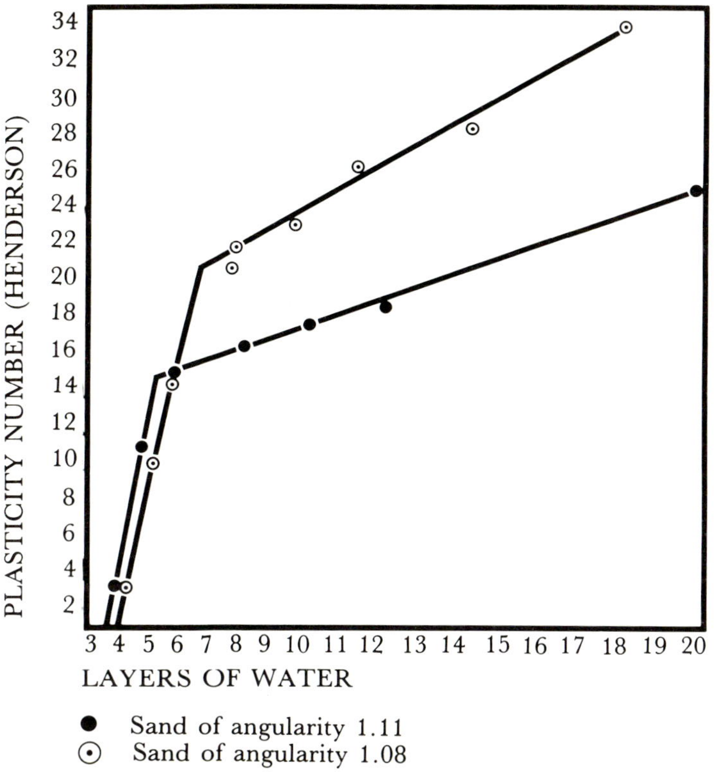

Fig. 92. (left). Plasticity number (Henderson) against nominal layers of water.

The FEU

One may safely infer that water is rigid or in an ordered state up to 4 water layers and perhaps up to nearly six. Angularity greatly reduces plasticity at high water contents but not at all at a water content which is suitable for greensand foundry work.

More familiar is the well-known Atterburg Plastic and Liquid Limits. The Plastic Limit is the lowest moisture content expressed as a percentage of the oven dried clay at which the clay can be rolled into threads ⅛ inch (3mm) in diameter without breaking into pieces. W. A. White's measurements may be taken from Grim's *Applied Clay Mineralogy*. The number of water layers is roughly obtained by multiplying the Plastic Limit by 0.0956. Thus Belle Fourche bentonite Ca^{2+}-saturated has 6.03 water layers.

A. Atterberg, Die Plastizität der Tone, *Intern. Mitt. Bodenkunde*, I, (1911), 4–37 and II, (1912).

W. A. White, *Amer. Mineral.* (1949), *34*, 508–512; Ph.D. Thesis, Univ. of Illinois (1955); and with E. Pichler, *Illinois State Geol. Surv. Circ.* 266 (1959).

Liquid Limit

THE Liquid Limit of the same clay was measured, and is the moisture content, expressed in the same way, at which the clay will just begin to flow when jarred in a standard way. The Ca^{2+}-bentonite had a water content equivalent to 16.93 'water layers', but at this water content most of the water is held between the clay crystals. (Sodium bentonite

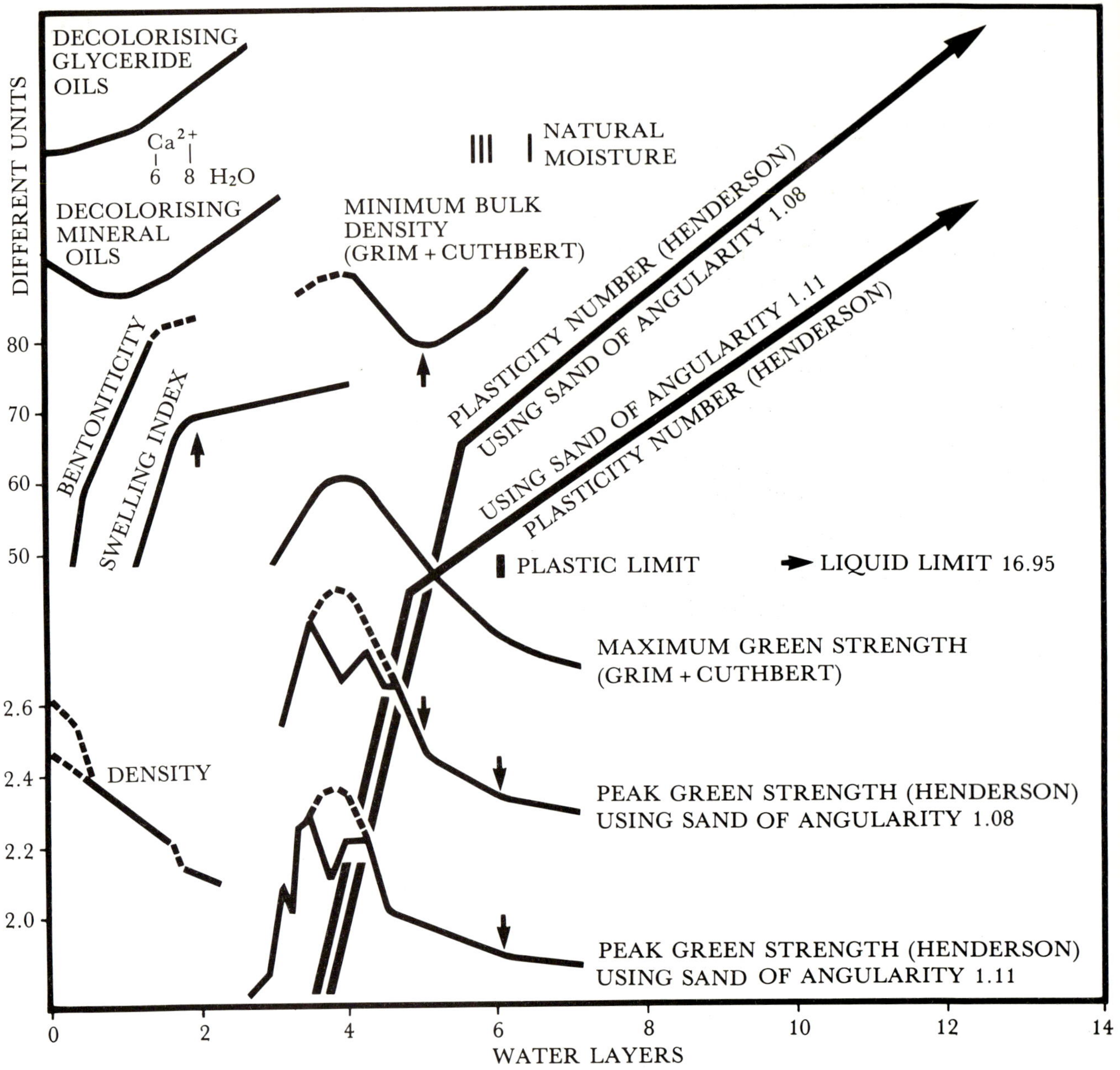

Fig. 93. Physical measurements on calcium montmorillonite against nominal water layers.

had its Plastic Limit at 9.27 'nominal water layers' and its Liquid Limit at 66.92!)

Conspectus and Conclusion

FIG. 93 shows diagrammatically how the physical measurements of industrial importance are related to the number of water layers present. The most striking coincidence perhaps is that natural moisture content and plastic limit both have close on six water layers or 24 H_2O in the formula. Up to that point at least the water molecules appear to have a certain degree of order amongst themselves. In any case all the uses of fuller's earth fall into a pattern and water is the link between them.

Fig. 94A *Dipsacus fullonum* (see Fig. 17, margin, p. 50). Cropping of the plant in South Somerset. The laden "pole" of about 25 teasel bunches ready for placing against the St. Andrew's Cross for drying.

Photograph by John L. Jones, Clevedon, Somerset

The Teasel in Modern Times

The teasel plant (see marginal note, page 50), as used in ancient times for "carding" is still being cultivated in England, as these photographs will show.

Fig. 94B. (left). The teasel crop, see Fig. 17, p. 50 and Fig. opposite.
Leaning against the racks to dry, the teasel "poles" hold an average of 25 bunches. In half light the rough St. Andrew's crosses against which they lean look like weird Arizonian cacti.
Photograph by John L. Jones, Clevedon, Somerset.

ACKNOWLEDGMENTS

to scholars

i

Naturally, in fifty years, I have been helped by innumerable people; and some were especially helpful:-

Sir Robert Annan	old mining books
Sir William (later Lord) Beveridge	access to library of the Master of University College, Oxford
C. E. N. Bromhead	Surrey, Assyria
Dr. John Broome	Midford
R. I. F. Dames	genealogy of George Dames' family
R. W. Darby	Darenth Roman Villa
V. A. and Joan Eyles	old mineralogy books, soils
Prof. D. E. Eichholz	ancient Greece
R. Goodburn	Chedworth Roman villa
J. J. Greenwood	Surrey
E. P. Harrison	the Elgin prices
S. G. Henderson	visits to recorded deposits
Dr. F. Hofmann	foundry test results, Switzerland
G. K. Jones	acid activation
O. F. Keevil	Bath and Arkansas
Prof. J. Konta	Czechoslovakia
Marian McKellar	for writing the first draft
Dr. R. C. Mackenzie	contributions over many years
Alice MacLeod (Mrs Cruickshank)	Gaelic
Dr. Jacques Martin	France, Nice
Prof. Dr.-Ing. R. Meldau	geophagy and folk medicine
Dr. B. S. Neumann	England and Hungary
Dr. Giovanni Novelli	Italy
Alec S. Palmer	Midford
Major R. Duncan Raikes	Aristophanes, Plato, Plautus, Plutarch, Theophrastus, Xenophon; Raikes pedigree
Lt. Col. W. T. Raikes	Dictionaries, 1797 Encyclopædia Britannica; Surrey
Sir Hugh Rankin, Bt.	fulling, sheep breeds, wool, etc
H. Leslie Salter	17th and 18th centuries; Algeria
Dr. Hugh S. Stannus	West of England
Dr. Hugh Torrens	Bath and Rhaetic clay
John Wills	collaboration; rough draft

My thanks are given to all of these, many of whom, alas, are no longer with us; and to hundreds of correspondents, museum curators and librarians. Of especial value were the libraries of the Athenaeum, the Geological Society and the Royal Institution of Great Britain, all in London,

ii for the illustrations

Association "Amour et Vie" 89
British Geological Survey 72, 77a, b
British Museum 3, 36
British Museum (Natural History) 10, 11, 44
Burns & Oates Ltd 19
Cementation Piling and Foundation Ltd 87
CIBA-GEIGY Ltd 1, 2, 21a, b, 22, 36, 37, 40, 57
Croydon Public Libraries 48
Davidson, Caroline 66, 69
Deribéré and Esme 80
Dover Publications 32, 33
Editions Albert René, Paris 51
Fowden, Sir Leslie, Dr. I. J. Graham-Bryce and The Royal Society 88
Geologists' Association, The 64
Goodburn, Roger 26
Heilerde-Gesellschaft LUVOS Just K. G. 90
Hofmann, Frank 81
Lambert, Uvedale 29
Laporte Industries Ltd 42b, 76, 78, 85a, b, 91, 92
Lucas, A. T. 67, 68
Matthews-Harlene Ltd 52
Naples Museum, Superintendent of 12
Oliver, Derek J. 59
Oxford: Clarendon Press 17, 24, 25, 37–40
Palmer, A. S. 61
Pelham, R. A. 34, 41a, b
Penny, Dorothy E. 78
Pevsner, N. (Penguin Books) 43
Rollins, M. B. 86
Science Museum 42b
Scott, E. Kilburn 20, 47, 49, 50
Seltman, Dr. Charles (Pan Books) 8a, b
Sudo, T. and S. Shimoda 79
Warner-Lambert (UK) Ltd 9

BIBLIOGRAPHY

ASHLEY, (Mrs) W. J. Early History of the Wool Industry of England, *Amer. Econom. Assocn*, (1887).

——— The English Manor, (1891).

——— An Introduction to English Economic History and Theory, Pt. I, The Middle Ages, 3rd ed. (1894).

BECKMANN, Johannes. Beiträge zur Geschichte der Erfindungen, Vol. 4, Leipzig (1799).

———, (John). A History of Inventions and Discoveries, translated by Wm. Johnston, 3rd ed., Vol. 3, pp. 224–253. London: Longmans etc. (1817).

——— A concise History of Ancient Institutions, Inventions and Discoveries. London: G. & W. B. Whittaker, (1823), Vol. 2, (Woollen Manufacture).

——— Bohn's 4th edition, London: Bohn, (1846), Vol. 2, pp. 92–106 has fewer notes.

CARUS-WILSON, E. The Woollen Industry. The Cambridge Economic History, (1952), Vol. 2, and other works referred to in the text.

DAVIDSON, Caroline. A Woman's Work is Never Done: A History of Housework in the British Isles 1650–1950, London: Chatto & Windus (1982).

DERIBÉRÉ, M. & A. Esme, La Bentonite, Paris: Dunod (1951).

GREENWOOD, J. J. Fullers earth in Surrey 1500–1900: an economic history, Reigate: Deering Press (50 Reigate Road), Surrey, (1983).

KAUSCH, O. Das Kieselsäuregel und die Bleicherden. Berlin: Springer (1927): Ergänzungsband (1935).

KRCZIL, F. Untersuchung und Verwertung technischer Adsorptionsstoffe, Leipzig: Akademische Verlagsges. (1931).

MACKENZIE, Colin. Five Thousand Receipts in all the useful and domestic arts, 5th ed. London: G. B. Whittaker, (1825), 788.

PELHAM, R. A. Fulling Mills. Publication No. 5 of the Society for the Protection of Ancient Buildings.

PONTING, Kenneth G. The Woollen Industry of South-West England, Bath: Adams & Dart, (1971).

SAVELLI, F. La Bentonite, Rome (1935).

SCHÖTTGEN, Antiquitates Triturae et Fulloniae, Traj. ad Rhen. (1727).

SCOTT, E. Kilburn. Early Cloth Fulling and its Machinery, *Trans. Newcomen Society*. (1931–2), *21*, 31–52.

——— Fulling and Milling Machinery, *J. Textile Inst.*, (1933), *24*, 247–263.

SINGER, Charles, E. J. Holmyard & A. R. Hall, Editors. A History of Technology. Oxford: Clarendon Press. Vol. 1 (1954): with T. L. Williams, Vol., 2 (1956); and Vol. 3 (1957).

SMITH, William, William Wayte & G. E. Marindin, Editors. A Dictionary of Greek and Roman Antiquities. London: John Murray (1890) (Fullo).

Cambridge Economic History.

CIBA Review 1937–1971.

Encyclopaedia Britannica, different editions.

Oxford English Dictionary.

Victoria County Histories.

INDEX

The index has been arranged according to people, places and subjects. Earths are put under places, companies under subjects, and gods under people; n means note; chapters are **bold**.

People

Places

Subjects

People

Places

Subjects

People

Places

Subjects

People

Places

Subjects

People	Places	Subjects

People

Places

Subjects

People

Places

Subjects

People

Places

Subjects

People

Places

Subjects

People

Places

Subjects

People

Places

Subjects

People

Places

Subjects